7

Topics in Organometallic Chemistry

Topics in Organometallic Chemistry

Recently Published and Forthcoming Volumes

Metal Carbons in Organic Synthesis
Volume Editor: K.-H. Dötz
Vol. 13, 2004

Theoretical Aspects of Transistion Metal Catalysis
Volume Editor: G. Frenking
Vol. 12, 2005

Ruthenium Catalysts and Fine Chemistry
Volume Editors: C. Bruneau, P. Dixneuf
Vol. 11, 2004

New Aspects of Zirconium Containing Organic Compounds
Volume Editor: I. Marek
Vol. 10, 2004

CVD Precursors
Volume Editor: R. Fischer
Vol. 9, 2005

Metallocenes in Stereoselective Synthesis
Volume Editor: T. Takahashi
Vol. 8, 2004

Transition Metal Arene π-Complexes in Organic Synthesis and Catalysis
Volume Editor: E. P. Kündig
Vol. 7, 2004

Organometallics in Process Chemistry
Volume Editor: R. D. Larsen
Vol. 6, 2004

Organolithiums in Enantioselective Synthesis
Volume Editor: D. M. Hodgson
Vol. 5, 2003

Organometallic Bonding and Reactivity Fundamental Studies
Volume Editors: J. M. Brown, P. Hofmann
Vol. 4, 1999

Activation of Unreactive Bonds and Organic Synthesis
Volume Editor: S. Murai
Vol. 3, 1999

Lanthanides: Chemistry and Use in Organic Synthesis
Volume Editor: S. Kobayashi
Vol. 2, 1999

Alkene Metathesis in Organic Synthesis
Volume Editor: A. Fürstner
Vol. 1,1998

Springer
Berlin
Heidelberg
New York
Hong Kong
London
Milan
Paris
Tokyo

Transition Metal Arene π-Complexes in Organic Synthesis and Catalysis

Volume Editor: E. Peter Kündig

With contributions by
A. Böttcher • A. Chlenov • B. Gotov • W. D. Harman
M. A. Kondratenkov • E. P. Kündig • K. Muñiz • A. Pape • J. H. Rigby
H.-G. Schmalz • M. F. Semmelhack • M. Uemura

Springer

The series *Topics in Organometallic Chemistry* presents critical overviews of research results in organometallic chemistry, where new developments are having a significant influence on such diverse areas as organic synthesis, pharmaceutical research, biology, polymer research and materials science. Thus the scope of coverage includes a broad range of topics of pure and applied organometallic chemistry. Coverage is designed for a broad academic and industrial scientific readership starting at the graduate level, who want to be informed about new developments of progress and trends in this increasingly interdisciplinary field. Where appropriate, theoretical and mechanistic aspects are included in order to help the reader understand the underlying principles involved.
The individual volumes are thematic and the contributions are invited by the volumes editors.
In references Topics in Organometallic Chemistry is abbreviated
Top. Organomet. Chem. and is cited as a journal

Springer WWW home page: springeronline.com
Visit the TOMC contents at springerlink.com

ISSN 1436-6002

ISBN 978-3-662-14422-0 ISBN 978-3-540-36967-7 (eBook)
DOI 10.1007/978-3-540-36967-7

Library of Congress Control Number: 2004104063

Bibliographic information published by Die Deutsche Bibliothek
Die Deutsche Bibliothek lists this publication in the Deutsche Nationalbibliographie; detailed bibliographic data is available in the Internet at <http:// dnb.ddb.de>.

Typesetting: Medio Technologies AG, Berlin
Production editor: Christiane Messerschmidt, Rheinau
Cover: design & production GmbH, Heidelberg

Printed on acid-free paper 02/3020 – 5 4 3 2 1 0

Volume Editor

Professor E. Peter Kündig
Département de Chimie Organique
Université Genève
30 Quai Ernest Ansermet
1211 Genève 4
Switzerland
peter.kundig@chiorg.unige.ch

Editorial Board

Topics in Organometallic Chemistry is also Available Electronically

For all customers who have a standing order to Topics in Organometallic Chemistry, we offer the electronic version via SpringerLink free of charge. Please contact your librarian who can receive a password for free access to the full articles by registration at:

www.springerlink.com

If you do not have a subscription, you can still view the tables of contents of the volumes and the abstracts of each article by going to the SpringerLink Homepage, clicking on "Browse by Online Libraries", then "Chemical Sciences", and finally choose Topics in Organometallic Chemistry.

You will find information about the

- Editorial Board
- Aims and Scope
- Instructions for Authors
- Sample Contribution

at www.springeronline.com using the search function.

Contents

Topics Organomet Chem (2004) 7: 1–2
DOI 10.1007/b94487

Introduction

E. Peter Kündig

Department of Organic Chemistry, University of Geneva, 30 Quai Ernest Ansermet, 1211 Geneva 4, Switzerland
E-mail: peter.kundig@chiorg.unige.ch

The unique bonding characteristics of aromatics, the stability of the benzene ring system, and the varied and often complex chemistry of arenes and heteroarenes have fascinated chemists for close to two centuries. Benzene rings are omnipresent in organic chemistry and they find important applications in the pharma, agrochemical, and polymer fields. New applications of aromatics include sectors such as functional materials and molecular machines.

Electrophilic aromatic substitution as a route to differentially substituted products is well established. The often forcing conditions, the incompatibility of this process with acid-sensitive functional groups, and the need for mild and selective syntheses have been the driving forces in the search for new methods of synthesis. A large range of methods has been developed over the past 20 years: they include the trimerization of alkynes, the directed *ortho*-metallation, the benzannellation via metal carbenes, and transition metal-catalyzed carbon-carbon and carbon-heteroatom bond formation. Aromatic C-H activation, while still in its beginning stages, is another area of promise.

The present monograph focuses on transition metal arene π complexes. Following the discovery of ferrocene and the determination of its sandwich structure, it did not take long before a large number of sandwich and half-sandwich complexes of benzene and its derivatives saw the light of day and became the subject of intense study. These events and the parallel development of transition metal catalyzed reactions were decisive in the vast interest that arose in the study and chemistry of compounds containing metal carbon bonds. Thus organometallics have become a major component in the chemistry field, a trend that has continued unabated to this day. Organometallics have strongly enriched the fields of homogeneous catalysis, coordination chemistry, and synthetic organic chemistry.

Metal-arene π-complexes show a rich and varied chemistry. The metal adds a third dimension to the planar aromatic compounds and the two faces of an arene with different *ortho* or *meta*-substituents are enantiotopic. Therefore, coordination of a metal to an arene not only alters the reactivity of ring-carbons and substituents as well as groups in benzylic positions but, in addition, also allows reactions with high stereoselectivities to be carried out. The aim of this book is to provide a coherent picture of the state-of-the-art in this field. It covers the entire spectrum of arene activation: from the electrophilic activation of a η^6-bound arene by the π-Lewis acids $Mn(CO)_3^+$, $Cr(CO)_3$, $FeCp^+$ and $RuCp^+$ to the activation of the reaction with electrophiles by η^2-coordination of the arene to the fragments $Os(NH_3)_5^{2+}$, TpRe(CO)(L), and TpMo(NO)(MeIm).

In this multi-author book, we document preparation, scope, limitations and challenges of reactions in contemporary π-arene metal chemistry with an emphasis on transformations of interest to organic synthesis and to their use in catalysis. The monograph is organized in nine chapters, written by leading scientists in the field. By focusing on the synthesis and transformations of arene complexes, as well as on their use as ligands and catalysts, the book provides the reader with an up-to-date treatise on the subject organized according to reaction type and use rather than according to the individual metal or each author's research focus.

We firmly hope that this book will provide an additional stimulus to the vigorous development of the chemistry of metal arene complexes, an area of research that we and our students and coworkers have all found so stimulating an area that again and again provides new reactions, selective methods, ligands and catalysts for future chemistry.

Topics Organomet Chem (2004) 7: 3–20
DOI 10.1007/b94489

Synthesis of Transition Metal ^{6}-Arene Complexes

E. Peter Kündig

Department of Organic Chemistry, University of Geneva, 30 Quai Ernest Ansermet, 1211 Geneva 4, Switzerland
E-mail: peter.kundig@chiorg.unige.ch

Abstract Methods of synthesis of η^6-arene complexes of $Cr(CO)_3$, $Mo(CO)_3$, $Mn(CO)_3^+$, $FeCp^+$, $RuCp^+$ are reviewed. These electrophilic transition metal complex fragments have found application in arene transformations. Critical comparison of the routes of access is made and methods of decomplexation and where possible methods of recovery of the activating group are also detailed. Excluded from the overview are methods involving arene transformations in the coordination sphere of the metal. These will be contained in subsequent chapters.

Keywords Arene complexes · Chromium · Molybdenum · Manganese · Iron · Ruthenium

1
(Arene)Cr(CO)$_3$ Complexes

Arene tricarbonyl chromium complexes are yellow to red, often crystalline compounds. They are stable to air in the solid state and can be stored for long periods provided that they are kept out of light. In solution, they are weakly to moderately air-sensitive. They are best purified by crystallization but other methods like sublimation, flash chromatography, and HPLC are generally applicable. This as well as the following sections focus on synthetic procedures. For a description of bonding- and structural characteristics, the reader is referred to the specialist literature [1].

1.1
Synthesis of (Arene)Cr(CO)$_3$ Complexes

The most common and most economic method for the synthesis of (arene)Cr(CO)$_3$ complexes is thermolysis of Cr(CO)$_6$ under an inert atmosphere (nitrogen or argon) in the presence of an excess of the arene. High boiling solvents need to be used. The solvent can be the arene itself, dibutyl ether/THF [2], 1,2-dimethoxyethane [3], diglyme/THF [4], heptane/diglyme [5], α-picoline [6], decalin [7], decalin/ethyl formate or decalin/butyl acetate [8, 9]. For aryl amino acids, a mixture of water and THF (80:20) has been successfully applied [10]. The polar ether and ester additives (or solvents) promote carbonyl dissociation, stabilize intermediates, and the vigorous reflux of lower boiling additives wash sublimed Cr(CO)$_6$ back into the reaction mixture. Prior to mixing and heating, solvents are degassed by several freeze/pump/thaw cycles or by bubbling inert gas through the solvent for 5–10 min. The most widely used solvent combination is dibutyl ether/THF (9:1) [2]. It allows the preparation of a wide range of complexes with useful functionalities in good yields with reaction times typically in the 1–4 day range (Scheme 1). Higher temperatures shorten reaction times but increase the risk of decomposition that, once started, can be autocatalytic and lead to rapid product loss [7]. The use of a wide-bore straight tube condenser is recommended and often sufficient. Special apparatus such as a double condenser system [8] or distillative recycling of Cr(CO)$_6$ [11] is advantageous or even required in some of the procedures. Complexes of condensed aromatics are unstable towards polar solvents (THF, DMSO, acetone) and their synthesis requires special attention [7, 12, 13] or the use of more labile Cr(CO)$_3$L$_3$ precursors (see below). A selection from the hundreds of mono- and polysubstituted chromium arene complexes made by the direct reaction of the arene with Cr(CO)$_6$ is shown in Scheme 1 [2, 7, 13–29].

Condensed aromatics coordinate the metal in a terminal ring (e.g. in the phenanthrene complex **23**) This bonding mode minimizes disruption of aromaticity [30]. Regioselectivity favors the arene over the heteroarene ring (e.g. in the indol complex **25**) and the non-substituted arene ring in 1- and 1,4 substituted naphthalenes (e.g. in **22**). Aromatic heterocycles can be complexed (e.g. **26**, **27**) though yields are not always high. Aryl bromides, aryl iodides, benzaldehyde, and arenes containing halides in the benzylic position cannot be complexed directly with Cr(CO)$_6$.

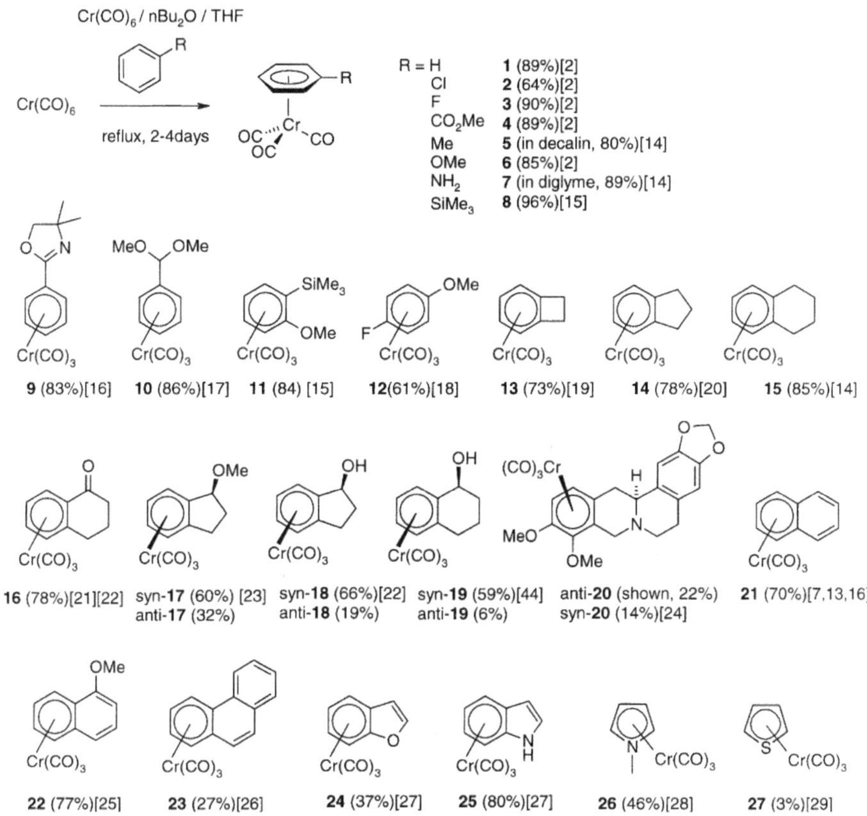

Scheme 1 Synthesis of (Arene)$Cr(CO)_3$ complexes by thermolysis of $Cr(CO)_6$

(Arene)$Cr(CO)_3$ complexes have also been prepared in low to moderate yields by photolysis of $Cr(CO)_6$ in the presence of the arene [31].

Milder complexation conditions are possible with suitable $M(CO)_3L_3$ precursors. These include the complexes with L=MeCN, NH_3, and pyridine. $Cr(MeCN)_3(CO)_3$ and $M(CO)_3py_3$ are prepared by refluxing $M(CO)_6$ in the appropriate solvent. $Cr(CO)_3(NH_3)_3$ is best prepared by treating $Cr(CO)_6$ with KOH in EtOH, followed by addition of NH_4OH (90% yield on a 10 g scale) [32, 33]. (Arene)$Cr(CO)_3$ complexes are obtained on refluxing the pyridine and NH_3 precursor complexes in the presence of an arene in dioxane. This allows thermal complexation typically in the temperature range of 80–100 °C. Still milder conditions (25–80 °C) apply to certain arene exchange reactions. The naphthalene complex **21** is labile [19] because it involves haptotropic slippage of the naphthalene ligand (change from η^6- to η^4- or η^2-coordination), thus facilitating the dissociation and coordination of the new arene [1a]. In situ generation of the naphthalene complex under conditions of arene exchange has also been used [34]. Facile arene exchange also occurs with the pyrrole complex **26** although this complex is not easy to handle [35]. Room temperature complexation of arenes is

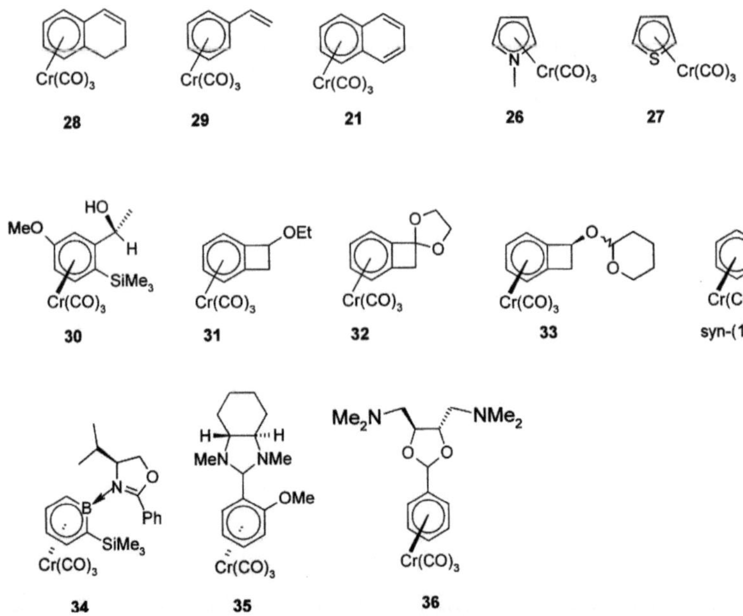

Scheme 2 Synthesis of (Arene)$Cr(CO)_3$ complexes from $Cr(CO)_3L_3$ precursors

Table 1 Synthesis of (Arene)$Cr(CO)_3$ complexes from $Cr(CO)_3L_3$ precursors

$Cr(CO)_3L_3$ starting complex	Product	Yield (%)	Ref
$Cr(CO)_3(NH_3)_3$, dioxane, Δ	**28**	85	[33]
$Cr(CO)_3(CH_3CN)_3$	**29**	35	[37]
$(C_{10}H_8)Cr(CO)_3$ (**21**)	**29**	90	[19]
$Cr(CO)_3(NH_3)_3/BF_3{\cdot}OEt_2$	**21**	70	[36]
$Cr(CO)_3(CH_3CN)_3$	**26**	74	[28] [35]
$Cr(CO)_3(py)_3/BF_3{\cdot}OEt_2$	**27**	70	[38]
$(C_{10}H_8)Cr(CO)_3$ (**21**)	**27**	70	[39]
$(C_{10}H_8)Cr(CO)_3$ (**21**)	**30**	97 (single diastereoisomer)	[40]
$(C_{10}H_8)Cr(CO)_3$ (**21**), THF	**31**	70 (1:1 mixture of diastereoisomers)	[41]
$Cr(CO)_3(NH_3)_3$, dioxane, Δ	**32**	83	[42]
$(C_{10}H_8)Cr(CO)_3$ (**21**), THF, 70 °C	**33**	76 (1:1 mixture of diastereoisomers)	[43]
$(C_{10}H_8)Cr(CO)_3$ (**21**), Et_2O, 70 °C	(1*S*,2a*S*)-**33**	61 (33:1 mixture of diastereoisomers)	[43]
$(C_{10}H_8)Cr(CO)_3$ (**21**)	(1*S*,4a*S*)-**19**	91 (single diastereoisomer)	[44]
$Cr(CO)_3(CH_3CN)_3$	**34**	72 (single diastereoisomer)	[45]
$(C_{10}H_8)Cr(CO)_3$ (**21**), THF, 25 °C, 4 days	**35**	80 (98:2 mixture of diastereoisomers)	[46]
$Cr(CO)_6$, naphthalene, *n*-Bu_2O, THF, reflux	**36**	90	[34]

accomplished by reaction of $Cr(CO)_3(NH_3)_3$ with $BF_3 \cdot OEt_2$ in the presence of an arene [36]. The advantages of lower temperatures for arene complexation are higher compatibility with arenes bearing functional groups and higher chemo- and diastereoselectivities (Scheme 2, Table 1). While these methods require an additional synthetic step to prepare the $M(CO)_3L_3$ complexes, the ease of the procedures and the high overall yield make them often the methods of choice.

Finally, a range of substituted phenol and naphthol complexes are accessible via the reaction of chromium carbene complexes with alkynes (Dötz annulation) [47]. On heating, the initially formed naphthol complex **37** undergoes haptotropic rearrangement to the isomer **38**, containing the $Cr(CO)_3$ fragment coordinated to the unsubstituted naphthalene ring. If an aminocarbene is used, CO insertion does not take place, and **39** is the sole product. In order to isolate the chromium complex with good yields, it is usually necessary to protect the phenol or naphthol function formed in the reaction [48] Scheme 3.

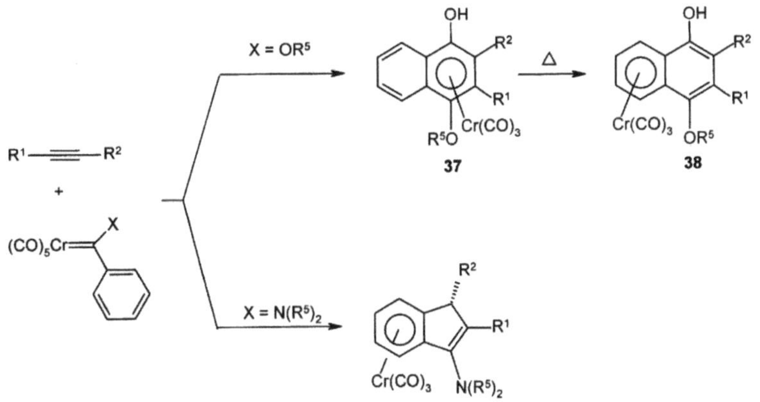

Scheme 3

1.2
Arene Decomplexation from Cr

The preceding section has shown that complexation of arenes to the $Cr(CO)_3$ complex fragment can be carried out in high yields and with often excellent selectivities. Purification by crystallization or chromatographic methods is also straightforward thanks to the stability of the uncharged complexes and their solubility pattern in organic solvents. These characteristics are complemented by the ease by which the metal can be removed at the end of a synthetic sequence. While inert to a large number of reaction conditions, the arene-metal bond in (arene)$Cr(CO)_3$ complexes is readily cleaved upon oxidation of the metal (Ce(IV), Fe(III), I_2, $h\nu/O_2$). The mildest procedure is the exposure of a solution of the complex in diethylether or acetonitrile to sunlight and air for a few hours. This allows the isolation of the arene in yields that are usually >80% and often considerably higher. Refluxing of an (arene)$Cr(CO)_3$ complex in pyridine cleaves the metal arene bond and allows recycling of the Cr(0) complex in the form of $Cr(CO)_3py_3$ [49]. In (naphthalene)$Cr(CO)_3$ (**19**) and substituted derivatives, the metal arene bond is readily cleaved by stirring the complex under an atmosphere of CO or by applying a few bars of pressure [50].

2
(Arene)$Mo(CO)_3$ Complexes

Arene tricarbonyl molybdenum complexes are yellow, often crystalline compounds. They are weakly air-sensitive in the solid state and have to be stored under inert atmosphere and out of light. They are best purified by crystallization. In solution, they are unstable to air. The trait that has most hampered development of the use of (arene)$Mo(CO)_3$ complexes in organic synthesis, however, is the lability of the arene metal bond. Lewis basic solvents such as THF, DMF, DMSO, acetone and acetonitrile rapidly displace benzene in (benzene)$Mo(CO)_3$. This lability of the arene-Mo bond, while making handling difficult, holds promise for the catalytic use of this class of compounds.

2.1
Synthesis of (Arene)$Mo(CO)_3$ Complexes

The direct synthesis of (arene)$Mo(CO)_3$ complexes from arene and $Mo(CO)_6$ is much more limited than for chromium (Scheme 4) [11, 51]. The long reaction times at elevated temperature (e.g., ten days for (benzene)$Mo(CO)_3$) and the high sensitivity to oxygen often results in low yields for substituted arenes. While (benzene)$Mo(CO)_3$ (**40**) has been reportedly obtained in near quantitative yield, the yield was based on liberated CO rather than isolated complex [11]. In the author's laboratory, an isolated yield of 50% is more realistic for this procedure. The reaction time can be shortened by reacting $Mo(CO)_6$ in benzene in the presence of pyridine in an autoclave [52]. Toma and coworkers have described a different procedure that uses a double condenser system, and decalin plus ethylformate as solvent [53]. With a bath temperature of 240 °C this cuts the preparation time of the aniline complex **42** to 1 h (55% yield) (Scheme 4). In the authors laboratory the method is used routinely for the synthesis of complex **40** (18 h, 60% yield).

Alternatively ligand substitution in $Mo(CO)_3L_3$ complexes can be used and a particularly useful method is the use of $Mo(CO)_3py_3$ and $BF_3 \cdot OEt_2$ in the presence of an excess arene (Scheme 5) [58–60]. Other sources of $Mo(CO)_3$ fragments that have been used in the synthesis of (arene)$Mo(CO)_3$ complexes are $Mo(CO)_3$(diglyme) [61], and $Mo(CO)_3(DMF)_3$ [62].

Thermochemical studies show the arene-Mo bond (68 kcal mol^{-1} in [(η^6-C_6H_6)$Mo(CO)_3$] (**40**)) to be stronger than the arene-Cr bond (53 kcal mol^{-1} in [(η^6-C_6H_6)$Cr(CO)_3$] (**1**)) [63, 64]. Kinetically, however, the situation is reversed. The metal arene bond in the Mo complex **40** is far more labile than that in the Cr complex **1**. In the absence of a Lewis base catalyst, arene exchange in (arene)$Mo(CO)_3$ complexes is measurable at temperatures as low as 60 °C (com-

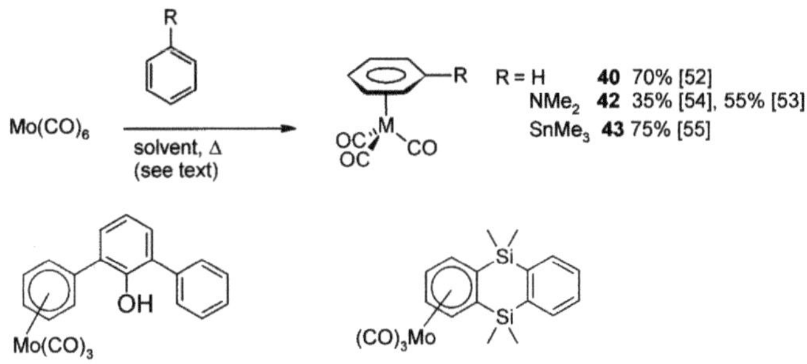

44 78% [56] **45** 70% [57]

Scheme 4

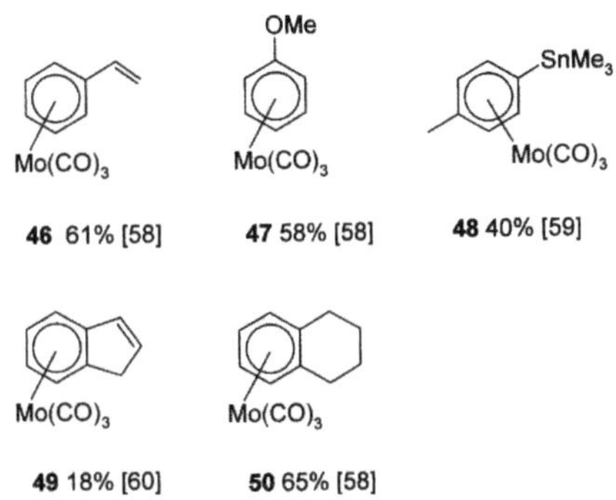

Scheme 5 Synthesis of (Arene)$Mo(CO)_3$ complexes from $Mo(CO)_3py_3$, arene, and $BF_3 \cdot OEt_2$

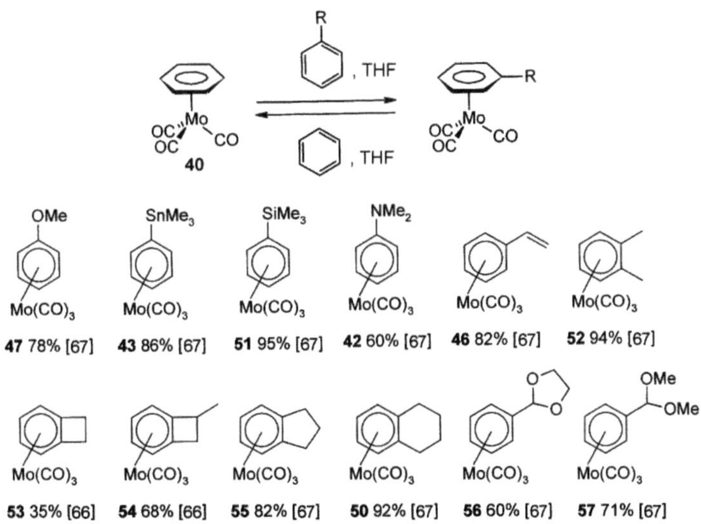

Scheme 6

pared to >150 °C in arene Cr complexes). In pioneering work, Muetterties determined equilibria of arene exchange in (arene)$Mo(CO)_3$ complexes (arene=benzene, toluene, xylene, mesitylene, ...). The acetone catalyzed reactions proceed at room temperature and equilibria are reached in less than three days. Stabilities of the arene complexes increases with the degree of Me substitution [65].

This lability can be used preparatively for functionalized arenes as shown by the examples in Scheme 6 [66, 67].

2.2 Arene Decomplexation from Mo

Given the lability of the Mo-arene bond, the removal of the activating group $Mo(CO)_3$ is very easy and is readily achieved by the addition of an excess of a coordinating solvent (MeCN, THF) or another Lewis base (PR_3, NR_3) to a solution of the arene complex.

3 (Arene)$Mn(CO)_3^+$ Complexes

Cationic arene tricarbonyl manganese complexes, usually isolated as their BF_4- or PF_6- salts, are yellow solids. They are stable to air in the solid state and can be stored for long periods provided that they are kept out of light. They are best purified by crystallization by slow diffusion of diethylether into acetone- or CH_2Cl_2-solutions of the complex. Only direct methods of complexation are reported here. For other methods, involving transformations of the arene in the complex, the reader is referred to Chap. 6, to reviews [68], and to the recent literature [69, 70].

3.1 Synthesis of (Arene)$Mn(CO)_3^+$ Complexes

Three procedures are based on the reaction of an arene with the complex fragment $[Mn(CO)_5]^+$. They vary in the way this coordinatively unsaturated fragment is generated. The classic method is the Fischer Hafner method that involves halide abstraction in $Mn(CO)_5Br$ (**58**) by the strong Lewis acid $AlCl_3$ upon heating in hydrocarbon solvents [71]. $[Mn(CO)_5]^+$ can also be generated directly by oxidative cleavage of $Mn_2(CO)_{10}$ under strongly acidic conditions (trifluoroacetic acid (TFA) in TFA anhydride) [72]. A milder route to $[Mn(CO)_5]^+$ is halide abstraction by $AgBF_4$ or $AgPF_6$ (room temperature, CH_2Cl_2) [73]. An excellent comparative study of the three methods has been published [74]. It shows that the harsh conditions of the TFA anhydride method is best for acid insensitive substrates and that the Fischer Hafner method ($AlCl_3$) is convenient for aryl ethers and nonconjugated carbonyl arenes but not for alkyl arenes that are susceptible to undergo retro Friedel-Crafts rearrangements. We note that $Mn(CO)_5OClO_3$ can be isolated from the TFA procedure and then used under neutral conditions in arene complexation [75]. This, and the mild silver(I) method are the methods of choice for acid sensitive arenes. The latter is best carried out under inert atmosphere in the dark.

The $Mn(CO)_3^+$ fragment is a powerful electrophile and strongly electron withdrawing groups such as nitro- and α-keto groups are not tolerated in the direct syntheses of the complexes. Scheme 7 gives a representative range of examples of complexes synthesized according to the three methods mentioned above.

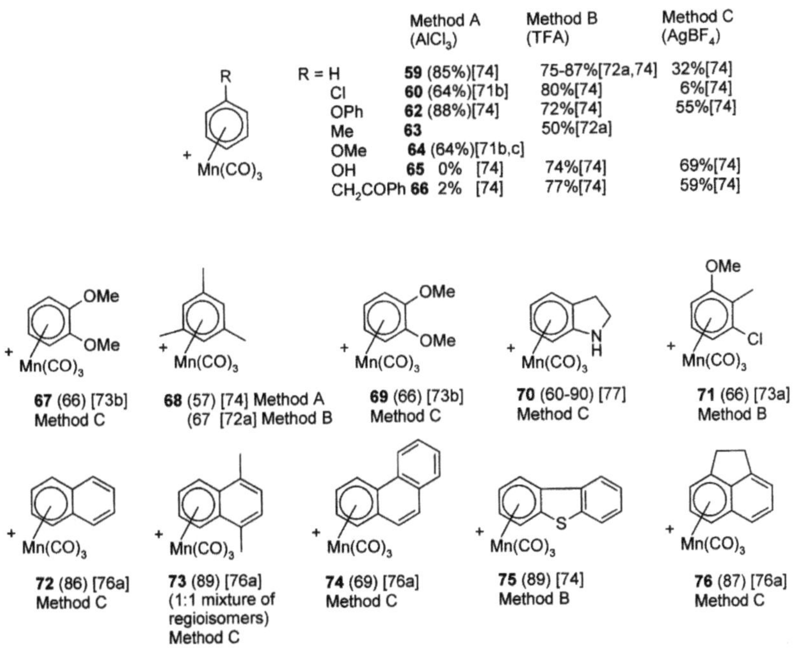

Scheme 7 Cationic arene manganese tricarbonyl complexes prepared from the $[Mn(CO)_5]^+$ fragment. $[Mn(CO)_5]^+$ is generated in situ via: Method A: $Mn(CO)_5Br$ (**58**)/$AlCl_3$/Δ; Method B: $Mn_2(CO)_{10}$/TFA/TFA-anhydride/reflux; Method C: Complex **58**/$AgBF_4$/CH_2Cl_2/reflux. In all cases the counteranion is either BF_4^- or PF_6^-

The most recent addition to the synthetic arsenal are arene exchange reactions in complexes containing condensed aromatics. Like in the $Cr(CO)_3$ chemistry detailed above, the most reactive complexes are those of naphthalene and its substituted analogues. They are accessible in high yield via the Ag(I) route but care needs to be taken to use rigorously dry solvents and avoid contact with donor solvents [76]. Examples are given in Scheme 8.

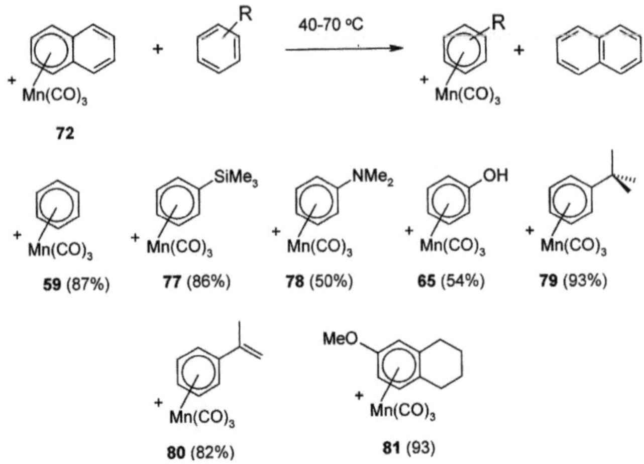

Scheme 8 Arene exchange reactions with (naphthalene)$Mn(CO)_3^+$. Arene exchange reactions are also successful with complexes **73–76** [76]

3.2
Arene Decomplexation from Mn

With the exception of the labile complexes of condensed aromatics, the metal-arene bond in (arene)$Mn(CO)_3^+$ complexes is very robust. It is more difficult to cleave than that in the neutral arene complexes of chromium and molybdenum and requires strongly oxidative conditions such as DDQ in refluxing acetonitrile.

4
(Arene)FeCp$^+$ Complexes

Cationic arene FeCp$^+$ complexes, usually isolated as their BF_4- or PF_6- salts, are yellow to red solids. They are stable in air in the solid state and can be stored for long periods provided that they are kept out of light. They are best purified by crystallization by slow diffusion of diethylether into acetone- or CH_2Cl_2-solutions of the complex. Only direct methods of complexation are reported here. For other methods, involving transformations of the arene in the complex, the reader is referred to Chap. 5, to reviews [78], and to the recent literature [79].

4.1
Synthesis of (Arene)FeCp$^+$ Complexes

Cationic arene cyclopentadienyl iron complexes were first isolated from the reaction of $CpFe(CO)_2Cl$ with arenes in the presence of $AlCl_3$ [80]. Subsequently, Nesmeyanov developed the ferrocene route (Scheme 9) and for most arenes this is more convenient and efficient and has become the standard synthesis for this class of compounds [81]. It has drawbacks in that it requires a stoichiometric amount of a strong Lewis acid and the method is not suited for alkyl arenes that are susceptible to undergo retro Friedel-Crafts rearrangements. Arenes with electron acceptor groups, e.g., acetophenone, benzonitrile, acetanilide, indole, etc., fail to form the corresponding $CpFe^+$ complex and dehalogenation occurs with halogenated arenes when Al powder is present [82]. Partial hydrogenation

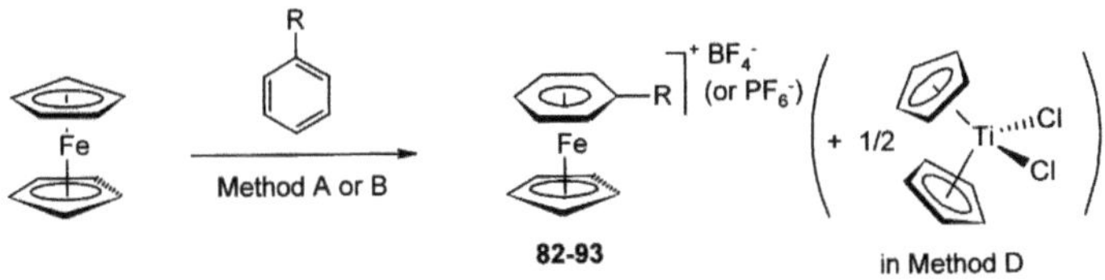

Method A: 1. $AlCl_3$, Al, arene(neat or in hydrocarbon solvent), 70 - 190 °C, 1-16h
2. aq. NH_4BF_4 or NH_4PF_6
ratio ferrocene/$AlCl_3$/Al = 1:2:1
Method B: 1. $AlCl_3$, Al (or Zn or Cu), arene(in trichlorobenzene solvent), Microwave heating, 1-16 min. 2. aq. NH_4BF_4 or NH_4PF_6
Method C: 1.1-Bu-3-Me-imidazolium chloride/$AlCl_3$ ([bmim]Cl-$AlCl_3$), [bmim][HCl_2], Al 80 °C, 8-18h, glove box. 2. H_2O extr. 3. aq. NH_4PF_6
Method D: 1. $AlCl_3$, Al, arene(neat or in Me-cyclohexane), 50-150 °C, 0.25-24h.
Ratio ferrocene/$AlCl_3$/Al/$TiCl_4$ = 2:2:1:1
2. aq. NH_4PF_6

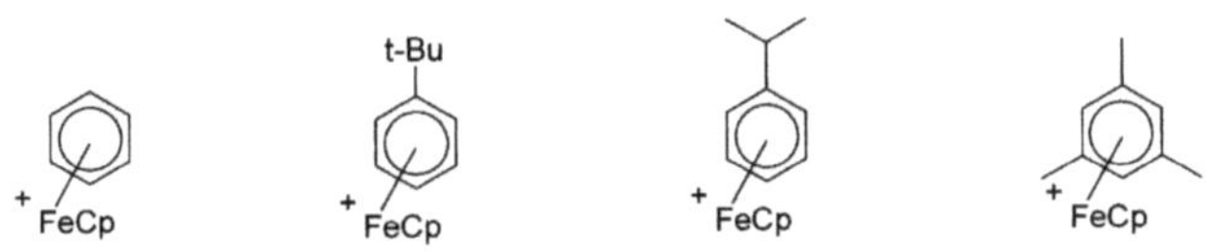

82
Meth. A: 30-50% [91a]
90% [84b]
Meth. B: 88%
Meth. C: 53% [86]

83
Meth. A: 20% [89]
Meth. B: 77% [85b]

84
Meth. D: 94% [87]

85
Meth. A: 42% [89]
90 [84c]
Meth. B: 80% [85a]
Meth. D: 66% [87]

86
Meth. B: 50% [85b]

87
Meth. A: 10-45% [88,90]
Meth. B: 47% [85a]

88
Meth. B: 65% [85a]

89
Meth. A: 40-60% [91a]

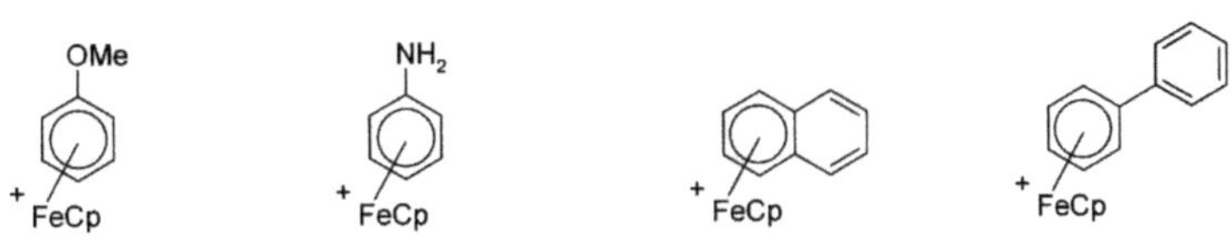

90
Meth. A: 30-50% [91b]
Meth. B: 17% [85a]
(phenol)

91
Meth. A: 22% [91a]
Meth. B: 53% [85a]
Meth. C: 53% [86]

92
Meth. A: 26% [83]
Meth. B: 55% [85a]
Meth. C: 53% [86]
Meth. D: 66% [87]

93
Meth. A: 40 [92]
Meth. B: 70% [85a]

Scheme 9

of fused arenes like anthracene, pyrene and naphthalene is a further problem [83]. To avoid side reactions, the preparation is best carried out at 70–100 °C in neat arene or in a hydrocarbon solvent. Addition of a Bronsted acid, e.g., in the form of 1 equiv. of water can strongly increase the yield in methyl arenes [84]. Despite these optimizations, yields of reactions with substituted arenes are often below 50% as shown in the examples in Scheme 9. The use of microwave heating has given good results and reaction times are much reduced [85]. Acidic ionic liquids as reaction medium have also been used successfully [86]. The patent literature presents another solution: addition of $TiCl_4$ to the reaction in the proportion ferrocene to $TiCl_4$=ca. 2:1. The titanium reagent traps the cyclopentadienyl ligands to form titanocene dichloride and this markedly increases the yield of the cationic arene complex [87]. The method has been scaled up to 1–10 mol quantities for several arenes (Scheme 9).

The metal arene bond in (arene)$FeCp^+$ complexes is photolabile [93], and this forms the basis of the use of these complexes as photoinitiators for cationic polymerizations of epoxides [94]. Arene exchange via this route has not been used extensively. On photolysis of (*p*-xylene)$FeCp^+$ in CH_2Cl_2 or acetone by a medium pressure Hg-arc or by 'bright sunlight', the complex undergoes arene exchange with hexamethylbenzene, paracyclophane and thiophene under mild conditions

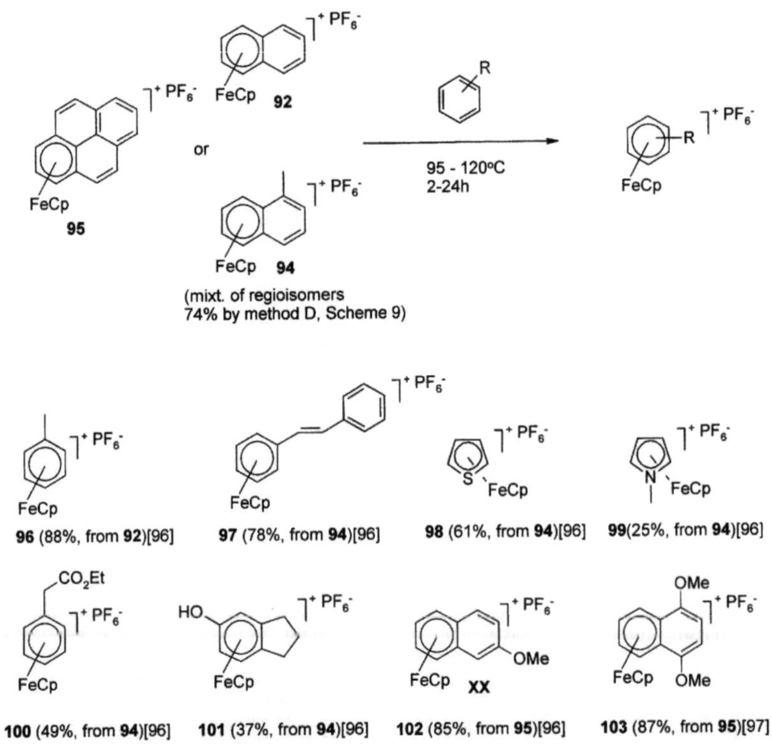

Scheme 10 Synthesis of (Arene)$FeCp^+$ complexes by arene exchange

(above room temperature) [93, 95]. An efficient route that uses neither Lewis acids nor light again stems from the patent-literature and involves thermal arene exchange in complexes of condensed aromatics [96, 97]. Examples are given in Scheme 10.

4.2 Arene Decomplexation from Fe

Cleavage of the arene-iron bond in (arene)FeCp complexes is not always facile. If applicable, the best way to remove the FeCp moiety from the arene is by photolytic cleavage. Irradiation in acetonitrile or another coordinating solvent can be efficient [78, 98]. Alternative methods involve oxidation by DDQ [99], pyrolytic sublimation [85], microwave heating [100], and electrochemical decomplexation [101]. The forcing conditions are not generally compatible with arenes bearing functional groups and this limits the scope of the use of these compounds in synthesis.

5 (Arene)RuCp+ Complexes

The (arene)RuCp+ complexes share many characteristics with the analogous iron compounds such as their robustness, stability to air and mild oxidants, and the photolytic cleavage of the arene metal bond. It has to be pointed out though that there are substantial differences in both the efficiency of the photolytic cleavage (Fe>Ru) and the stability of the resulting cationic $CpM(solvate)_3^+$ complexes (Ru>Fe). As will be detailed below the latter will form the basis for the best synthetic routes to the title compounds.

5.1 Synthesis of (Arene)RuCp+ Complexes

In contrast to the iron system, ligand exchange in ruthenocene is not a major route to (arene)RuCp+ complexes. The $AlCl_3$ mediated cleavage of a Ru-Cp bond gives only low yields (<10%) of the arene complex (arene: benzene, biphenyl, mesitylene, chlorobenzene) under the conditions of the ferrocene reactions [102]. At 165 °C, ruthenocene, in the presence of the arene, $AlCl_3$ and Al afforded, after work up and anion exchange [(mesitylene)RuCp][PF_6] in the better, yet still unsatisfactory yield of 30% [102b].

A better route is the one shown in Scheme 11. The benzene complex **104** forms readily in near quantitative yield upon heating Ru(III) chloride and 1,3- or 1,4-cyclohexadiene in EtOH. The introduction of the Cp ligand is by halide/Mcp exchange. The classic procedure to **105** required a stoichiometric quantity of thallium or silver [103], but a very recent, improved procedure shows that cyclopentadiene/K_2CO_3/ abs. EtOH can be used successfully [104]. Although beyond the scope of this review, it is interesting to note that the much less electrophilic complexes [(arene)Ru(C_5Me_5)][PF_6] are accessible in a one-pot procedure from $RuCl_3$ [105]. The sequence Birch reduction/complexation/halide substitution can also be applied to substituted arenes (e.g., cymene, anisol) [103]. Yet another access to analogues of **104** is by arene exchange in the cymene analogue to **104**. The high temperatures required (180 °C) limits the scope of this procedure but the hexamethyl benzene complex was made this way [107].

Photolytic cleavage of the benzene-Ru bond of **105** in acetonitrile gives a quantitative yield of the pale orange, crystalline complex $[CpRu(CH_3CN)_3][PF_6]$ (**106**). The acetonitrile ligands in **106** are labile and the compound therefore serves as a convenient transfer agent for the $CpRu^+$ fragment. A selection of $(arene)RuCp^+$ complexes made by this route is assembled in Scheme 11.

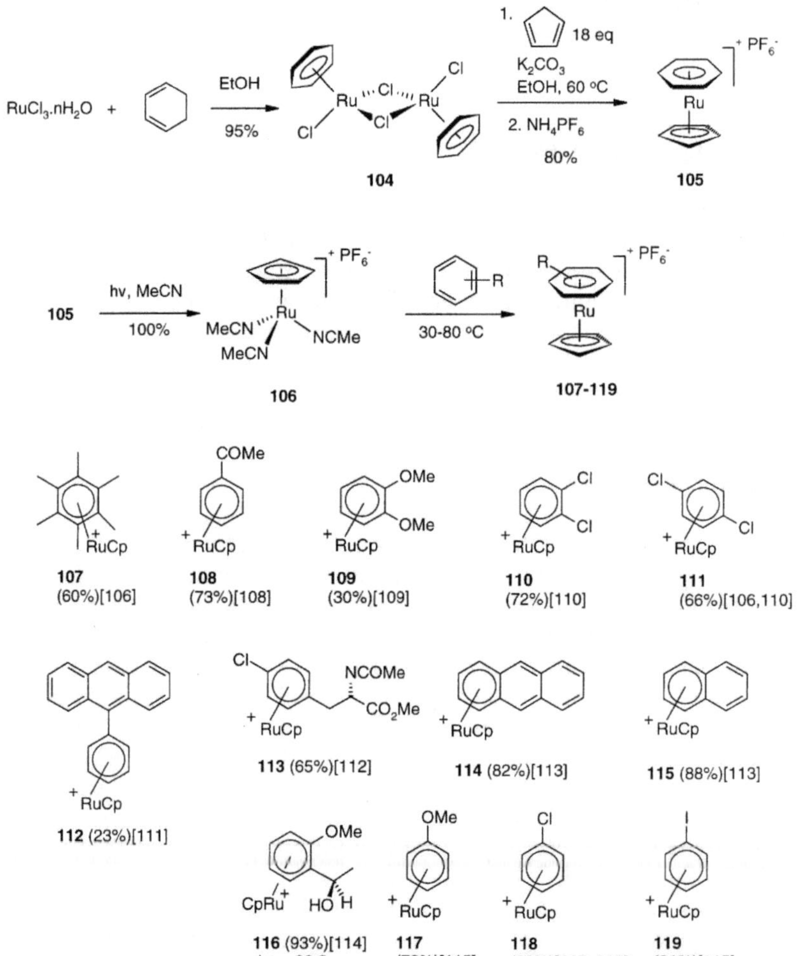

Scheme 11 Synthesis of $(arene)RuCp^+$ complexes from $(arene)Ru(CH_3CN)_3{}^+$

5.2
Arene Decomplexation from Ru

The best way to remove the RuCp moiety from the arene is by photolytic cleavage. Irradiation in acetonitrile is efficient and $[(CH_3CN)_3RuCpl\ [PF_6]$ (**106**) can be recovered from these reactions in often good yield (>80%), making up for the disadvantage of using ruthenium in stoichiometric quantity [115, 116].

References

1. (a) Muetterties EL, Bleeke JR, Wucherer EJ, Albright TA (1982) Chem Rev 82:499; (b) Hunter AD, Mozol V, Tsai SD (1992) Organometallics 11:2251; (c) Morris MJ (1993) In: Abel EW, Stone FGA, Wilkinson G (eds) Comprehensive organometallic chemistry II, vol 5. Pergamon, New York, p 471
2. Mahaffy CAL, Pauson PL (1990) Inorg Synth 28:136
3. Schöllkopf K, Stezowski JJ, Effenberger F (1985) Organometallics 4:922
4. Nagelberg SB, Reinhold CE, Willeford BR, Bigwood MP, Molloy KC, Zuckerman JJ (1982) Organometallics 1:851
5. Kaganovich VS, Rybinskaya MI (1991) J Organomet Chem 407:215
6. Pruett RL (1965) Perp Inorg React 2:187
7. Deubzer B, Fischer EO, Fritz HP, Kreiter CG, Kriebitzsch N, Simmons HD, Willeford BR (1967) Chem Ber 100:3084
8. Hudecek M, Toma S (1991) J Organomet Chem 406:147
9. Hrnciar P, Hudecek M, Magomedov GKI, Toma S (1991) Collect Czech Chem Commun 56:1477
10. Segheraert C, Brunet JC, Tartar A (1982) J Chem Soc, Chem Commun 1417
11. Strohmeier W (1961) Chem Ber 94:2490
12. Desobry V, Kündig EP (1981) Helv Chim Acta 64:1288
13. Hudecek M, Gadja V, Toma S (1991) J Organomet Chem 413:155
14. Nicholls B, Whiting MC (1959) J Chem Soc 551
15. Effenberger F, Schoellkopf K (1985) Chem Ber 118:4356
16. Kündig EP, Ripa A, Liu R, Amurrio D, Bernardinelli G (1993) Organometallics 12:3724
17. Davies SG, Goodfellow CL (1990) J Chem Soc Perkin Trans 1:393
18. Gilday JP, Negri JT, Widdowson DA (1989) Tetrahedron 45:4605
19. Kündig EP, Perret C, Spichiger S, Bernardinelli G (1985) J Organomet Chem 286:183
20. Gracey DF, Jackson WR, Jennings WB, Rennison SC, Spratt R (1969) J Chem Soc B 1210
21. Jackson WR, Mitchell TRB (1969) J Chem Soc B 1228
22. Davies SG, Goodfellow CL (1988) J Organomet Chem 340:195
23. Ohlsson B, Ullenius C, Jagner S, Grivet C, Wenger E, Kündig EP (1989) J Organomet Chem 365:243
24. Baird PD, Blagg J, Davies SG, Sutton KH (1988) Tetrahedron 44:171
25. Kündig EP, Inage M, Bernardinelli G (1991) Organometallics 10:2921
26. King RB, Stone FGA (1960) J Am Chem Soc 82:4557
27. Fischer EO, Goodwin HA, Kreiter CG, Simmons HD, Sonogashira K, Wild S-B (1968) J Organomet Chem 14:359
28. Oefele K, Dotzauer E (1971) J Organometal Chem 30:211
29. Fischer EO, Oefele K (1958) Chem Ber 91:2395
30. Zhang S, Shen JK, Basolo F, Ju TD, Lang RF, Kiss G, Hoff CD (1994) Organometallics 13:3692
31. Kostermans GBM, Bobeldijk M, Kwakman PJ, De Wolf WH, Bickelhaupt F (1989) J Organomet Chem 363:291
32. Moser GA, Rausch MD (1974) Syn Reactiv Inorg Metal-Org Chem 4:37
33. Vebrel J, Mercier R, Belleney J (1982) J Organomet Chem 235:197

34. Kondo Y, Green JR, Ho J (1993) J Org Chem 58:6182
35. Goti A, Semmelhack MF (1994) J Organomet Chem 470:C4
36. Morley JA, Woolsey NF (1992) J Org Chem 57:6487
37. Knox GR, Leppard DG, Pauson PL, Watts WE (1972) J Organometal Chem 34:347
38. Oefele K (1966) Chem Ber 99:1732
39. Struharik M, Toma S (1994) J Organomet Chem 464:59
40. Uemura M, Kobayashi T, Isobe K, Minami T, Hayashi Y (1986) J Org Chem 51:2859
41. Kündig EP, Bernardinelli G, Leresche J (1991) J Chem Soc, Chem Commun 1713
42. Brands M, Wey HG, Kroemer R, Krueger C, Butenschön H (1995) Liebigs Ann 253
43. Kündig EP, Leresche J, Saudan L, Bernardinelli G (1996) Tetrahedron 52:7363
44. Schmalz HG, Millies B, Bats JW, Duerner G (1992) Angew Chem, Int Edn Eng 31:631
45. Tweddell J, Hoic DA, Fu GC (1997) J Org Chem 62:8286
46. Alexakis A, Mangeney P, Marek I, Rose-Munch F, Rose E, Semra A, Robert F (1992) J Am Chem Soc 114:8288
47. (a) Wulff WD (1995) In: Abel EW, Stone FGA, Wilkinson G (eds) Comprehensive organometallic chemistry II, vol 12. Pergamon, New York, p 469; (b) Dötz KH, Tomuschat P (1999) Chem Soc Rev 28:187
48. Chamberlin S, Wulff WD, Bax B (1993) Tetrahedron 49:5531
49. Carganico G, Del Buttero P, Maiorana S, Riccardi G (1978) J Chem Soc Chem Commun 989
50. (a) Kündig EP, Desobry V, Grivet C, Rudolph B, Spichiger S (1987) Organometallics 6:1173; (b) Kündig EP, Pache SH (2003) Sci Synth 2:153
51. Fischer EO, Oefele K, Essler H, Fröhlich W, Mortensen JP, Semmlinger W (1958) Chem Ber 91:2763
52. Magomedov GKI (1990) J Organomet Chem 385:113
53. Hudecek M, Toma S (1990) J Organomet Chem 393:115
54. Pidcock A, Smith JD, Taylor BW (1967) J Chem Soc A 872
55. Poeth TP, Harrison PG, Long TV, Willeford BR, Zuckerman JJ (1971) Inorg Chem 10:522
56. Kerschner JL, Torres EM, Fanwick PE, Rothwell IP, Huffman JC (1989) Organometallics 8:1424
57. Chernyshev EA, Afanasova OB, Frenkel AS, Zubarev YE, Magomedov GK (1987) Zh Obshch Khim 57:2531
58. Nesmeyanov AN, Krivykh VV, Kaganovich VS, Rybinskaya MI (1975) J Organomet Chem 102:185
59. Alyea EC, Jain VK (1996) Polyhedron 15:433
60. Novikova LN, Ustynyuk NA, Zvorykin VE, Dneprovskaya LS, Ustynyuk YA (1985) J Organomet Chem 292:237
61. Chatt J, Guy RG, Watson HR (1961) J Chem Soc 2332
62. Pasquali M, Leoni P, Sabatino P, Braga D (1992) Gazz Chim Ital 122:275
63. Hoff CD (1985) J Organomet Chem 282:201
64. Mukerjee SL, Lang RF, Ju T, Kiss G, Hoff CD, Nolan SP (1992) Inorg Chem 31:4885
65. Muetterties EL, Bleeke JR, Sievert AC (1979) J Organomet Chem 178:197
66. (a) Butenschön H (1999) Synlett 680; (b) Wey H (1990) PhD Thesis, Bochum
67. Kündig EP, Fabritius CH, Grossheimann G, Romanens P (submitted for publication)
68. Pike RD, Sweigart DA (1999) Coord Chem Rev 187:183
69. Auffrant A, Prim D, Rose-Munch F, Rose E, Vaissermann J (2001) Organometallics 20:3214
70. Seo HM, Lee SG, Shin DM, Hong BK, Hwang SG, Chung DS, Chung YK (2002) Organometallics 21:3417
71. (a) Winkhaus G, Pratt L, Wilkinson G (1961) J Chem Soc 3807; (b) Pauson PL, Segal JA (1975) J Chem Soc Dalton Trans 1677; (c) Kane-Maguire LAP, Sweigart DA (1979) Inorg Chem 18:700

72. (a) Rybinskaya MI, Kaganovich VS, Kydinov AR (1984) Izv Akad Nauk SSR Ser A Khim 885; (b) Pearson AJ, Shin DM (1992) Tetrahedron 48:7527
73. (a) Basin KK, Balkeen WG, Pauson PL (1981) J Organomet Chem 204:C25; (b) Pearson AJ, Richards IC (1983) J Organomet Chem 258:C41
74. Jackson JD, Villa SJ, Bacon DS, Pike RD, Carpenter GB (1994) Organometallics 13:3972
75. Chaffee SC, Sutton JC, Babbitt CS, Maeyer JT, Guy KA, Pike RD, Carpenter GB (1998) Organometallics 17:5586
76. (a) Sun SH, Yeung LK, Sweigart DA, Lee TY, Lee SS, Chung YK, Switzer SR, Pike RD (1995) Organometallics 14:2613; (b) Sun SH, Dullaghan CA, Carpenter GB, Sweigart DA, Lee SS, Chung YK (1997) Inorg Chim Acta 262:213
77. Ryan WJ, Peterson PE, Cao Y, Williard PG, Sweigart DA, Baer CD, Thompson CF, Chung YK, Chung TM (1993) Inorg Chim Acta 211:1
78. (a) Abd-El-Aziz AS, Bernardin S (2000) Coord Chem Rev 203:219; (b) Pape AR, Kaliappan KP, Kundig EP (2000) Chem Rev 100:2917; (c) Pike RD, Sweigart DA (1999) Coord Chem Rev 187:183
79. (a) Manzur C, Millan L, Figueroa W, Boys D, Hamon JR, Carrillo D (2003) Organometallics 22:153; (b) Ruhland T, Bang KS, Andersen K (2002) J Org Chem 67:5257; (c) Moulines F, Kalam-Alami M, Martinez V, Astruc D (2002) J Organomet Chem 643:125; (d) Storm JP, Andersson CM (2000) J Org Chem 65:5264
80. (a) Coffield TH, Sandel V, Closson RD (1957) J Am Chem Soc 79:5826; (b) Green MLH, Pratt L, Wilkinson G (1960) J Chem Soc 989
81. (a) Nesmeyanov AN, Volkenau NA, Bolesova IN (1963) Dokl Akad Nauk SSSR 149:615; (b) Nesmeyanov AN, Vol'kenau NA, Bolesova IN (1963) Tetrahedron Lett 1725; (c) Nesmeyanov AN, Volkenau NA, Bolesova IN (1966) Dokl Akad Nauk SSSR 166:607
82. Sutherland RG, Chen SC, Pannekoek WJ, Lee CC (1976) J Organomet Chem 117:61
83. (a) Sutherland RG, Chen SC, Pannekoek J, Lee CC (1975) J Organomet Chem 101:221; (b) Sutherland RG, Pannekoek WJ, Lee CC (1978) Can J Chem 56:1782; (c) Guerchais V, Astruc D (1986) J Organomet Chem 312:97
84. (a) Nesmeyanov AN, Vol'kenau NA, Petrakova VA (1974) Izv Akad Nauk SSSR Ser Khim 9:2159; (b) Nesmeyanov AN, Vol'kenau NA, Bolesova IN, Polkovnikova LS (1975) Koord Khim 1:1252; (c) Roman E, Astruc D (1979) Inorg Chim Acta 37:L465
85. (a) Dabirmanesh Q, Fernando SIS, Roberts RMG (1995) J Chem Soc Perkin Trans 1:743; (b) Dabirmanesh Q, Roberts RMG (1997) J Organomet Chem 542:99
86. Dyson PJ, Grossel MC, Srinivasan N, Vine T, Welton T, Williams DJ, White AJP, Zigras T (1997) J Chem Soc Dalton Trans 3465
87. Doggweiler HO, Desobry V (Ciba Geigy AG) (1988) Eur Pat Appl EP 270490, (1988) Chem Abstr 109:129310z
88. Astruc D (1983) Tetrahedron 39:4027
89. Astruc D, Dabard R (1971) CR Acad Sci Paris Ser C 272:1337
90. Sirotkina EI, Nesmeyanov AN, Vol'kenau NA (1969) Izv Akad Nauk SSSR Ser Khim 1524
91. (a) Khand IU, Pauson PL, Watts WE (1968) J Chem Soc C 2261; (b) Khand IU, Pauson PL, Watts WE (1969) J Chem Soc C 116
92. Helling JF, Hendrickson WA (1979) J Organomet Chem 168:87
93. (a) Gill TP, Mann KR (1980) Inorg. Chem 19:3007
94. (a) Roloff A, Meier K, Riediker M (1986) Pure Appl Chem 58:1267; (b) Park KM, Schuster GB (1991) J Organomet Chem 402:355
95. (a) Boekelheide V (1983) Top Curr Chem 113:87; (b) Schumann H (1984) Chem Ztg 108:239
96. Meier K (Ciba-Gigy AG) (1987) Eur Pat Appl EP 207889, (1987) Chem Abstr 107:23505k
97. (a) Kündig EP, Jeger P, Bernardinelli G (1995) Angew Chem Int Edn Eng 34:2161; (b) Kündig EP, Jeger P, Bernardinelli G (2004) Inorg Chim Acta in press
98. (a) Pearson AJ, Park JG, Zhu PY (1992) J Org Chem 57:3583; (b) Storm JP, Ionescu RD, Martinsson D, Andersson CM (2000) Synlett 975

99. Storm JP, Andersson CM (2000) J Org Chem 65:5264
100. Abd-El-Aziz AS, Lee CC, Piorko A, Sutherland RG (1988) J Organomet Chem 348:95
101. Sutherland RG, Abdelaziz AS, Piorko A, Baranski AS, Lee CC (1989) Synth Commun 19:189
102. (a) Nesmeyanov AN, Volkenau NA, Bolesova IN, Shulpina LS (1979) J Organomet Chem 182:C36; (b) Roman E, Astruc D (1979) Inorg Chim Acta 37:L465
103. (a) Zelonka RA, Baird MC (1972) J Organomet Chem 44:383; (b) Zelonka RA, Baird MC (1972) Can J Chem 50:3063; (c) Bennett MA, Smith AK (1974) J Chem Soc Dalton Trans 233; (d) Oshima N, Suzuki H, Morooka Y (1986) Inorg Chem 3407
104. Trost BM, Older CM (2002) Organometallics 21:2544
105. Schmid A, Piotrowski H, Lindel T (2003) Eur J Inorg Chem 2255
106. Gill TP, Mann KR (1982) Organometallics 1:485
107. Bennett MA, Huang TN, Matheson TW, Smith AK (1982) Inorg Synth 21:74
108. Becker E, Sligovc C, Rueba E, Standfest-Hauser C, Mereiter K (2002) J Organomet Chem 649:55
109. Glueck DS, Brough AR, Mountford P, Green MLH (1993) Inorg Chem 32:1893
110. Cambie RC, Coulson SA, Mackay LG, Janssen SJ, Rutledge PS, Woodgate PD (1991) J Organomet Chem 409:385
111. Koefod RS, Mann KR (1991) Inorg Chem 30:541
112. Pearson AJ, Park JG, Yang SH, Chuang YH (1989) J Chem Soc Chem Commun 1363
113. McNair AM, Mann KR (1986) Inorg Chem 25:2519
114. Kamikawa K, Norimura K, Furusyo M, Uno T, Sato Y, Konoo A, Bringmann G, Uemura M (2003) Organometallics 22:1038
115. Cambie RC, Clark GR, Coombe SL, Coulson SA, Rutledge PS, Woodgate PD (1996) J Organomet Chem 507:1
116. Pearson AJ, Lee K (1994) J Org Chem 59:2304
117. Nesmeyanov AN, Vol'kenau NA, Shul'pina LS, Bolesova IN (1981) Dokl Akad Nauk SSSR 258:120

Topics Organomet Chem (2004) 7: 21–42
DOI 10.1007/b94490

(Arene)$Cr(CO)_3$ Complexes: Arene Lithiation/Reaction with Electrophiles

Martin F. Semmelhack · Anatoly Chlenov
Princeton University, Princeton NJ 08540, USA
E-mail: mfshack@princeton.edu

Abstract Coordination of an arene with a transition metal changes the reactivity of the arene pi system and the substituents. The ring hydrogens are made more acidic compared to the free arene, and can be removed with strong bases such as LDA and *n*-BuLi to give (ArLi)ML_n complexes. This effect has been demonstrated with several arene-metal systems but investigated systematically only with the arene-$Cr(CO)_3$ complexes and these are the focus of this review. Other methods for producing (ArLi)$Cr(CO)_3$ complexes such as halogen-metal and metal-metal exchange are also known. In general, the metal-complexed aryl lithium derivatives are formed more readily and react like a stabilized version of the free aryl lithiums. Many examples of the effect of substituents on the regioselectivity of proton abstraction have been reported, including alkyl, halo, alkoxy, and alkylamino, and benzylic heteroatoms. The (ArLi)$Cr(CO)_3$ complexes react as strong nucleophiles with a variety of electrophiles to provide substituted (arene)$Cr(CO)_3$ species. Even (*o*-chlorophenyl lithium)$Cr(CO)_3$ gives substantial yields of addition instead of benzyne formation, the primary pathway in the absence of the chromium stabilization.

Keywords Lithiation · Arene-chromium complexes · Aryl-lithium · Regioselectivity in arene lithiation

1 Introduction and Scope

The coordination of a transition metal to an arene ligand in a pi complex modifies the reactivity of the arene ligand in several distinct ways. One strong effect is the acidification of the ring protons, which allows direct proton abstraction from the arene ligand or, less commonly, halogen-lithium exchange to give an aryl lithium derivative (**1**) coordinated to the metal (Eq. 1). The influence of substituents (R in Eq. 1) on the regioselectivity of proton abstraction is an important consideration. Trapping with electrophiles provides the substituted arene complex (**2**). In cases with substituted arene ligands, the complexes have molecular asymmetry and there can be an issue of stereoselectivity in the creation of a new stereogenic center during lithiation or electrophilic trapping. Most of the general methodology related to this process is based on the $Cr(CO)_3$ system and these complexes will be the focus of this review.

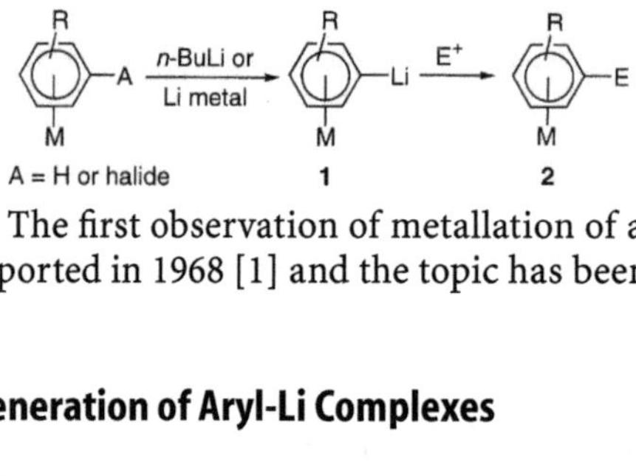

1 2

(1)

The first observation of metallation of arene in (arene)$Cr(CO)_3$ complex was reported in 1968 [1] and the topic has been included in several reviews [2–10].

2 Generation of Aryl-Li Complexes

2.1 Direct Proton Abstraction with Base

There are three electrophilic sites in (benzene)$Cr(CO)_3$: the aromatic ring, the CO ligand, and the ring protons. Most common nucleophiles/bases will add to the ring; a few organo-lithium reagents are known to add to the CO ligand [11, 12]. Selective proton abstraction is highly demanding of the properties of the base, requiring high kinetic basicity and low nucleophilic reactivity. Typical bases include alkyllithiums, among which *n*-BuLi is the most widely used, and hindered amides such as LDA and LiTMP [13]. Using three equivalents of LiTMP and electrophile, a symmetrical trisubstituted arene can be obtained [14].

2.2
Halogen-Metal Exchange of Halobenzene-Cr(CO)$_3$ Complexes

The complexes (PhX)Cr(CO)$_3$ are known [15] for X=F, Cl, Br, and I, but only the fluoro and chloro examples are easily prepared. Reaction with *n*-BuLi might be expected to bring about halogen-Li exchange, but the halide atom (F and Cl) strongly favors *ortho* proton abstraction (see below) instead. In one example, (PhI)Cr(CO)$_3$ was converted to (PhLi)Cr(CO)$_3$ and the latter was methylated in 78% yield overall [16].

2.3
Metal-Metal Exchange

One of the earliest techniques for preparation of (PhLi)Cr(CO)$_3$ is the reaction of diphenylmercury bis-Cr(CO)$_3$ with *n*-BuLi [17]. Quenching with CO_2 gave the benzoic acid complex in 50% yield. Analogously, a lithium derivative can be obtained from trialkyl tin compound (**3**) via transmetallation with *n*-BuLi. The resulting anion was quenched with a variety of electrophiles with retention of stereochemistry [18].

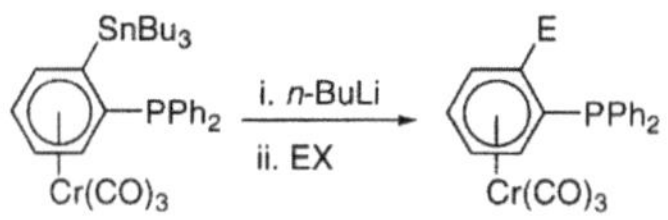

3

EX	yield (%)	EX	yield (%)
Me_3SiCl	80	MeI	67
PhCHO	83	EtI	53
c-$C_6H_{12}CHO$	70	$PhCH_2Br$	31
PhC(O)Cl	70	$BrCF_2CF_2Br$	61

(2)

The aryl anion complex can also be generated in situ from a trimethylsilyl derivative **4** via nucleophile-assisted desilylation. Nucleophiles used for this purpose include KO*t*-Bu, CsF [19–21] and KH [22].

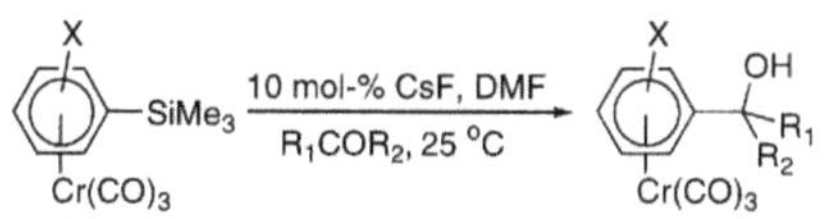

4 (3)

X = o-Me, m-Me, p-Me, o-Cl, m-Cl, p-Cl, H
R_1 = H, Me, Ph; R_2= Ph, nPr, Me

Internal alkoxide **5** generated after MeLi addition to the carbonyl group of **6** can serve the same purpose. The resulting anion **7** can be trapped with a variety of electrophiles [23].

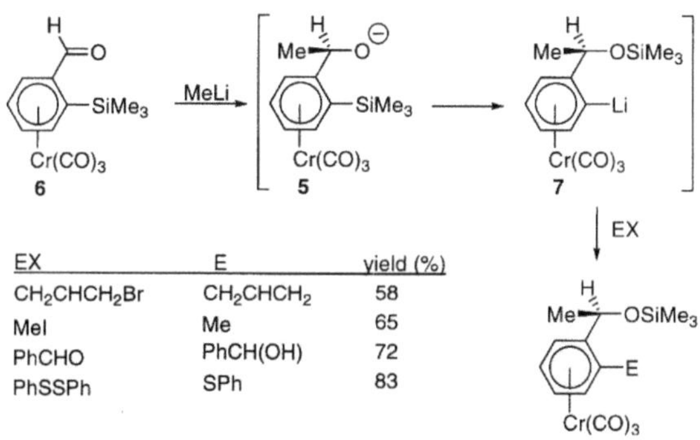

EX	E	yield (%)
CH_2CHCH_2Br	CH_2CHCH_2	58
MeI	Me	65
PhCHO	PhCH(OH)	72
PhSSPh	SPh	83

(4)

However, both the tin and the silicon precursors are generally synthesized from the corresponding lithium compounds and therefore these methods have limited applicability.

3 Trapping with Electrophiles

The main requirement for the electrophile in this reaction is high electrophilicity and low kinetic acidity. Several types have been reported (see Table 1) [16, 17, 24].

Electrophilic halide sources include iodine [25–27], diiodoethane (Eq. 4) [28–29], *N*-bromosuccinimide [30, 31, 25], dibromoethane [32–34], dibromotetrachloroethane [35], dibromotetrafluoroethane [23, 36], and *N*-chlorosuccinimide [30, 31, 25]. Dihaloethane derivatives are typically preferred since any excess iodine and *N*-halosuccinimides oxidize the chromium which leads to a reduction in yield of the substituted arene complex [16, 25, 31].

Table 1 Quenching of $(PhLi)Cr(CO)_3$ with electrophiles

Entry	Electrophile	Product (yield %)	$(PhH)Cr(CO)_3$ Yield %	Reference
1	I_2	E=I (26)	-	16
2	$ClSiMe_3$	$E=SiMe_3$ (94, 50)	0	16, 24
3	MeI	E=Me (71, 78)	-	16
4	$MeOSO_2F$	E=Me (91)	2	24
5	Ph_2PCl	$E=PPh_2$ (54)	22	17
6	CO_2	$E=CO_2H$ (50, 71)	0	16, 24
7	MeCOCl	E=COMe (18)	22 (39%)[a]	16
8	PhCHO	E=CH(OH)Ph (60)	0	24
9	PhCOPh	$E=CPh_2(OH)$ (63)	0	16
10	MeCOMe	$E=CMe_2(OH)$ (29)	60	24
11	CD_3COCD_3	$E=C(CD_3)(OH)$ (68)	(8)[b]	24
12	EtCHO	E=CH(OH)Et (76)	17	24
13	*n*-PrCHO	E=CH(OH)*n*-Pr (65)	0	24
14	*i*-PrCHO	E=CH(OH)*i*-Pr (90)	0	24
15	MeSSMe	E=SMe	45	52
16	EtSSEt	E=Set	80	52
17	$(MeOCH_2CH_2S)_2$	$E=SCH_2CH_2OMe$	70	52
18	$PhCH_2SSCH_2Ph$	$E=SCH_2Ph$	26	52
19	$(4\text{-}MeC_6H_4S)_2$	$E=SC_6H_4\text{-}4\text{-}Me$	39	52
20	PhSSPh	E=SPh	54	52

[a] Yield of bis(η^6-tricarbonylchromiumphenyl)methylcarbinol
[b] Yield of $(PhD)Cr(CO)_3$

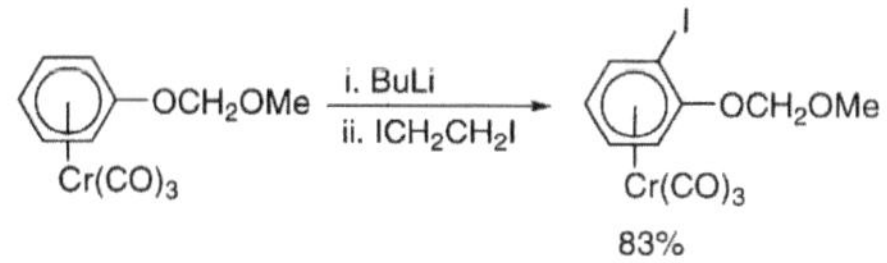

(5)

Among other heteroatom nucleophiles, trimethylsilyl chloride is widely used [36–41] and tri(isopropyl)silyl chloride was also reported [42–44]. Tin derivatives include trimethyl- [33, 45–48] and tributyltin chloride [34, 49–51]. Several sulfur transfer reagents have been used (Table 1) [52], including a wide range of disulfides [23, 29, 34, 48, 53–55], succinimide derivatives [52], phenylsulfinyl chloride, [45, 46, 56–58], and elemental sulfur [52, 59].

Several sulfenyl group sources are also known [60, 61]. Diphenyl selenide was reported [52]. Diphenylphosphorus chloride was used extensively in the synthesis of new phosphines [38, 47, 51, 62]. A chiral version is shown in Eq. (6) [51].

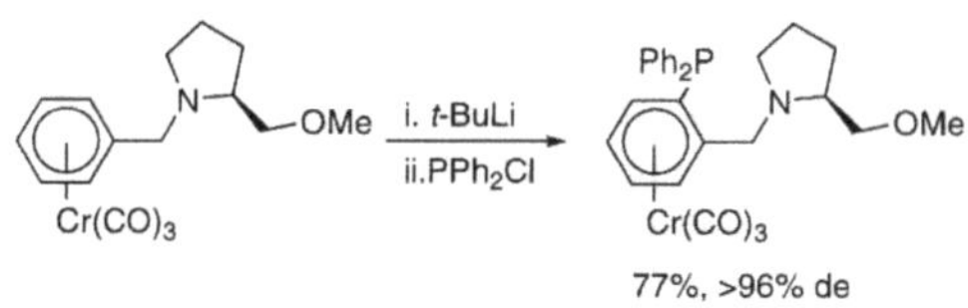

(6)

A molybdenum peroxide derivative was used as a source of "OH$^+$" [63]. Deuteration of (arene-Li)Cr(CO)$_3$ is typically achieved using D_2O [54, 58, 64, 65], CF_3CO_2D [66, 67] or MeOD [60, 68].

Carbonyl electrophiles can suffer competing enolization (Table 1). With the methoxymethyl directing group, reactive carbonyl compounds including aldehydes [24, 28, 49, 56, 69–72], acid chlorides [17, 24, 34, 44, 56, 57, 73–76], anhydrides [77, 78], esters [22, 24, 79–81], amides [22, 28, 82–84] and several ketones [19, 22, 28, 70, 85, 86] were reported (Eq. 7). Others include phenyl isocyanate [54, 78, 87], benzonitrile [55], and imine derivatives [28].

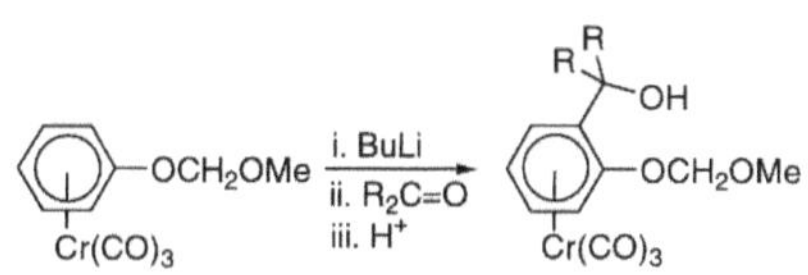

(7)

R_2CO	yield (%)	R_2CO	yield (%)
DMF	83	PhCHO	73
$(CH_2O)_n$	53	4-MeOC$_6$H$_4$CHO	83
Ph_2CO	94	(2,6-MeO)$_2$C$_6$H$_3$CHO	80
(4-MeOC$_6$H$_4$)$_2$CO	92	4-O$_2$NC$_6$H$_4$CHO	69

Among methylating agents, methyl iodide is the most popular [23, 88–90], although dimethyl sulfate [91] and methyl fluorosulfonate [68, 70, 85] have also been reported. Other alkylating agents include ethyl iodide [18, 88] (propyl iodide was unsuccessful [92]), allyl bromide [23, 88], and benzyl bromide [18, 23, 47, 88, 93]. Benzyl chloride was shown to be insufficiently reactive [54].

Arene metal complexes of Cr and Mn can also serve as electrophiles (Eq. 8) [42, 49, 66, 67, 94].

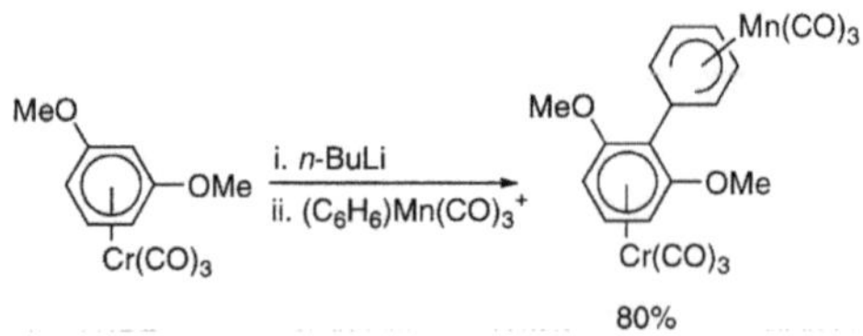

(8)

4
Ortho Lithiation with Intramolecular Substitution

When an electrophile possesses a nucleophilic site, ring formation becomes possible, employing the activation of an arene ligand toward *o*-lithiation and activation toward nucleophile addition. Both begin in a manner closely related to the applications immediately above: lithiation is followed by addition of a carbon electrophile, in this case one which bears an anion-stabilizing group in the side chain. In a second step, anion generation can initiate ring formation via addition/oxidation, substitution for H [70] or via S_NAr with a leaving group. The latter pathway is known only with X=F, and with heteroatom nucleophiles. Starting from (anisole)$Cr(CO)_3$, lithiation followed by trapping with 5-oxo-hexanenitrile and methylation of the alkoxide generates an anisole derivative with a side chain ready for activation as a nucleophile (in **8**); the cyclized product **9** was obtained in 50% yield as a mixture of diastereoisomers [70].

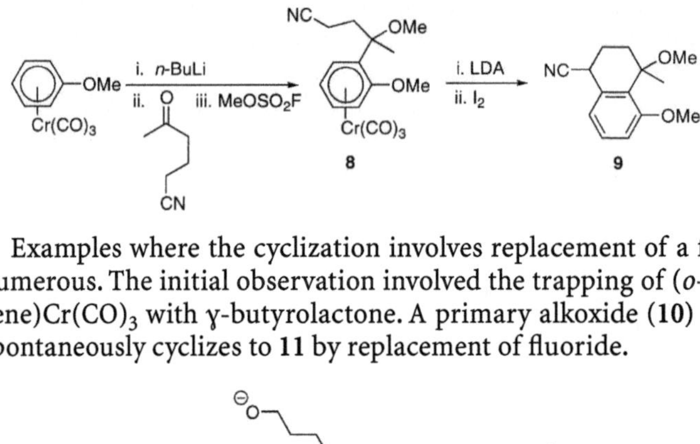

(9)

Examples where the cyclization involves replacement of a fluoride are more numerous. The initial observation involved the trapping of (*o*-lithiofluoro-benzene)$Cr(CO)_3$ with γ-butyrolactone. A primary alkoxide (**10**) is released which spontaneously cyclizes to **11** by replacement of fluoride.

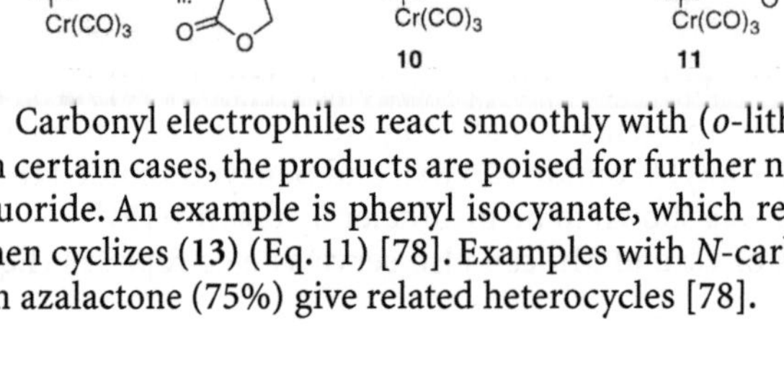

(10)

Carbonyl electrophiles react smoothly with (*o*-lithiofluorobenzene)$Cr(CO)_3$. In certain cases, the products are poised for further nucleophilic substitution for fluoride. An example is phenyl isocyanate, which reacts twice (to give **12**) and then cyclizes (**13**) (Eq. 11) [78]. Examples with *N*-carboxyanhydrides (59%) and an azalactone (75%) give related heterocycles [78].

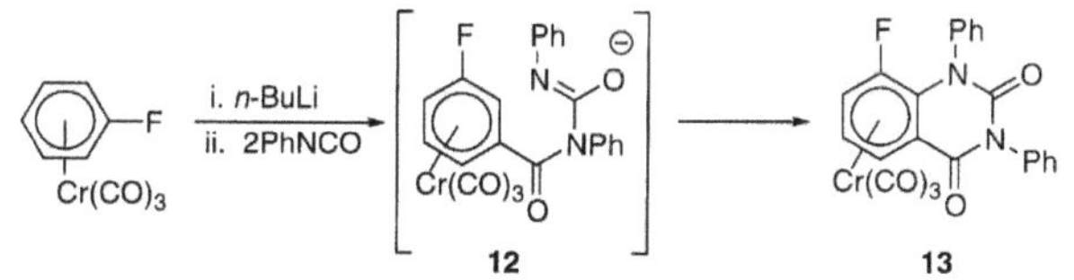

(11)

5
Regioselectivity in the Deprotonation of (Arene)Cr(CO)$_3$ Complexes

5.1
Alkyl Substituents

Deprotonation of (alkylbenzene)Cr(CO)$_3$ with *n*-BuLi can occur at three different ring H positions as well as at the benzylic position. In free toluene, the kinetic acidity of the methyl protons is higher than the ring protons by a factor of 150 [95]. Under equilibration conditions, the benzylic position of (PhMe)Cr(CO)$_3$ is strongly favored. However, kinetic deprotonation with *n*-BuLi at low temperature followed by quenching with CO_2 produces a mixture of toluic acids, along with phenyl acetic acid (Eq. 12). The apparent retardation in kinetic acidity for the benzylic position is attributed to a steric effect of the Cr(CO)$_3$ group, blocking the Me substituent [96].

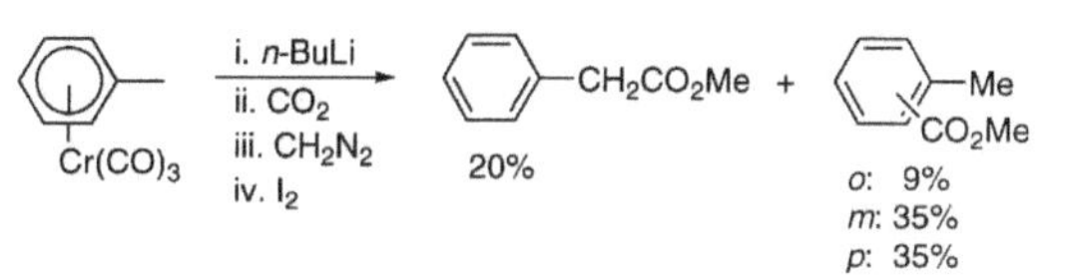

(12)

In a systematic study of lithiation of alkylarene ligands [96], steric inhibition by *ortho* substitution is clear. The unhindered positions tend to be comparably reactive. With mesitylene, only benzylic (α) lithiation/alkylation is observed. The cyclobutabenzene complex **14**, on the other hand, is lithiated on the arene ring; quenching with Me_3SiCl results in primarily the *ortho* (**15**) product [97–99].

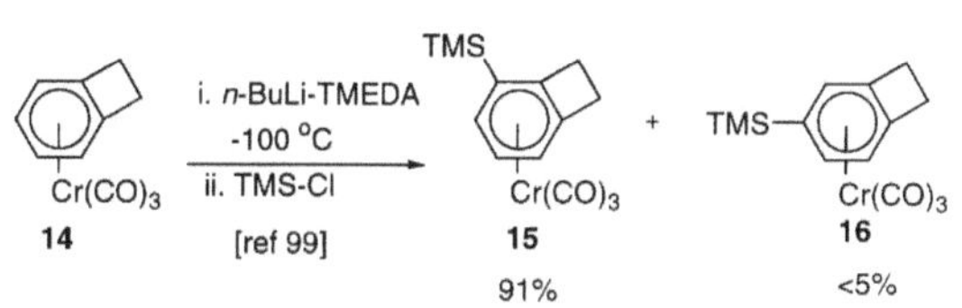

(13)

5.2
Halo Substituents

Treatment of uncomplexed haloaromatics using strong base produces selective *ortho* metallation [100], and opens a pathway to *o*-benzyne intermediates. The overall conversion to an *o*-lithiohaloarene and quenching with an electrophile is generally successful only with fluoroarenes [101]. For chloride and the heavier halides, halogen-metal exchange competes with proton abstraction, and benzyne formation for the *o*-lithiohaloarene is too fast to allow efficient electrophile trapping [102].

The effect of coordination of the haloarene to the $Cr(CO)_3$ group is to stabilize the *ortho*-lithiated adduct. While *o*-lithiochlorobenzene (uncomplexed) has a half-life time of minutes at -78 °C [102], the concentration of (*o*-lithiochlorobenzene)-$Cr(CO)_3$ shows a unimolecular decay with a $t_{1/2}$ of 136 min at 0 °C [103]. While the kinetics are consistent with formation of a coordinated benzyne, trapping experiments failed to show more than traces of an appropriate adduct [103]. If the solution is kept at low temperature prior to quenching, high yields of adducts with electrophiles are obtained (Table 2). The reaction with CO_2 (98% yield) shows that (*o*-lithiochlorobenzene)$Cr(CO)_3$ was produced efficiently. Reaction with acetone suffers from significant protonation of the lithioarene, presumably due to enolization of the acetone. In these examples, there is no evidence for deprotonation at a site other than *ortho*.

The initial proton abstraction from $(PhF)Cr(CO)_3$ is complete within 0.5 h at -78 °C to give a clear yellow solution; the mixture turns red with obvious decomposition when heated above -20 °C [70, 96]. The results of quenching with representative simple electrophiles are summarized in Table 2. Evidence for double deprotonation appears in entry 2. The enolization problem in reaction with acetone is less significant (entry 5).

The two *ortho* protons in a monosubstituted (arene)$Cr(CO)_3$ complex are enantiotropic and can be distinguished with a chiral base. (Halobenzene)$Cr(CO)_3$ was deprotonated with (**17**) and quenched in situ with Me_3SiCl to afford enantiomerically enriched *o*-substituted products [49, 104]. Small ees for X=F were attributed to the small difference in size between F and H [49, 104].

Table 2 *o*-Deprotonation/electrophile quenching of halobenzene-$Cr(CO)_3$

Entry	X	Electrophile	Product (yield %)	$(PhF)Cr(CO)_3$ Yield %	Reference
1	F	PhCHO	E=CH(OH)Ph (57)	0	24
2	F	$ClSiMe_3$	E=$SiMe_3$ (46)	0 (10–20%)[a]	24
3	F	MeI	E=Me (71)	0	96
4	F	$MeOSO_2F$	E=Me (68)	30	24
5	F	MeCOMe	E=CMe_2(OH) (85)	15	24
6	F	CO_2	E=CO_2H (99)	0	24
7	Cl	PhCHO	E=CH(OH)Ph (71)	0	96
8	Cl	$ClSiMe_3$	E=$SiMe_3$ (49)	0	24
9	Cl	$MeOSO_2F$	E=Me (81)	6	24
10	Cl	MeCOMe	E=CMe_2(OH) (67)	28	24
11	Cl	CO_2	E=CO_2H (98)	0	24

[a] Yield of the 2,6-bis(trimethylsilyl) derivative

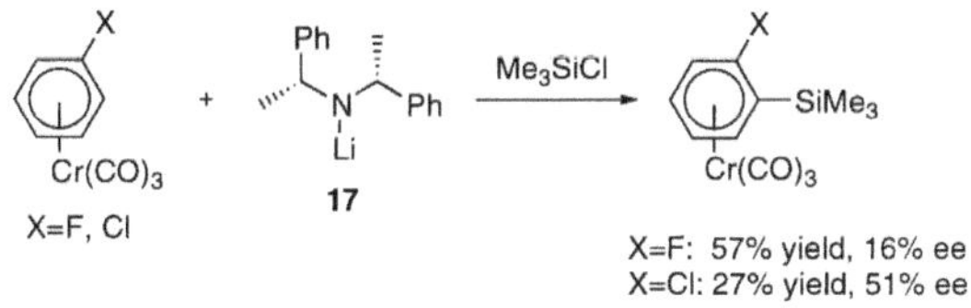

(14)

5.3 Lithiation of Anisole and Aniline Ligands

The usual [100] *o*-directing effect of a methoxy group is at least as powerful in the $Cr(CO)_3$ complexes of alkoxyarenes. (Anisole)$Cr(CO)_3$ shows a pKa of 33, compared to 39 for anisole itself [13]. Proton abstraction is fast with either *n*-BuLi or $LiNR_2$ bases. Deprotonation of (anisole)$Cr(CO)_3$ and trapping with simple electrophiles (PhCHO, Me_3SiCl, CO_2, MeI, $MeOSO_2F$, Me_2CO, nPrCHO) gives substitution only at the *ortho* positions, and 2,6-disubstitution was significant in several of the cases. Even with acetone, enolization is not significant (<10% recovered anisole complex).

(Anisole)$Cr(CO)_3$ can be deprotonated stereoselectively with a chiral base (**17**). In situ quench with Me_3SiCl affords (**18**, R=Me) in 83% yield and 84% ee [49, 104, 105]. Increasing the size of the alkoxy group leads to a slight improvement in the stereoselectivity at the expense of yield [49].

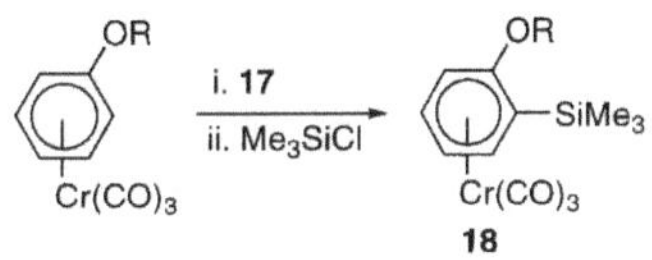

R	yield (%)	ee (%)
Me	83	84
Et	82	81
i-Pr	65	90
t-Bu	0	-

(15)

The procedure is complicated by a relatively fast proton exchange between the starting material and the lithiated intermediate [105]. This problem can be partially overcome via addition of LiCl which accelerates the lithiation or by quenching with an electrophile in situ (typically limited to Me_3SiCl) [49, 105].

The presence of the additional ether functionality on the arene (in **19**) does not improve the yield of substitution [28, 49]. Results of a systematic study are shown in Eq. (16) [28].

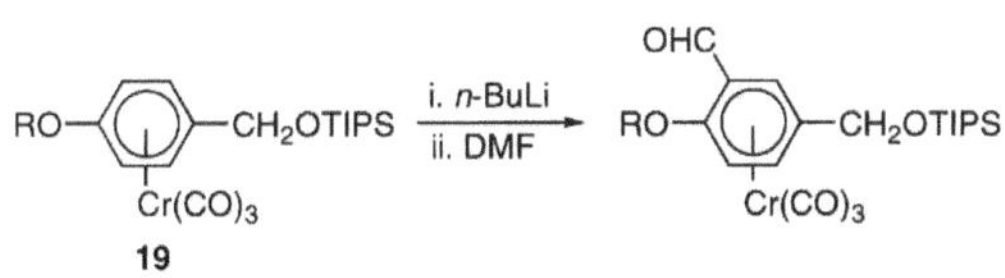

R	yield (%)
Me	87
$MeOCH_2$	61
$EtOCH_2$	62
$MeOCH_2CH_2$	74
$MeOCH_2CH_2OCH_2$	73

(16)

Chiral information present in the ether **20** allows formation of a single diastereomer **21** using an achiral base [75].

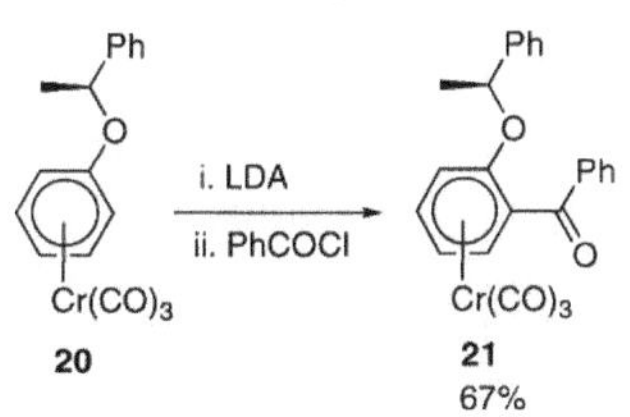

(17)

When one *ortho* position is blocked, lithiation occurs at the other *ortho* position [49, 106]. When both *ortho* positions are blocked by methyl groups, lithiation occurs predominantly *para* to the MeO group [92]. The *ortho* directing ability of the MeO group was used in the modification of the steroid **22** [63] and a podocarpic acid derivative [107].

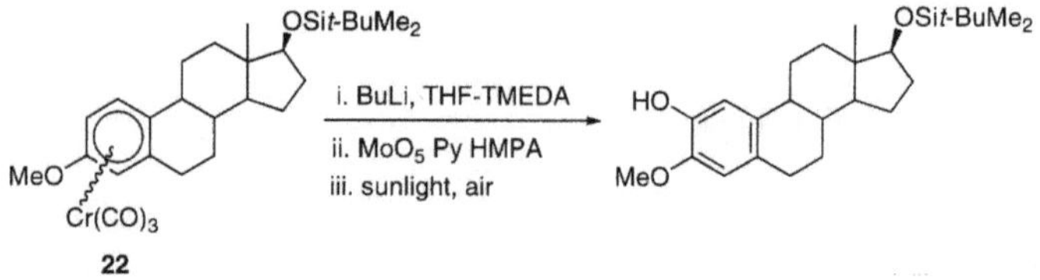

(18)

When two methoxy groups are present *ortho* to each other, the 3-position is lithiated [39, 42, 49, 66, 67, 47, 108–113]. Both 1,4-benzodioxole [76, 82] and 1,4-benzodioxin **23** [107] gave analogous selectivities.

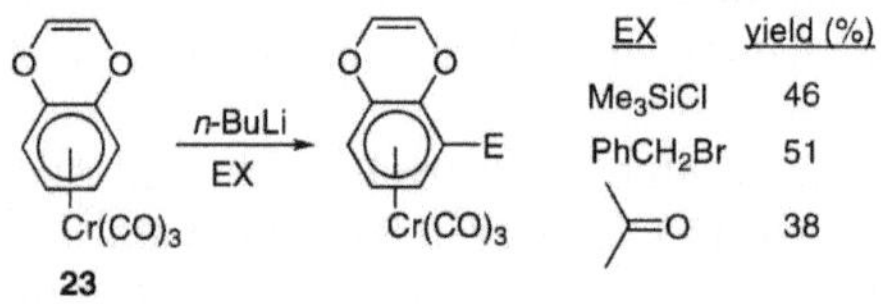

(19)

When two methoxy groups are *meta* to each other, lithiation first occurs between them at the 2-position [114–116]. When the 2-position of the *m*-dimethoxybenzene complex is blocked (as in **24**), lithiation occurs at the 5-position (in **25**) when lithium amide base is used, and at the 4-position (in **26**) when *n*-BuLi is used [25, 31, 114]. The 5-lithio- species was shown to be a kinetic product that isomerizes to a 4-lithio- intermediate at 0 °C [114] (Scheme 1).

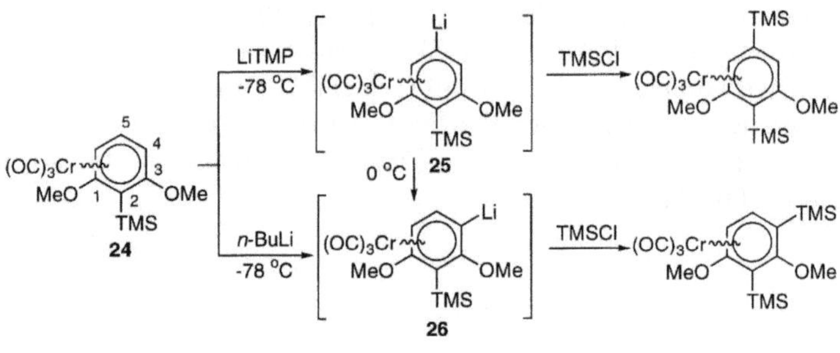

Scheme 1 Lithiation of (1,3-dimethoxy-2-trimethylsilylbenzene)Cr(CO)$_3$

The directing effect of the alkoxy group can be altered by use of a sufficiently large protecting group. For example, the complex from phenol tri(isopropyl)silyl ether was lithiated with *t*-BuLi and then methylated to give predominantly the *meta* substitution product [29, 117]. When the TBDMS or TBDPS protecting groups were used instead, the results were inferior due to a competing *o*-lithiation followed by silyl migration (see below) [55]. Additional examples are given in Fig. 1, showing selectivity for an unhindered position [117].

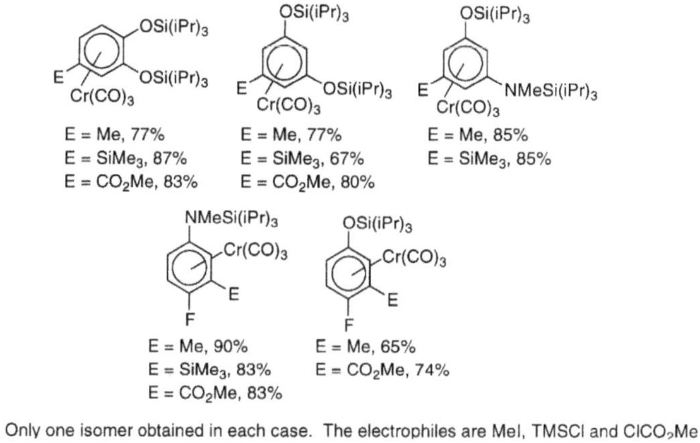

Fig. 1 Selective lithiation and quenching of heteroatom-substituted (arene)Cr(CO)$_3$

A surprising contrast is lithiation/quenching with the Cr(CO)$_3$ complex of 3-(tri(isopropyl)siloxy)anisole (**27**); the final product (**28**) has a silane unit at C-2 and the pathway is shown to be initial lithiation at C-2 in **29** (instead of the expected lithiation at C-6) followed by silyl migration and then allylation of the phenoxide anion **30** [118].

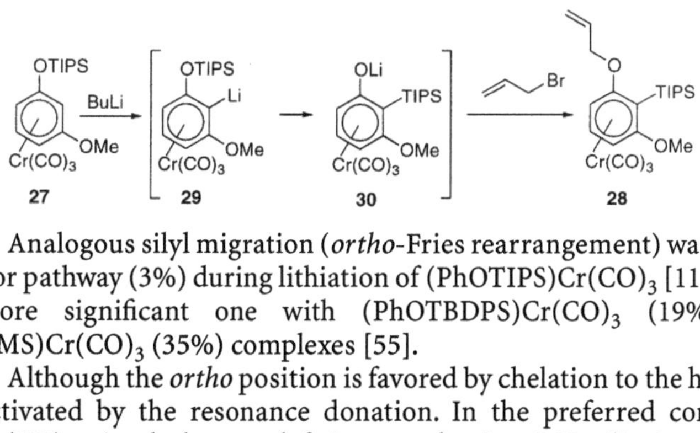

27 29 30 28 (20)

Analogous silyl migration (*ortho*-Fries rearrangement) was observed as a minor pathway (3%) during lithiation of (PhOTIPS)Cr(CO)$_3$ [117] and a somewhat more significant one with (PhOTBDPS)Cr(CO)$_3$ (19%) and (PhOTBDMS)Cr(CO)$_3$ (35%) complexes [55].

Although the *ortho* position is favored by chelation to the heteroatom, it is deactivated by the resonance donation. In the preferred conformation of the Cr(CO)$_3$ tripod, electron deficiency at the ring C-H eclipsing a CO ligand results in *meta* activation (Fig. 2) [55, 86, 117].

This effect is also apparent in lithiation of aniline complexes. While the free ligand *N,N*-dimethylaniline undergoes deprotonation selectively at the *ortho*

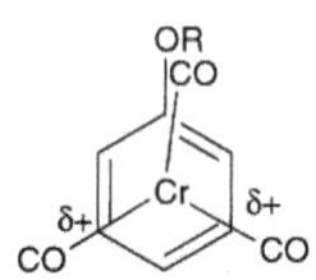

Fig. 2 Preferred eclipsed conformation for (anisole)Cr(CO)$_3$ and resulting electron deficiency

position, presumably largely due to specific coordination effects [100], the corresponding $Cr(CO)_3$ complex gives a mixture with *meta* substitution as the major product (*o*:*m*:*p* 31:52:17) [96]. In a related system, (**31**), a dimethylamino group directs lithiation *meta* to itself [82].

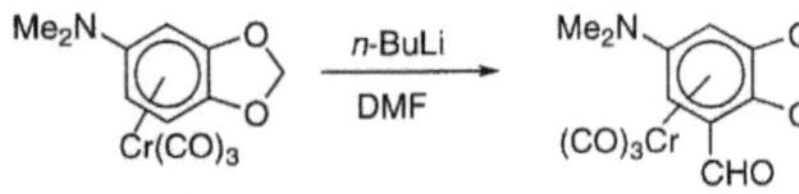

31

(21)

Larger alkyl substituents on the nitrogen favor meta lithiation [55]. The extreme case of a *t*-$BuMe_2Si$ or TIPS group further disfavor *ortho*-metallation, and the process becomes very selective for *meta* [55, 117].

Positioning a carbonyl substituent on the nitrogen results in a change of selectivity in favor of *o*-metallation [49, 77]. With a Boc substituent (in **32**), *ortho*-substitution is the sole product and use of a chiral amine allows modest chiral induction as shown in [77].

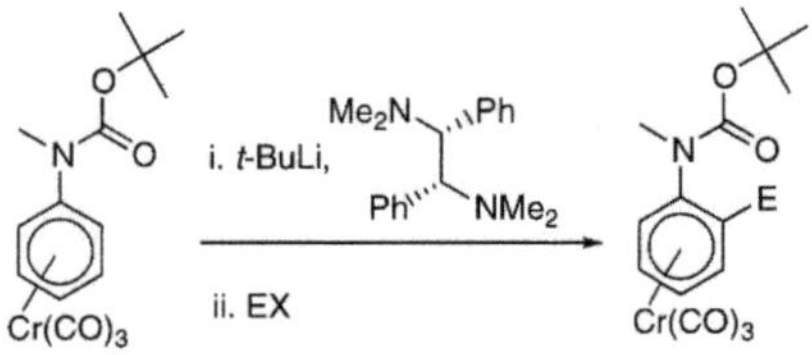

EX	yield (%)	ee (%)
$ClCO_2Me$	59	59
DMF	14	61
Ac_2O	48	65
Me_3SiCl	50	57
Ph_2PCl	96	54

32

(22)

A more complex system is the indoline complex **33** [71, 119]. The selectivity again favors the position *meta* to the nitrogen substituent; the two positions (5,7) *meta* to N account for 90% of the product mixture.

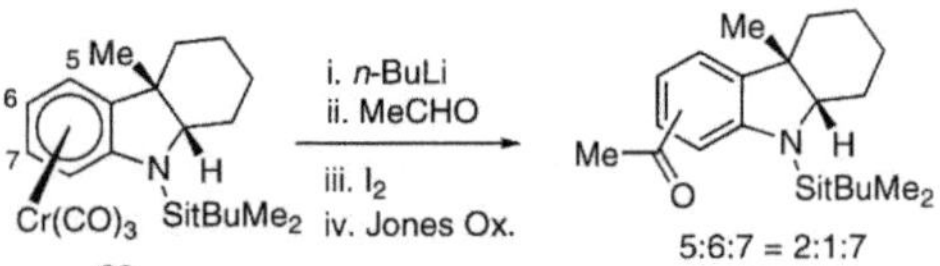

33

5:6:7 = 2:1:7
59%

(23)

In the absence of a substituent at the 3-position of indoline (in **34**), lithiation can be directed to the 4-position using a bulky TIPS protecting group on the nitrogen [44].

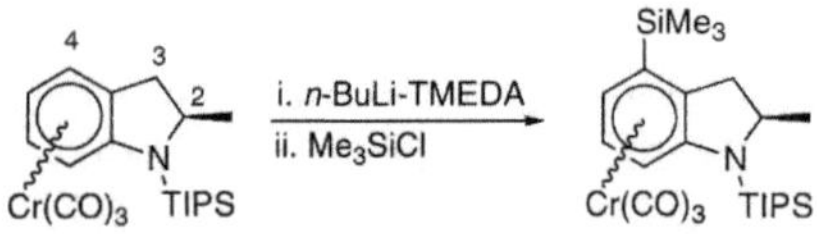

34

45%

(24)

5.4
o-Lithiation Directed by Benzylic Nitrogen and Oxygen Substituents

Lithiation of free benzylic alcohols and amines leads to efficient and selective *o*-lithiation [100]. Lithiation of (benzyl alcohol)$Cr(CO)_3$ with *n*-BuLi leads to a directed nucleophilic addition to the ring, rather than *o*-lithiation [120]. When the benzylic oxygen is part of an ether or an acetal, *ortho* lithiation competes with benzylic deprotonation [49, 58, 74, 104, 121, 122]. The Cr(CO)$_3$ complex of *N,N*-dimethyl-1-phenylethylamine (**35**) is deprotonated smoothly by *t*-BuLi and the *o*-lithio derivative can be coupled with various electrophiles (Scheme 2) [29, 53, 58, 59, 68, 86, 123, 124]. Protons H_a and H_b in the starting complexes are diastereotopic. Coordination of the R-Li base with the benzylic heteroatom is assumed to form a five-membered ring in the plane of the arene (Scheme 2). Then the benzylic Me in **36** can be protruding into the space occupied by the Cr(CO)$_3$ (disfavored diastereoisomer) or positioned on the opposite face of the arene (favored diastereoisomer, **37**). The products are formed with high diastereoselectivity and in good yield [124].

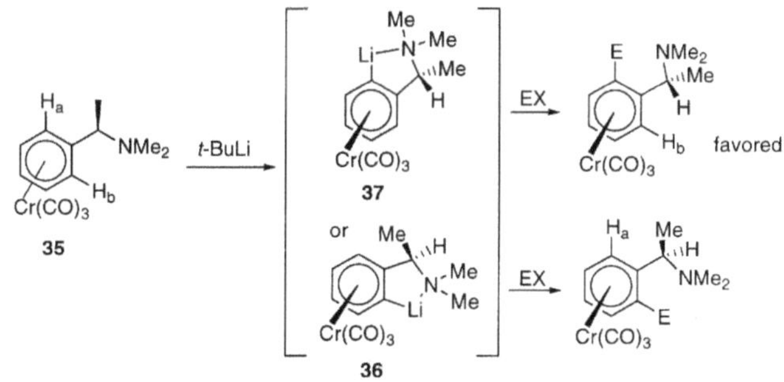

Scheme 2 Diastereoselective lithiation of (*N,N*-dimethyl-1-phenylamine)Cr(CO)$_3$

This methodology was utilized in the synthesis of the chiral sulfur nucleophile **38** used for asymmetric ring opening of *meso*-epoxides [59].

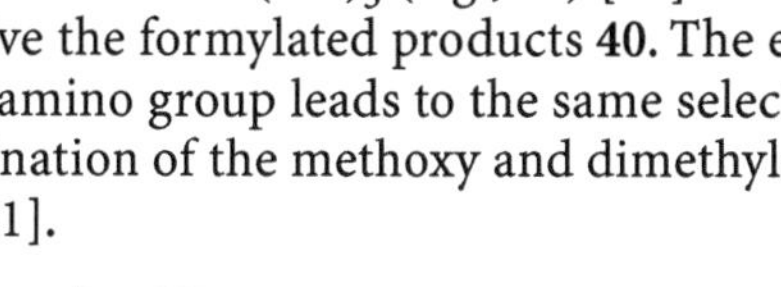 (25)

In the benzyl ether series, enantiopure methylated amino alcohols were activated with Cr(CO)$_3$ (e.g., **39**) [81]. The diastereoselectivity can be very high to give the formylated products **40**. The epimer at the carbon bearing the dimethylamino group leads to the same selectivity. A rationale based on internal coordination of the methoxy and dimethylamino groups accounts for the selectivity [81].

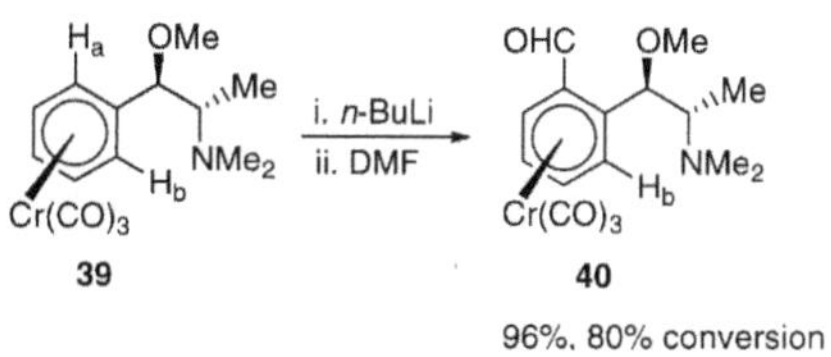 (26)

The second coordinating atom can also be an oxygen instead of nitrogen, as in **41** [38].

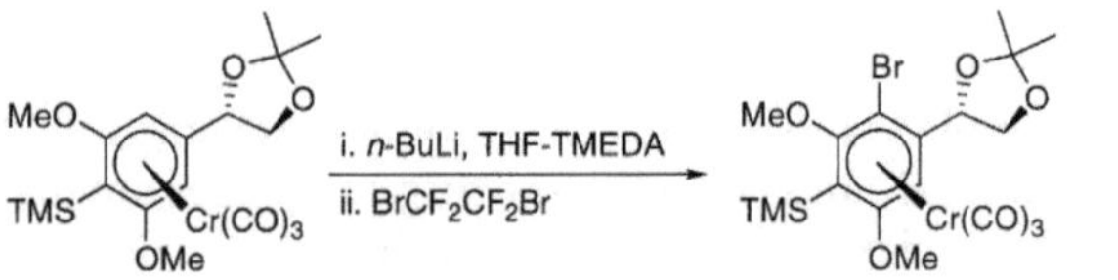

(27)

41 98%, single diastereomer

Diastereoselective deprotonation has been observed with enantiopure complexes (e.g., **42**) prepared by acetalization of benzaldehyde with asymmetric 1,2-diols (Fig. 3) [32, 34, 36, 50, 85, 116, 122, 125]. A series of experiments with *S*-butane-1,2,4-triol as a source of chiral acetal auxiliary failed to demonstrate synthetic utility [122].

42

yields 62-77%
ee's (after hydrolysis) 86- >94%

R	X	yield (%)	de (%)
Me	OMe	74-82	28-58
Me	NMe_2	61-85	78-96
H	NMe_2	64	12

Fig. 3 Chiral acetal auxiliaries employed for diastereoselective lithiation

The main limitation of the acetal methodology is competitive benzylic deprotonation and low efficiency of the acetal removal. Blocking the benzylic position with a methyl group was reported [85]. Aminals can be removed easily but the lithiation was shown to generate significant amounts of *meta* and *para* substitution [32, 47]. An oxazolidine auxiliary (in **43**), on the other hand, gave quantitative conversions to *ortho*-substituted products diastereospecifically [47]. Removal of this auxiliary was not reported, however. Oxazoline auxiliary was also used [126].

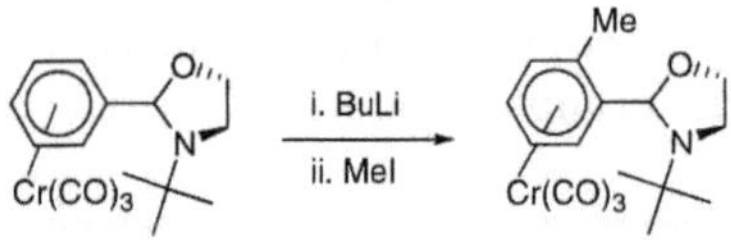

(28)

43 98%, 100% de

Another approach utilized temporary formation of a chiral α-aminoalkoxide derivative (e.g., **44**) of the benzaldehyde complex [127, 49]. Several chiral amines were investigated and although the yields were good, stereoselectivity did not exceed 72% ee [127].

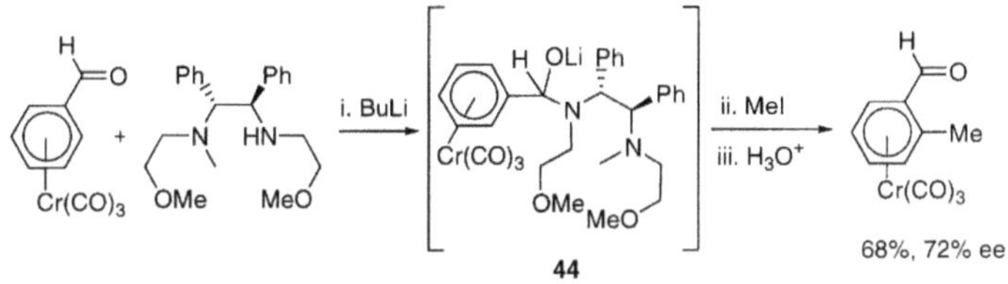

(29)

An achiral acetal can be *ortho*-lithiated enantioselectively using a chiral base. A dialkylamide base gives large amounts of benzylic deprotonation [37], whereas a hydrocarbon base derived from menthol (**45**) selectively deprotonated the benzene ring with good stereoselectivity [90].

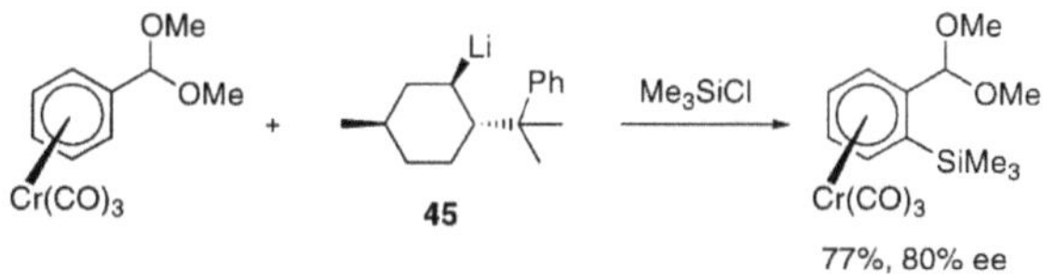

(30)

An interesting competition is presented by the complex **46** from 3-methoxybenzyl alcohol. Lithiation is the predominant reaction with *n*-BuLi, and the major product (**47**) is due to proton abstraction adjacent to the methoxy group and away from the benzyl alcohol unit [79]. In the absence of Cr(CO)$_3$ coordination, lithiation occurs exclusively at the 2-position [79].

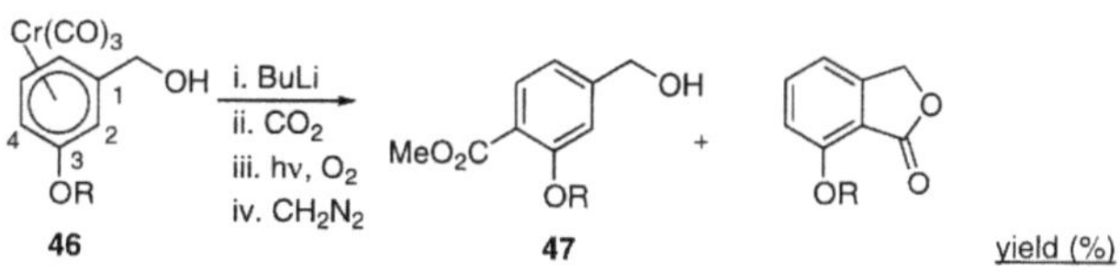

(31)

47				yield (%)	
77	:	23	*n*-BuLi	71	R=Me
85	:	15	*s*-BuLi	55	R=Me
95	:	5	*t*-BuLi	48	R=Me
98	:	2	*n*-BuLi	45	R=MeOCH$_2$

The same effect is active with a 7-methoxytetralin (**48**), leading to one product isomer (**49**) in generally high yields with six representative electrophiles and several substituents on the saturated ring [79, 80]. The selectivity is the same with either epimer of the benzylic hydroxyl group.

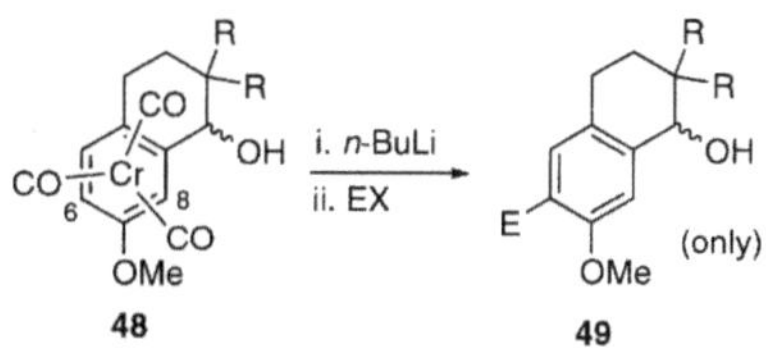

(32)

R=H, Me

EX= CO_2, Me_3SiCl, *p*-$MeOC_6H_4CHO$, *o*-$C_6H_4(CO_2Me)_2$, $MeCH{=}C(Me)C(O)Cl$, DMF

In this case, an X-ray structure of a representative substrate verified that the Cr(CO)$_3$ tripod is situated with one CO group eclipsed with C-6 and one CO eclipsed with C-8 in **48**. Therefore, both C-6 and C-8 should be activated according to the conformational argument. The lack of reactivity at C-8 (or C-2 with the 3-methoxybenzyl alcohol case) has not been satisfactorily explained. A suggestion [79] of a repulsive interaction between the benzyl alkoxide and the incoming base which might disfavor deprotonation at C-8 (or C-2) is not compelling. That repulsion should also operate during the corresponding reaction with the uncomplexed benzyl alcohols, but it is not apparent. However, the repulsive interaction for an alkoxide side chain *is* consistent with the observation that the corresponding benzyl methyl *ethers* undergo lithiation and electrophile trapping at the (expected) site between the functionality [79]. Surprisingly, the benzyl methyl ethers of the 7-methoxytetraline system (**50**) favor the same lithiation selectivity (**51**) seen with the free benzyl alcohol [79].

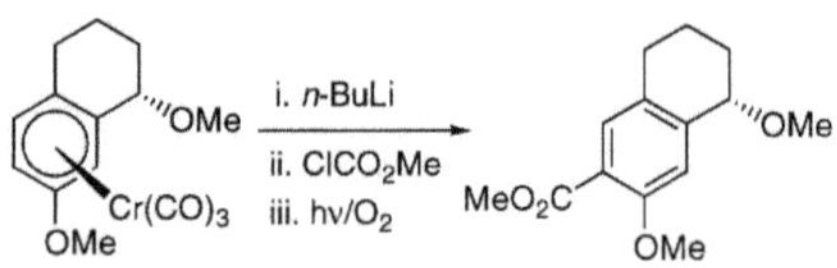

50 **51** (33)

5.5
Correlation of Heteroatom Directing Effects

A very useful study demonstrated that the substituent effects on regioselectivity in deprotonation of free arenes is distinctly different from that for $Cr(CO)_3$-complexed arenes [4, 128]. A series of *p*-disubstituted arenes were complexed and subjected to deprotonation conditions. From the ratio of the two possible products in each case, a reactivity series was generated. For example, the complex of *p*-methoxy-*N,N*-dimethylaniline (**52**) provides a comparison of the directing effects of OMe vs CH_2NMe_2. In this case, the effects are almost exactly balanced. Figure 4 displays two cases, complexed and uncomplexed, with the yields shown for lithiation/electrophile trapping [32, 50, 85, 125].

From these and related data, the following sequences of *ortho* lithiation reactivity can be defined:

Free arenes

-$CONR_2$ > -SO_2NR_2 > -NHCOR > -CH_2NR_2 > -OMe > -NMe_2 = F

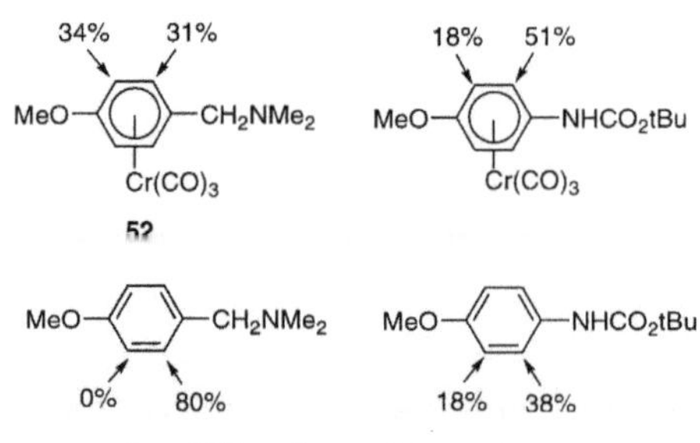

Fig. 4 Competition between pairs of directing groups

Cr(CO)$_3$ Complexes

-F > -CONHR > -NHCOR > -CH_2NR_2 = -OMe >> -CH_2OMe > -NMe_2 = -SR

A most dramatic difference is the case of fluoride [4, 129]. In a competition between carboxamide and fluoro substituents arranged ortho on a benzene ligand, LDA rapidly removes the proton *ortho* to F with complete selectivity [4]. In another example, amide dominates over a *m*-methoxide but with low selectivity [24]. The collective results suggest that conformational effects are significant (best conformation of the Cr(CO)$_3$ tripod), and that inductive effects are relatively more important than specific coordination of the base to a ring *ortho* substituent (compared to the uncomplexed arene). The high inductive effect of F is critical in directing deprotonation of the fluorobenzene complexes.

5.6 Miscellaneous Examples of Regioselective Deprotonation of (Arene)Cr(CO)$_3$ Complexes

Other complexes that selectively lithiate *ortho* to the substituent are shown in Fig. 5 [18, 35, 37, 38, 49, 60, 74, 77, 89, 104, 130]. Complex **53** undergoes a competing rearrangement upon *ortho* lithiation analogous to the silyl migration described earlier [89, 49].

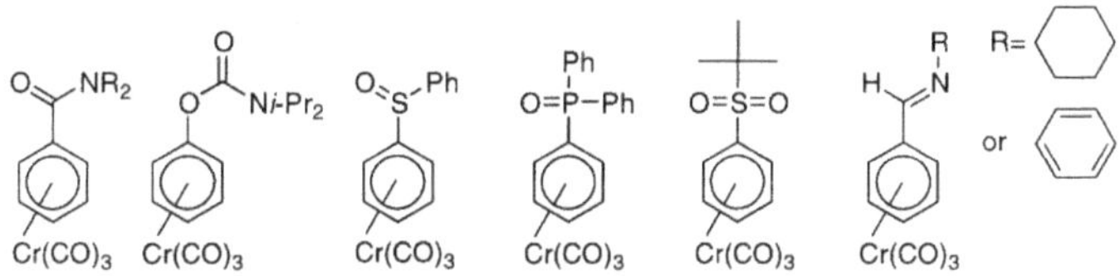

R=Et, *i*-Pr **53**

Fig. 5 Examples of complexes that can be lithiated selectively at the *ortho* position

The lithiation of fused aromatic polycycle ligands has received relatively little attention. In the first report, (naphthalene)Cr(CO)$_3$ (**54**) was metallated by *n*-BuLi giving a mixture of α- and β-substituted products [64]. In a more carefully designed experiment, **54** was metallated with LDA in THF at -95 °C and shows evidence of equilibration (as well as decomposition) above -45 °C with kinetic deprotonation favoring C-1 and thermodynamic deprotonation favoring C-2 [131].

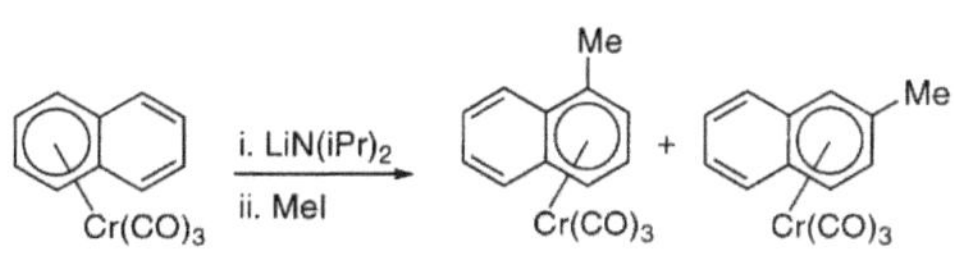

54 (34)

-95 °C/15 min	22	:	78
-95 °C/10 hr	61	:	39
-45 °C/5 min	87	:	13

A more extensive study reported a few months later gave conflicting results, showing that *n*-BuLi (presumably kinetic control) gave substitution at C-2/C-1= 70:30, after trapping with three different electrophiles, and also showed that treatment with TMPLi gives *exclusive* deprotonation at C-2. A series of simple electrophiles was used and the yields ranged from 65 to 95% [132].

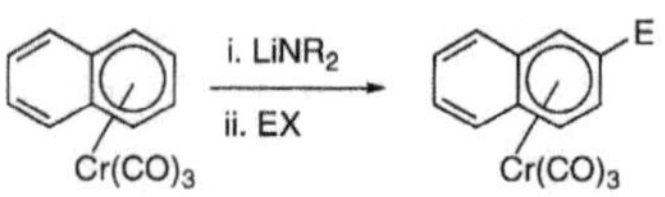

EX	yield
D_2O	93%
$ClSiMe_3$	96%
$EtOSO_2F$	62%
MeI	97%
$ClCO_2Me$	68%

(35)

In a clarification of the situation, the same authors subsequently showed that the changing ratios of 1- and 2-substituted products in Eq. (36) can be ascribed to selective decomposition of the less stable 2-Li-intermediate [133].

The selectivity with (phenanthrene)$Cr(CO)_3$ (**55**) favors C-3 on the coordinated ring (Fig. 6), and shows minor changes upon warming possibly due to equilibration [131]. Coordination of 1-methoxynaphthalene positions $Cr(CO)_3$ on the unsubstituted ring (**56**), and metallation proceeds at both C-2 and C-3 (Fig. 6) [65]. It is interesting that the effect of $Cr(CO)_3$ overwhelms the usual selectivity for metallation adjacent to the methoxy group [65]. Metallation of (2,6-dimethoxynaphthalene)$Cr(CO)_3$ (**57**) also favors positions *ortho* to the methoxy group on the coordinated ring (C-1 and C-3, Fig. 6).

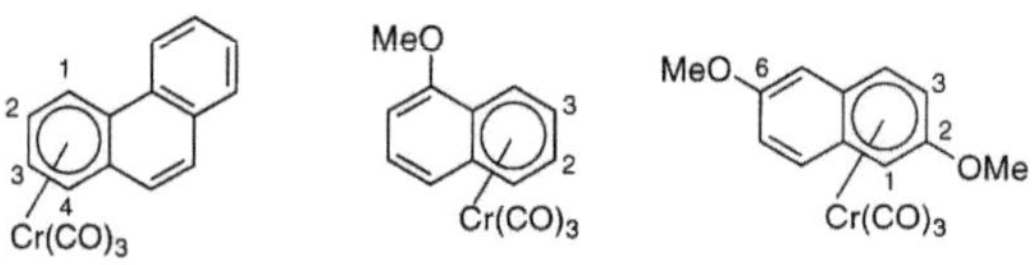

55 1:2:3:4 = 0:14:83:3

56 2:3 = 26:74

57 1:3 = 81:19

Fig. 6 Summary of site of lithiation of phenanthrene and substituted naphthalene complexes

The complex of *N*-methylindole with $Cr(CO)_3$ is efficiently deprotonated at C-2 by *n*-BuLi in the presence of TMEDA. Quenching with $ClCO_2Me$ produces the 2-carboxylate ester in 78% yield [46, 56]. This selectivity is the same as for the uncomplexed indole, but the reaction is faster (lower temperature) for the complexed version. Blocking C-2 by in situ silylation (**58**) allows alternate deprotonation, and providing a chelating unit at the indole N (2-trimethylsilylethoxymethyl) directs the deprotonation to C-7 with high selectivity (**59**, Eq. 37) [46]. The silyl groups can be removed to give the C-7 substituted indole. Electrophiles which were successful include: MeI (95%), DMF (70%), and allyl bromide (69%).

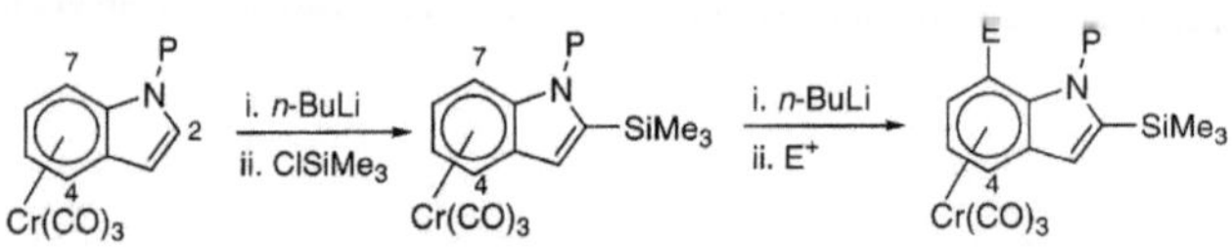

58 P = $CH_2OCH_2CH_2SiMe_3$ **59**

(36)

When the C-3 position is substituted (e.g., with Me), deprotonation is selective for C-7 even in the presence of a proton at C-2. At the same time, a large blocking group at the indole N (**60**) generates steric hindrance to deprotonation of both C-2 and C-7; selective lithiation at C-4 results [29, 45, 52, 57]. The *N*-(tri(isopropyl)-silyl)indole complex (**60**) can be deprotonated with 2 mol-eq. of *n*-BuLi at -78 °C for 3 h, and then trapped with a variety of electrophiles to give **61**.

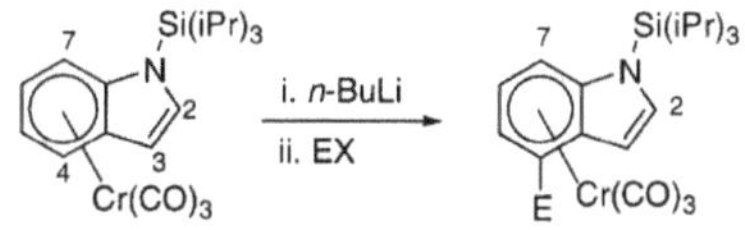

60 **61**

EX	yield
Me_3SiCl	56 %
EtO_2CCl	60 %
MeO_2CCl	59 %
PhSCl	26 %
$Me_2C{=}CHCH_2Br$	36 %

(37)

From these results, it is clear that convenient protecting group strategies allow selective functionalization of the indole ligand at C-4 or C-7, a selectivity and control not available for uncomplexed indole. The Cr(CO)$_3$ unit can be removed by gentle oxidation, as usual, and good yields of the substituted, uncomplexed indole are reported [45, 57].

The (thiophene)Cr(CO)$_3$ complex (**62**) is unusual in undergoing highly efficient double deprotonation at the positions adjacent to the sulfur (to give **63**, Eq. 39) [134]. The first and second deprotonation steps appear to proceed with comparable rates, making selective formation of the mono-lithiation product difficult. When both α-positions are occupied, lithiation occurs in the β-position [54].

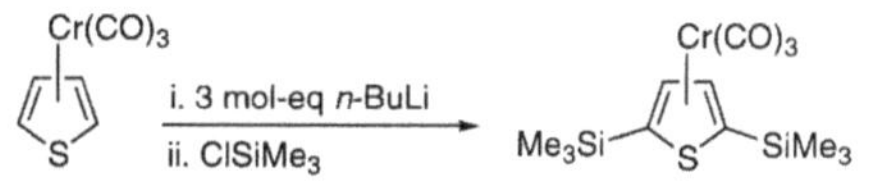

62 **63**

(38)

6
Summary

The lithiation of arenes activated by Cr(CO)$_3$ extends the general concept of arene lithiation and generally proceeds under milder conditions. Certain *o*-lithiations, such as that of chlorobenzene, can be done only with the metal activation. In other cases, the regioselectivity is different for the complexed and uncomplexed cases, attributed to a direct effect of the orbital interactions of the arene with the Cr(CO)$_3$ unit. The Cr(CO)$_3$ complexes provide new features which are particularly interesting in synthesis methodology, such as introduction of a center of asymmetry and promoting further elaboration through nucleophilic addition to the arene ring.

References

1. Nesmeyanov AN, Kolobova NE, Anisimov KN, Makarov YV (1968) :2665
2. Semmelhack MF (1995) Transition metal arene complexes: ring lithiation. In: Abel EW, Stone FGA, Wilkinson G (eds) Comprehensive organometallic chemistry II. Pergamon
3. Berger A, Djukic J-P, Michon C (2002) Coord Chem Rev 225:215

4. Dickens PJ, Gilday JP, Negri JT, Widdowson DA (1990) Pure Appl Chem 62:575
5. Rose-Munch F, Rose E (1999) 3:445
6. Gibson SE, Reddington EG (2000) Chem Commun 989
7. Kalinin VN (1987) Uspehi Khimii 46:1190
8. Uemura M (1991) Adv Met Org Chem 2:195
9. Gill US, Moriarty RM, Ku YY, Butler IR (1991) J Organomet Chem 417:313
10. Berger A, Dukic J-P, Michon C (2002) Coord Chem Rev 225:215
11. Beck H-J, Fischer EO, Kreiter CG (1971) J Organomet Chem 26:C41
12. Fischer EO, Stuckler P, Beck H-J, Kressl FR (1976) Chem Ber 109:3089
13. Fraser RR, Mansour TS (1986) J Organomet Chem 310:C60
14. Gibson SE, Saladin SA, Sur S (2000) Chem Commun 2011
15. Silverthorn WE (1975) Adv Organomet Chem 13:47
16. Card RJ, Trahanovsky WS (1980) J Org Chem 45:2555
17. Rausch MD, Gloth RE (1978) J Organomet Chem 153:59
18. Ariffin A, Blake AJ, Ewin RA, Li W-S, Simpkins NS (1999) J Chem Soc Perkin Trans 1:3177
19. Effenberger F, Schollkopf K (1981)Angew Chem Int Edn Eng 20:266
20. Effenberger F, Shollkopf K (1985) Chem Ber 118:4356
21. Effenberger F, Shollkopf K (1985) Chem Ber 118:4377
22. Mandal SK, Sarkar A (1998) J Org Chem 63:1901
23. Moser WH, Endsley KE, Colyer JT (2000) Org Lett 2:717
24. Uemura M, Nishikawa N, Hayashi Y (1980) Tetrahedron Lett 21:2069
25. Rose-Munch F, Chavignon R, Tranchier J-P, Gagliardini V, Rose E (2000) Inorg Chim Acta 300/302:693
26. Mann IS, Widdowson DA, Clough JM (1991) Tetrahedron 47:7981
27. Rausch MD, Moser GA, Lee WA (1979) Synth React Inorg Metal-Org Chem 9:357
28. Sebhat IK, Tan Y-L, Widdowson DA, Wilhelm R, White AJP, Williams DJ (2000) Tetrahedron 56:6121
29. Christian PWN, Gil R, Muniz-Fernandez K, Thomas SE, Wierzchleyski AT (1994) J Chem Soc Chem Commun 1569
30. Beswick PJ, Leach SJ, Masters NF, Widdowson DA (1984) 46
31. Gagliardini V, Tranchier J-P, Chavignon R, Rose-Munch F, Rose E (1998) C R Acad Sci 137
32. Alexakis A, Kanger T, Mangeney P, Rose-Munch F, Perrotey A, Rose E (1995) Tetrahedron: Assymetry 6:47
33. Kondo Y, Green JR, Ho J (1991) J Org Chem 56:7199
34. Kondo Y, Green JR, Ho J (1993) J Org Chem 58:6182
35. Nelson SG, Hilfiker MA (1999) Org Lett 1:1379
36. Kamikawa K, Watanabe T, Daimon A, Uemura M (2000) Tetrahedron 56:2325
37. Pache S, Botuha C, Franz R, Kündig EP (2000) Helv Chim Acta 83:2436
38. Gibson SE, Steed JW, Sur S (2001) J Chem Soc Perkin Trans 1:636
39. Horstermann D, Schmalz H-G, Kociok-Kohn G (1999) Tetrahedron 55:6905
40. Schmalz H-G, Schellhaas K (1996) Angew Chem Int Ed Engl 35:2146
41. Schellhaas K, Schmalz H-G, Bats JW (1998) Chem Eur J 4:57
42. Rose-Munch F, Rose E, Semra A (1989) J Organomet Chem 377:C9
43. Davies SG, Goodfellow CL (1990) J Chem Soc Perkin Trans I 393
44. Costa MF, da Costa MRG, Curto MJM (2001) J Organomet Chem 626:233
45. Beswick PJ, Greenwood CS, Mowlem TJ, Nechwatal G, Widdowson DA (1988) Tetrahedron 44:7325
46. Masters NF, Methews N, Nechvatal G, Widdowson DA (1989) 45:5955
47. Woodgate PD, Singh Y, Pickard CEF (1998) J Organomet Chem 560:197
48. Gibson SE, Saladin SA, Sur S (2000) Chem Commun 2011
49. Ewin RA, MacLeod AM, Price DA, Simpkins NS (1997) J Chem Soc Perkin Trans 1:401
50. Han JW, Son SU, Chung YK (1997) J Org Chem 62:8264

51. Bolm C, Muniz K, Ganter C (1998) New J Chem 1371
52. Dickens MJ, Gilday JP, Mowlem TJ, Widdowson DA (1991) Tetrahedron 47:8621
53. Christian PWN, Gibson SE, Gil R, Jones PCV, Marcos CF, Prechtl F, Wierzchleyski AT (1995) Recl Trav Chim Pays-Bas 114:195
54. Loft MS, Mowlem TJ, Widdowson DA (1995) J Chem Soc Perkin Trans I 97
55. Fukui M, Ikeda T, Oishi T (1982) Tetrahedron Lett 23:1605
56. Nechvatal G, Widdowson DA, Williams DJ (1981) J Chem Soc Chem Commun 1260
57. Nechvatal G, Widdowson DA (1982) J Chem Soc Chem Commun 467
58. Heppert JA, Aube J, Thomas-Miller ME, Milligan ML, Takusagawa F (1990) 9:727
59. Fukuzawa S, Karo H, Ohtaguchi M, Hayashi Y, Yamazaki H (1997) J Chem Soc Perkin Trans 1:3059
60. Davies SG, Loveridge T, Clough JM (1995) J Chem Soc Chem Commun 817
61. Davies SG, Loveridge T, Teixeira MFCC, Clough JM (1999) J Chem Soc Perkin Trans 1:3405
62. Hayashi Y, Sakai H, Kaneta N, Uemura M (1995) J Organomet Chem 503:143
63. Gill JC, Marples BA, Traynor JR (1987) Tetrahedron Lett 28:2643
64. Oprunenko YF, Malygina SG, Ustynyuk YA, Ustynyuk NA (1985) Izv Akad Nauk SSSR Seriya Khim 2405
65. Kirss RU, Treichel PM (1986) J Am Chem Soc 108:853
66. Boutonnet J-C, Levisalles J, Rose-Munch F, Rose E (1985) J Organomet Chem 290:153
67. Rose-Munch F, Bellot O, Mignon L, Semra A (1991) J Organomet Chem 402:1
68. Heppert JA, Thomas-Miller ME, Milligan ML, Velde DV, Aube J (1988) 7:2581
69. Rose-Munch F, Gagliardini V, Perrotey A, Tranchier J-P, Rose E, Mangeney P, Alexakis A, Kagner T, Vaisermann J (1999) Chem Commun 2061
70. Semmelhack MF, Bisaha J, Czarny M (1979) J Am Chem Soc 101:768
71. Fukui M, Endo Y, Yamada Y, Asakura A, Oishi T (1981) Heterocycles 15:86
72. Uemura M, Take K, Hayashi Y (1983) J Chem Soc Chem Commun 858
73. Breimar J, Wieser M, Beck W (1992) J Organomet Chem 441:429
74. Kündig PE, Quattropani A (1994) Tetrahedron Lett 35:3497
75. Davies SG, Hume WE (1995) J Chem Soc Chem Commun 251
76. Schmalz H-G, Schellhaas K (1995) Tetrahedron Lett 36:5515
77. Uemura M, Hayashi Y, Hayashi Y (1994) Tetrahedron: Assymetry 5:1427
78. Ghavshou M, Widdowson DA (1983) J Chem Soc Perkin Trans I 3065
79. Uemura M, Nishikawa N, Take K, Ohnishi M, Hirotsu K, Higuchi T, Hayashi Y (1983) J Org Chem 48:2349
80. Uemura M, Take K, Isobe K, Minami T, Hayashi Y (1985) Tetrahedron 41:5771
81. Coote SJ, Davies SG, Goodfellow CL, Sutton KH, Middlemiss D, Naylor A (1990) Tetrahedron: Asymmetry 1:817
82. Costa MF, da Costa MRG, Curto MJM, Magrinho M, Damas AM, Gales L (2001) J Organomet Chem 632:27
83. Uemura M, Isobe K, Take K, Hayashi Y (1983) J Org Chem 48:3855
84. Uemura M, Isobe K, Hayashi Y (1985) Chem Lett 91
85. Aube J, Heppert JA, Milligan ML, Smith MJ, Zenk P (1992) J Org Chem 57:3562
86. Uemura M, Miyake R, Nakayama K, Shiro M, Hayashi Y (1994) J Org Chem 58:1238
87. Gilday JP, Widdowson DA (1986)J Chem Soc Chem Commun 1235
88. Koide H, Uemura M (2000) Chirality 12:352
89. Quattropani A, Bernardinelli G, Kundig EP (1999) Helv Chim Acta 82:90
90. Uemura M, Koboyashi T, Isobe K, Minami T, Hayashi Y (1986) J Org Chem 51:2859
91. Siwek MJ, Green JR (1996) Chem Commun 2359
92. Author names in Japanese (1993) Nippon Kagaku Kaishi 1353
93. Lee TV, Leigh AJ, Chapleo CB (1989) Tetrahedron Lett 30:5519
94. Renard C, Valentic R, Rose-Munch F, Rose E (1998) Organometallics 17:1587
95. Streitweiser A Jr, Boerth DW (1978) J Am Chem Soc 100:755
96. Card RJ, Trahanovsky WS (1980) J Org Chem 45:2560

97. Wey HG, Butenschon H (1988) J Organomet Chem 350:C8
98. Wey HG, Betz P, Butenschon H (1991) Chem Ber 124:465
99. Kundig EP, Perret C, Rudolph B (1990) Helv Chim Acta 73:1970
100. Gshwend HW, Rodrigez HR (1979) Org React 26:1
101. Gilman H, Soddy TS (1957) J Org Chem 22:1715
102. Nefedov OM, Dyachenko AI (1971) Dokl Akad Nauk SSSR 198:593
103. Semmelhack MF, Ullenius C (1982) J Organomet Chem 235:C10
104. Price DA, Simpkins NS (1994) J Org Chem 59:1961
105. Price DA, Simpkins NS (1994) Tetrahedron Lett 35:6159
106. Uemura M, Minami T, Hayashi Y (1984) J Chem Soc Chem Commun 1193
107. Clark GR, Kuipers B, Metzler MR, Nguyen MH, Woodgate PD (1997) J Organomet Chem 545/546:225
108. Schmalz H-G, Schwarz A, Durner G (1994) Tetrahedron Lett 35:6861
109. Schmalz H-G, Arnold M, Hollander J, Bats JW (1994) Angew Chem Int Ed Engl 33:109
110. Majdalani A, Schmalz H-G (1997) Tetrahedron Lett 38:4545
111. Schmalz H-G, Majdalani A, Geller T, Hollander J, Bats JW (1995) Tetrahedron Lett 36:4777
112. Majdalani A, Schmalz H-G (1997) Synlett 1303
113. Geller T, Schmalz H-G, Bats JW (1998) Tetrahedron Lett 39:1537
114. Schmalz H-G, Volk T, Bernicke D (1997) Tetrahedron 53:9219
115. Mann IS, Widdowson DA, Clough JM (1991) Tetrahedron 47:7991
116. Watanabe T, Shakadou M, Uemura M (1999) Inorg Chim Acta 296:80
117. Masters NF, Widdowson DA (1983) J Chem Soc Chem Commun 955
118. Clough JM, Mann IS, Widdowson DA Synlett :469
119. Fukui M, Endo Y, Yamada Y, Asakura A, Oishi T (1981) Heterocycles 15:415
120. Blagg J, Davies SG, Goodfellow CL, Sutton KH (1986) J Chem Soc Chem Commun 1283
121. Nirchio PC, Wink DJ (1991) Organometallics 10:2499
122. Kendall JD, Woodgate PD (1998) Aust J Chem 51:1083
123. Blagg J, Davies SG, Goodfellow CL, Sutton KH (1987) J Chem Soc Perkin Trans I 1805
124. Uemura M, Miyake R, Shiro M, Hayashi Y (1991) Tetrahedron Lett 32:4569
125. Uemura M, Daimon A, Hayashi Y (1995) J Chem Soc Chem Commun 1943
126. Overman LE, Owen CE, Zipp GG (2002) Angew Chem Int Edn 41:3884
127. Alexakis A, Kanger T, Mangeney P, Rose-Munch F, Perrotey A, Rose E (1995) Tetrahedron: Assymetry 6:2135
128. Gilday JP, Negri JT, Widdowson DA (1989) Tetrahedron 45:4605
129. Gilday JP, Widdowson DA (1986) Tetrahedron Lett 27:5525
130. Gibson SE, Guillo N, White AJP, Williams DJ (1996) J Chem Soc Perkin Trans 1:2575
131. Treichel PM, Kirss RU (1987) Organometallics 6:249
132. Kündig EP, Desbory V, Grivet C, Rudolph B, Spichiger S (1987) Organometallics 6:1173
133. Kündig EP, Grivet C, Spichiger S (1987) J Organomet Chem 332:C13
134. Nefedova MN, Setkina VN, Kursanov DN (1983) J Organomet Chem 244:C21

Topics Organomet Chem (2004) 7: 43–69
DOI 10.1007/b94491

(Arene)Cr(CO)$_3$ Complexes: Aromatic Nucleophilic Substitution

Martin F. Semmelhack · Anatoly Chlenov
Princeton University, Princeton NJ 08540, USA
E-mail: mfshack@princeton.edu

Abstract Transition metals coordinated to the pi system of arenes function as strong electron-withdrawing groups and favor nucleophilic addition to the arene to give cyclohexadienyl-metal complexes. The regioselectivity of addition is influenced in subtle and indirect ways, compared to the powerful effect on both reactivity and regioselectivity of electron-withdrawing groups attached to the sigma bond system of the arene. The intermediate cyclohexadienyl complexes can be processed in several ways, leading to several distinct synthesis methods. Useful reactivity is known for several metal systems and each is discussed for specific reaction types. The presence of a fluoro or chloro substituent on the arene allows an effective analog of classical S_NAr reactivity, with a wide range of nucleophiles including simple amines, alkoxides, and carbon anions. Another general process is the addition/protonation protocol, where a hydrogen substituent on the arene is replaced by a nucleophile. This opens new questions of regioselectivity since the typical arene ligand has several hydrogens which are candidates for substitution. The formation of five- and six-membered rings by both general mechanisms are known. Less general indirect mechanisms (e.g., tele substitution) have also been defined.

1 Introduction

This chapter covers reactions in which coordination of a transition metal to the pi system of an arene ring activates the ring toward addition of nucleophiles, to give η^5-cyclohexadienyl-metal complexes (e.g., **1** in Scheme 1) as an intermediate. If an electronegative atom is present in the *ipso* position, elimination of that atom (X in **1**) leads to nucleophilic aromatic substitution (**2**, path a). If a hydrogen occupies the *ipso* position, oxidation of the intermediate gives formal nucleophilic substitution for hydrogen (**3**, path b). General reviews have appeared [1–4] as well as others with an emphasis on η^6-arene-$Cr(CO)_3$ complexes [5, 6], on η^6-arene-$FeCp^+$ ($Cp=\eta^5$-cyclopentadienyl) complexes [7–9], η^6-arene-$RuCp^+$ [10, 11], and on η^6-arene-$Mn(CO)_3$ complexes [12]. The purpose of this chapter is to summarize the established or potential synthesis methodology available via nucleophile addition to a metal-coordinated arene ligand.

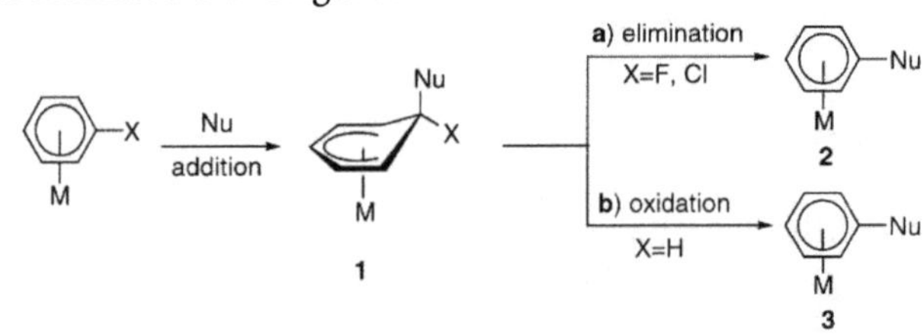

Scheme 1 Nucleophile addition to metal-coordinated arene ligand

Five types of arene π complexes have seen significant development in synthesis methodology (Fig. 1): neutral η^6-arene-$Cr(CO)_3$ (**4**), cationic (η^6-arene)(η^5-cyclopentadienyl)Fe(II) (**5**) and (η^6-arene)(η^5-cyclopentadienyl)Ru(II) (**6**), cationic η^6-arene-$Mn(CO)_3$ (**7**), and the dicationic (η^6-arene)(η^5-ethyltetramethyl-cyclo-pentadienyl)Rh(III) complexes (**8**).

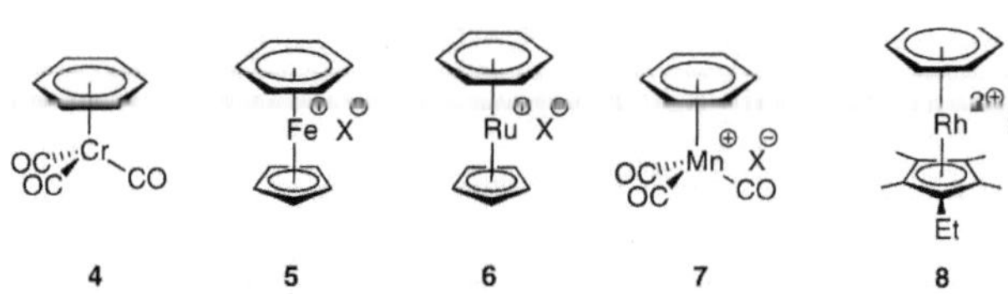

Fig. 1 Arene-metal complexes discussed in this chapter

For the series Cr, Mo, and W, the nature of the metal has little effect on rate of nucleophile addition, while for Fe, Ru, and Os, the reactivity decreases with increasing atomic number [4].

2 Nucleophilic Substitution for Heteroatoms on Arene Ligands, S_NAR Reaction

The smooth replacement of a heteroatom (usually halide) via **9** and **10** requires reversible addition of the nucleophile, since the kinetic site of addition is usually at a position bearing a hydrogen substituent (**11** in Scheme 2; charges omitted for clarity).

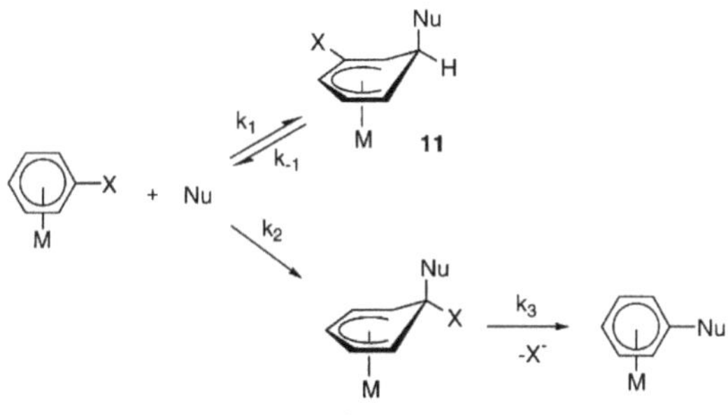

Scheme 2 Nucleophile addition to metal-coordinated haloarene

The relative rates of each step depend critically on the nature of M and the nucleophile. More reactive nucleophiles and more reactive complexes disfavor equilibration ($k_1 >> k_{-1}$), and the process can stop with formation of the first cyclohexadienyl intermediate (**11**) [2].

Higher charge on the metal leads to higher reactivity of the arene. Chromium tricarbonyl activation is comparable to a *p*-nitro group, whereas manganese tricarbonyl activates like two nitro groups, and Rh(III) activation is equal to that of three nitro groups [2].

2.1 $(Halobenzene)Cr(CO)_3$ Complexes

Stabilized nucleophiles (e.g., alkoxides, enolates, etc.) add to the *ipso* carbon of $(haloarene)Cr(CO)_3$ providing an intermediate (**9:** $M=Cr(CO)_3$, Scheme 2) which upon loss of the leaving group gives a substituted complex **10:** $M=Cr(CO)_3$. Unstabilized nucleophiles (e.g., sulfur-stabilized alkyllithium reagents) give a kinetic addition product (e.g., **11:** $M=Cr(CO)_3$-) which does not rearrange to **9** and therefore can give substitution only by loss of a leaving group by a different mechanism, as will be discussed later. The rate-determining step depends on the leaving group and the nucleophile. A good leaving group such as chloride makes the second step (k_3) fast and the initial addition rate determining. In case of a poor leaving group such as fluoride, the situation is more complex. With alkoxides the first step (k_2) remains rate-determining [13], whereas with amines the rates (k_2, k_3) become comparable [14]. After nucleophilic substitution is complete, the substituted arene can be liberated by mild oxidation, e.g., with excess iodine at 25 °C. This process releases the free arene, CO, chromium(III), and iodide anion [2]. The overall reaction is shown in Eq. (1)

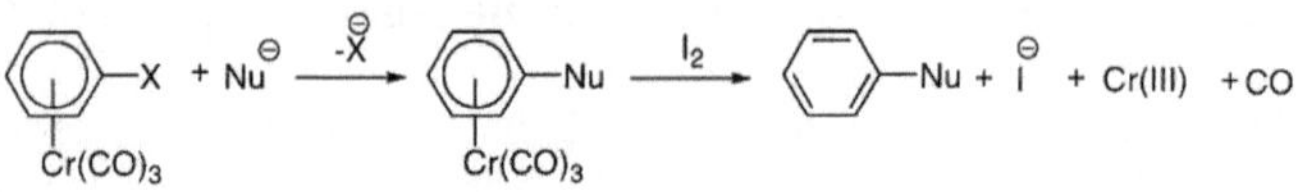

(1)

Replacing a CO ligand with a phosphine donor ligand lowers the electrophilicity of the arene (resulting in low S_NAr reactivity), whereas when an isocyanide [15] or an electron poor phosphine (e.g., tris(pyrrolyl)phosphine) [16] is used as a ligand, the arene ring remains very reactive towards nucleophiles. Isocyanide (Eq. 2) and tris(pyrrolyl)phosphine ligands can be modified with a side chain to allow attachment of the activating Cr unit to the solid support [15–17].

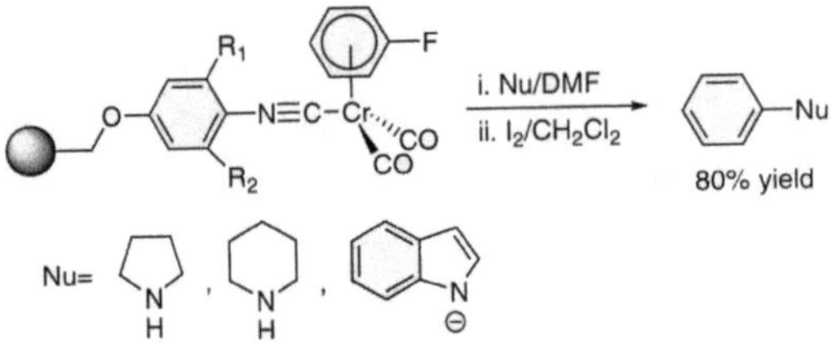

(2)

2.1.1
Intermolecular Substitution with Heteroatom Nucleophiles

A patent issued in 1965 claims substitution for fluoride on fluorobenzene-$Cr(CO)_3$ in DMSO by a long list of nucleophiles [18]. Chloroarene complexes are typically less reactive [2]. The bromo- and iodoarene complexes are known, but generally are not effective in the S_NAr reaction. The high reactivity of the fluorobenzene complex allows nucleophilic substitution under mild reaction conditions. A variety of alkoxides [19] including chiral versions [20, 21] were used successfully. Alkyl sulfides replace fluoride in an analogous process [2, 15, 22]. Several difluoroarene complexes are known and both fluorides can be displaced by a methoxide nucleophile [23].

Hydroxylamines deprotonated by KOH substitute for chloride in (chloroarene)$Cr(CO)_3$ in high yields [24]. Reaction of a (fluoroarene)$Cr(CO)_3$ complex with dialkyl phosphite **12** resulted in formation of a phosphate **13** [25]. The arylation of oximes under phase transfer conditions leads to a simple process for benzofuran formation [26].

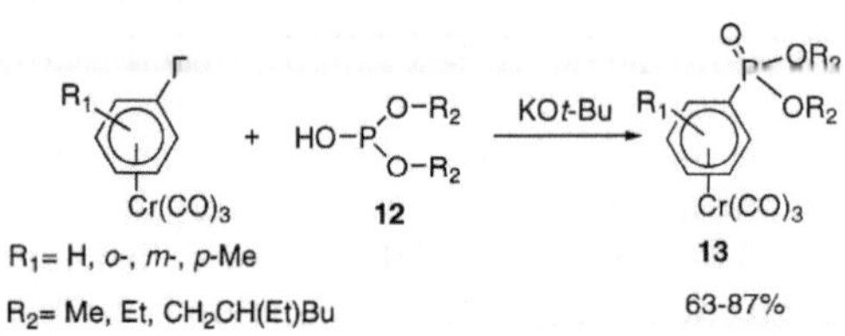

(3)

Simple primary and secondary amine nucleophiles react smoothly in the absence of added base (excess of amine nucleophile is used to remove HX), in a very general and efficient process to produce aniline derivatives [2,16]. Piperazine adds to fluoroanisoles in the presence of potassium carbonate [27] whereas indole has to be first deprotonated with sodium hydride; the resulting anion then adds to variously substituted fluorobenzenes [15,28]. To add an -NH_2 group, an effective procedure is reaction of the trifluoroacetamide anion with a (fluorobenzene)Cr(CO)$_3$ complex, (e.g., **14** R=H, X=F), followed by gentle base treatment [29].

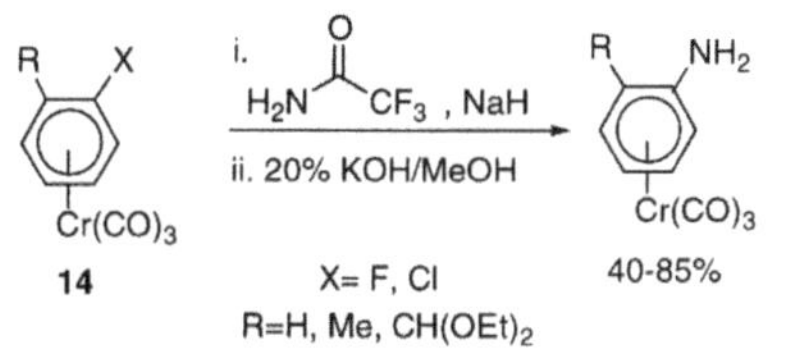

(4)

Cyanide has also been used as a nucleophile. Although in some cases, nucleophilic attack at Cr with concomitant loss of the arene ligand was observed, recent work showed that halide substitution can be achieved in DMSO [30].

Direct replacement of sulfonate groups, such as tosylate, is not observed, perhaps due to steric retardation of the departure of the leaving group from the *endo* direction. A few special examples of oxygen leaving groups are known. Lithium diphenyl phosphide replaced a carbamate leaving group to generate an aryl phosphine **15** in good yield [31].

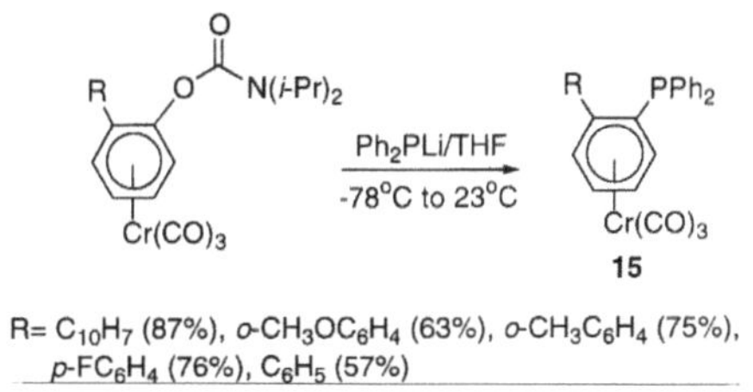

(5)

2.1.2
Cyclizations via Heteroatom Substitution for Halide

Intramolecular substitution for chloride or fluoride is particularly effective. Oxygen heterocycles with fused benzo rings are obtained from Cr(CO)$_3$ complexes of fluorobenzene with an *o*-(hydroxyalkyl) side chain [2]. Using a Cr(CO)$_3$ unit to direct a [2+2] cycloaddition to an imine, a stereoselective synthesis of a β-lactam was achieved [32].

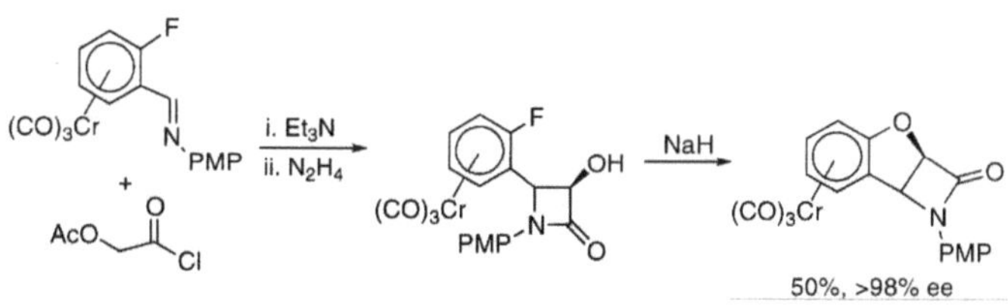

(6)

From the *o*-dichlorobenzene complex, reaction with a dialkoxide can produce a cyclic bis-ether. This idea has been applied in the preparation of benzo-crown ethers [33], their bisthia analog [33], and several dioxin derivatives [34].

Coordination of arenes with $Cr(CO)_3$ also activates the ring hydrogens toward abstraction with strong base (metallation) as discussed in Chap. 3 of this volume. Simple arene ligands can be metallated with alkyllithium reagents; alkoxy, amino, and halo substituents on the arene direct the metallation to the *ortho* position with rates and regioselectivity higher than with the corresponding free arene. This allows a strategy for annulating aromatic rings via *ortho*-lithiation (**16**), trapping with a bifunctional electrophile (**17**), and finally nucleophilic substitution for the electronegative substituent (usually F), (**18**).

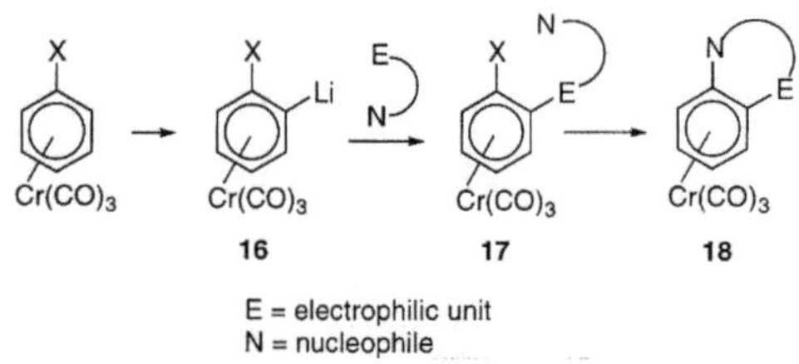

(7)

Following an initial report [2] including carbon nucleophiles for the cyclization, a series of papers have defined useful possibilities with heteroatom nucleophiles, especially N and O derivatives [2]. The *ortho*-lithio haloarene complexes (e.g., **16**) are highly basic, and the electrophilic trapping species is restricted to those with low kinetic acidity and high electrophilicity. Good results are obtained with isocyanates, ketenes, and acyl derivatives with α-protons of low acidity.

2.1.3
Nucleophilic Substitution with Carbon Nucleophiles

2.1.3.1
S_NAr Addition/Elimination, ipso Substitution

While carbon nucleophiles were suggested to be efficient in substitution for fluoride in the early patent, the first examples in the primary literature appeared only in 1974 [2]. It is now clear that there are three reactivity classes of carbon nucleophiles (Scheme 3): (a) stabilized carbanions (from carbon acids with pK_a<ca. 18) which give unfavorable and easily reversible addition to the arene ligand ($k_1 << k_{-1}$), (b) more reactive carbanions (pK_a>20) which give complete conversion to cyclohexadienyl addition products (**19**) at low temperatures (-78 °C) and rearrange among other regioisomers only when the reaction mixture is warmed ($k_1 > k_{-1}$), and (c) very reactive carbanions (pK_a>30) which give irreversible addition to the arene ligand and do not rearrange ($k_1 >> k_{-1}$) [2]. In general, addition to the unsubstituted position (k_1; **19a, 19b**) is preferred kinetically over addition to a substituted position (k_2; **19c**).

Sodio diethylmalonate is an example of type (a). Reaction with fluorobenzene-$Cr(CO)_3$ proceeds to completion after 20 h at 50 °C in HMPA to give the diethyl phenylmalonate complex in over 95% yield [2]. There is no evidence for an

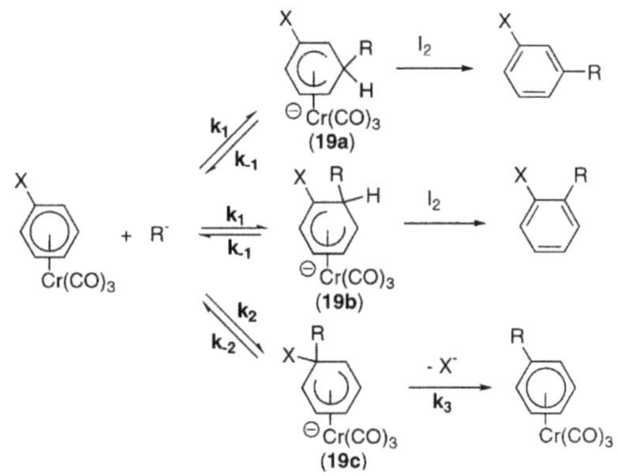

Scheme 3 Equilibration possibilities in nucleophile addition to haloarene-$Cr(CO)_3$ complexes

intermediate (e.g., the cyclohexadienyl anion complexes **19**); interruption of the reaction by addition of iodine at less than 20 h gave significant amounts of unreacted fluorobenzene. A satisfactory picture assumes that the anion adds reversibly and unfavorably ($k_1 << k_{-1}$, Scheme 3), slowly finding itself at the *ipso* position ($k_2 << k_1$); then irreversible loss of fluoride gives the substitution product [2]. Lithio isobutyronitrile is an example of anion type (b). The initial addition to chlorobenzene-$Cr(CO)_3$ is over within minutes at -78 °C ($k_1 > k_{-1}$), but the substitution product does not appear until the mixture is warmed to 25 °C [2]. Quenching with iodine after short reaction time leads to a mixture of phenylisobutyronitrile and *o*- and (*m*-chlorophenyl)-isobutyronitrile. This appears to be a case of fast addition *ortho* and *meta* to the chloride to give cyclohexadienyl anionic complexes (e.g., **19a,b**) followed by slow rearrangement to the *ipso* intermediate (**19c**). Quenching with iodine before equilibration is complete leads to oxidation of the intermediates (**19a,b**) and formal substitution for hydrogen.

Carbon nucleophiles of type (c) add to the arene ligand and do not rearrange except in special cases; examples include very reactive anions such as 2-lithio-2-methyl-1,3-dithiane and phenyl lithium. In these cases, the anion adds to an unsubstituted position (mainly *ortho* or *meta* to Cl, (e.g., **19a,b**, Scheme 3) and does not rearrange. Then iodine quenching, even after a long period at 25 °C, gives almost exclusively the products from formal substitution for hydrogen.

Reversibility in the addition of the carbanion also depends on the structure of the arene ligand, the nature of the counterion and the solvent [35]. A lithium counterion allows faster equilibration compared to potassium, whereas addition of HMPA slows down equilibration up to fourfold in some cases.

Hydrocarbon anions such as fluorenyl, indenyl, and cyclopentadienyl substitute for fluoride, leading, for example, to (phenylindene)$Cr(CO)_3$ complex (**20**) [36, 37].

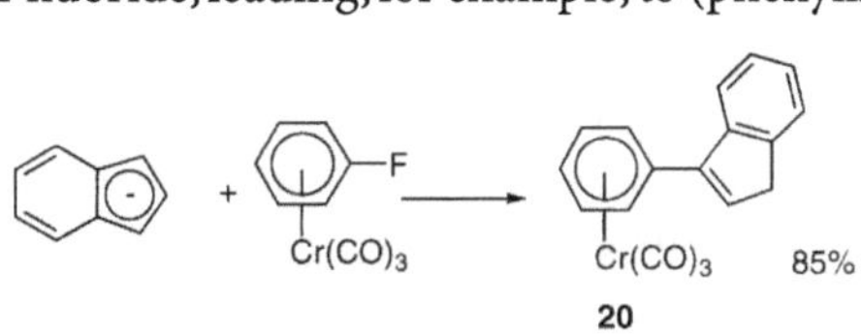

(8)

Other successful nucleophiles include α-ketoesters [38], diesters [39], iminoesters, and iminonitriles [40, 41]. The carbanion derived from benzonitrile reacted with (chloroarene)$Cr(CO)_3$ complex **21** [42].

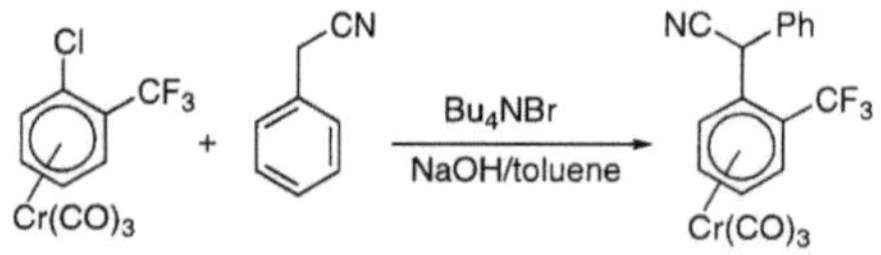

(9)

21

Amino acid derivatives possessing a chiral auxiliary can be successfully arylated using (fluorobenzene)$Cr(CO)_3$ as an electrophile [43, 44].

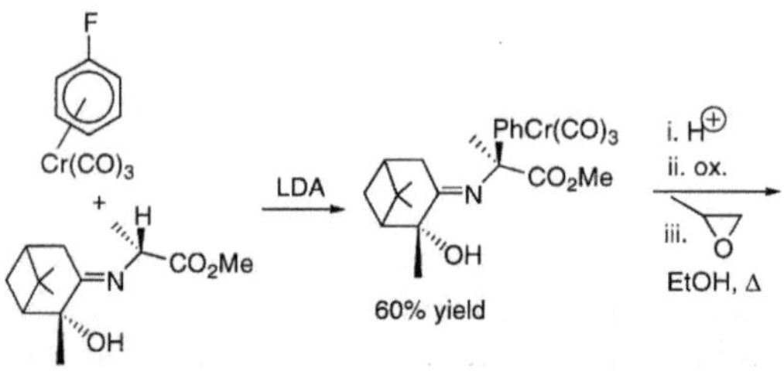

(10)

Application of Grignard reagents is limited to aryl Grignards [45]. Axially chiral biaryls were obtained when an *ortho* substituted aryl Grignard **22** replaced a MeO leaving group in complex **23** [45].

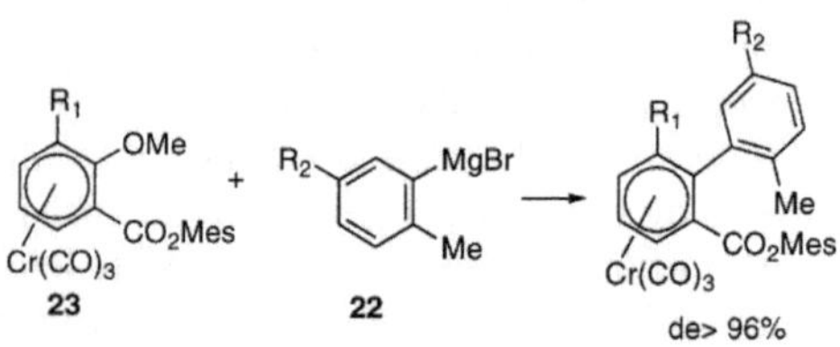

23 **22**

R_1	R_2	yield(%)
OMe	H	68
Me	H	68
Me	Me	81

(11)

2.1.3.2
Cine/Tele Substitution

There is an alternate mechanism for substitution, following the sequence of nucleophile addition, protonation, and elimination of HX. In this pathway, the addition of the nucleophile need not be at the *ipso* position; it can be *ortho* to halide leading to "*cine*" substitution or it can be at the *meta* (in **24**) or *para* positions, leading to "*tele*" substitution via elimination of HX from an intermediate such as **25** [2, 46]. The processes depend on the formation of the cyclohexadienyl anion intermediates (**24**) in a favorable equilibrium (carbon nucleophiles from carbon acids with pKa>22 or so), followed by protonation (which can occur at low temperature with even weak acids such as acetic acid), and hydrogen shifts in the proposed diene-chromium intermediates (**26**) [46]. Hydrogen shifts lead to an isomer (**25**) which allows elimination of HX and regeneration of an arene-chromium complex, now with the carbanion unit indirectly substituted for X.

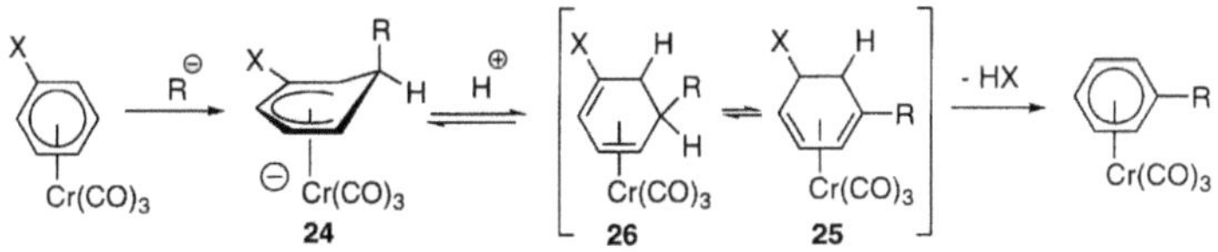

(12)

A potentially useful example of indirect substitution utilizes alkoxy [47] or aryloxy [48, 49] leaving groups; the intermediate (**27**) from the reaction of the diphenylether-Cr(CO)$_3$ complex (**28**) and the 2-phenyl-1,3-dithiane anion can be induced to eliminate phenol after protonation at low temperature. The result is *tele-meta* substitution.

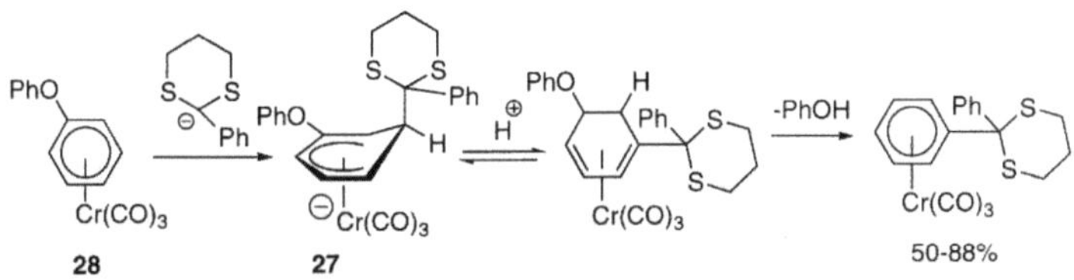

(13)

Halogen leaving groups can be used with type (c) nucleophiles [49] including alkynyl lithium derivative [50]. An impressive example is the reaction of (2,6-dimethyl-1-fluorobenzene)Cr(CO)$_3$ (**29**) with 2-lithio-2-methyl-1,3-dithiane at -78 °C followed by treatment with trifluoroacetic acid at -78 °C [51]. Loss of HF leads to the 1,2,4 (*tele*) substitution product, (**30**), in 62% yield.

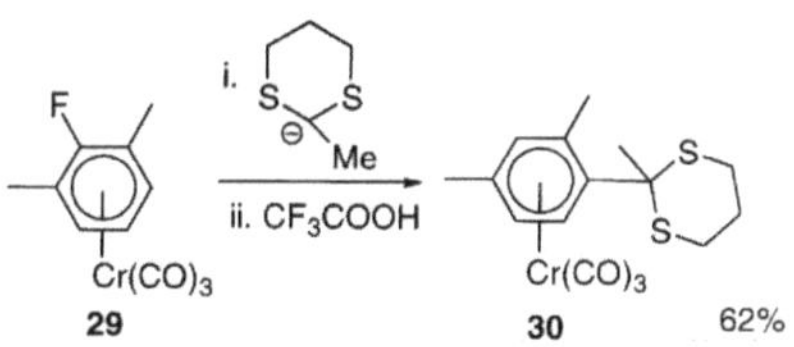

(14)

The directing effects of F (strong *meta*) vs Cl (weak, *o*/*m*; discussed below) allow control over the site of *tele* substitution; the chloro analog of **29** leads to the 1,3,5-substitution product [51]. Type (b) nucleophiles can also be used if low temperature is maintained throughout the reaction to avoid equilibration [52].

Hydride nucleophiles can give substitution directly, via reversible addition or by an indirect (addition-protonation) mechanism. In the tetrahydronaphthalene complex (**31**), replacement of a methoxy group by hydride gave (**32**), using LiAlH$_4$ followed by oxidative demetallation [53].

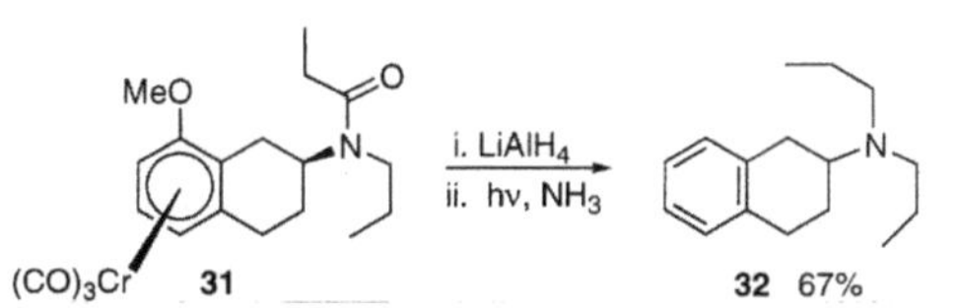

(15)

Addition of LiEt$_3$BH to (haloarene)Cr(CO)$_3$ complexes leads, after further treatment with acid, to the product from cleavage of the heteroatom bond and overall hydride substitution for the heteroatom via *cine* or *tele* pathways [54].

Examples include cleavage of carbon-oxygen [55], carbon-halogen [56], and carbon-nitrogen [57] bonds.

2.2
Halobenzene CpFe+

The activating ability of the FeCp+ unit is higher than $Cr(CO)_3$ and the inorganic starting materials are somewhat less expensive and less toxic. The main limitations are the severe conditions necessary for the preparation of the starting complexes and sometimes inefficient demetallation of the products.

2.2.1
S_NAr with Heteroatom Nucleophiles

Phenoxides undergo substitution for chloride [58–60]. The first substitution in a dichlorobenzene complex occurs at -20 °C [61]. This allowed synthesis of tyrosine derivative **33** without the loss of optical purity [62a].

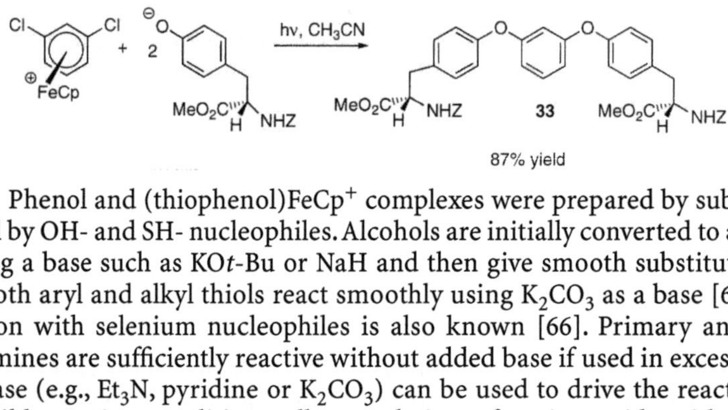

(16)

Phenol and (thiophenol)FeCp+ complexes were prepared by substitution for Cl by OH- and SH- nucleophiles. Alcohols are initially converted to alkoxides using a base such as KO*t*-Bu or NaH and then give smooth substitution [63, 64]. Both aryl and alkyl thiols react smoothly using K_2CO_3 as a base [65]. Substitution with selenium nucleophiles is also known [66]. Primary and secondary amines are sufficiently reactive without added base if used in excess, although a base (e.g., Et_3N, pyridine or K_2CO_3) can be used to drive the reaction [67–69]. Mild reaction conditions allow arylation of amino acids with (PhF)FeCp+ (Eq. 17) [69]. The process is of limited value, however, since decomplexation of the arene led to decomposition.

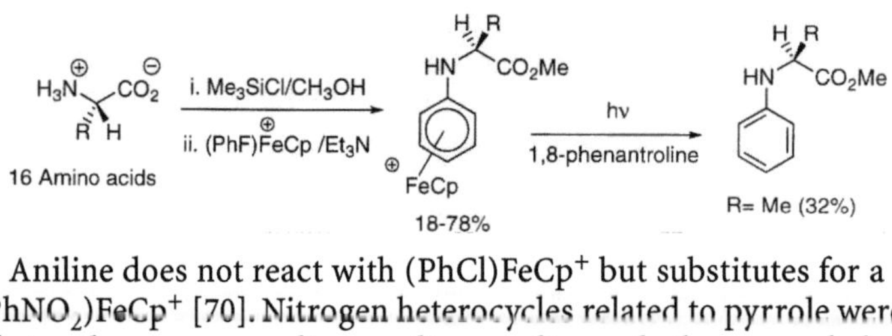

(17)

Aniline does not react with (PhCl)FeCp+ but substitutes for a nitro group in ($PhNO_2$)FeCp+ [70]. Nitrogen heterocycles related to pyrrole were first converted to the corresponding sodium salts and then coupled with (chloroarene)FeCp+ [71, 72].

Azide can be used as a nucleophile, although attempts at decomplexation resulted in rearrangement of the azidobenzene and yielded a mixture of ring contraction products, cyanoferrocene and the aniline complex [73]. Hydrazine was successful in substituting for chloride, but demetallation was not attempted [74].

Selective nucleophilic substitution of one of the chlorides in (dichlorobenzene)FeCp$^+$ complexes can be achieved with a variety of nucleophiles. The first substitution is fast while the resulting complex is typically less electrophilic so that the second substitution is slower [75]. In the case of a primary amine nucleophile, the product is deprotonated by excess of amine and thereby made completely unreactive towards a second substitution [66, 76]. A second substitution can be achieved after neutralization with an equivalent amount of acid. If the first nucleophile is attached to the solid phase, a library of substituted arenes can be generated [77]. With η^6-(*o*-dichlorobenzene)FeCp$^+$ and a bis-heteroatom nucleophile, benzo-fused heterocycles (e.g., **34**) are obtained after pyrolytic demetallation [78].

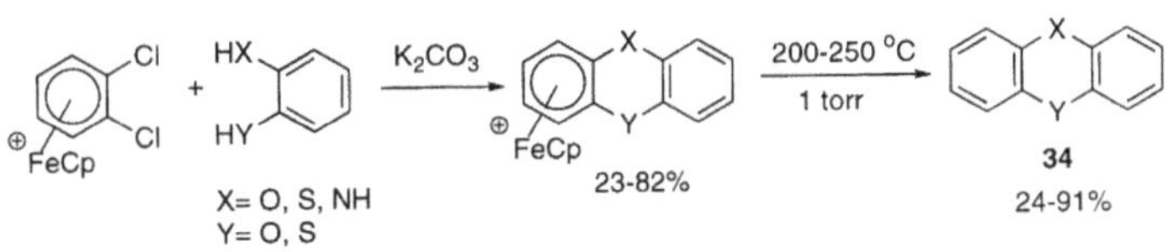

(18)

The advantage of a fluoride leaving group was clear in reaction of a hydroxycarbamate anion with (fluorobenzene)FeCp$^+$ complex, where the yield was 72% for F and 41% for Cl [79]. The (1,2-difluorobenzene)FeCp$^+$ complex underwent smooth double substitution with phenethyl amine, whereas a dichloro complex led to monosubstitution. Replacement of a nitro group in η^6-nitrobenzene-FeCp$^+$ is possible with O, S, and N-nucleophiles [70, 80, 81]. The nitroarene complexes, which are not known in the Cr(CO)$_3$ series, are prepared by oxidation of the corresponding aniline complexes.

Decomplexation of the arene in (arene)FeCp$^+$ is typically not easy. Pyrolysis in high vacuum or using microwave radiation and graphite flakes [82] is limited to arenes stable at temperatures around 200 °C. Strong base (e.g., KO^{t}Bu) in DMF [72] allows decomplexation only of the base-stable arenes. Electrochemical reduction is useful for ketones and esters but not convenient on large scale [83]. Irradiation in acetonitrile is the most popular procedure, phenanthroline is sometimes added to improve decomplexation [72], although for delicate substrates, such as amino acid derivatives, no reliable procedure was found so far.

2.2.2
Nucleophilic Substitution with Carbon Nucleophiles

Similar to Cr(CO)$_3$ complexes, less stabilized carbon nucleophiles such as MeLi and acetone enolate add irreversibly at the position *ortho* to chloride in (chlorobenzene)FeCp$^+$ [84]. Nucleophiles stabilized by a nitro, β-dicarbonyl, or related groups add reversibly with ultimate formation of a substitution product [66, 85]. Ketoarene complexes **35** can be prepared by addition of a malonate derivative followed by acid-induced decarboxylation [86].

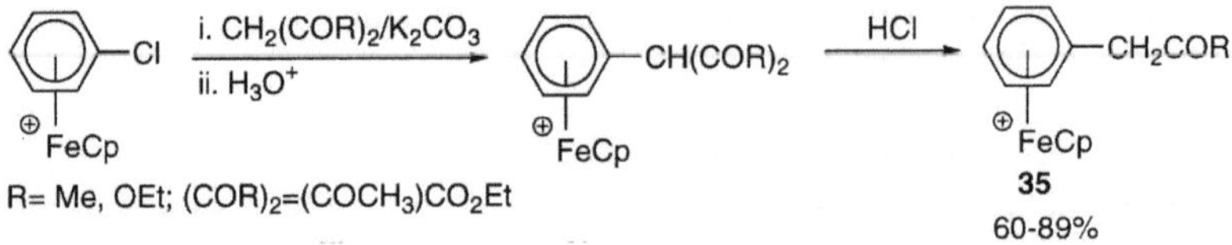

(19)

(Dichlorobenzene)FeCp$^+$ reacts with diethylmalonate only once due to in situ deprotonation of the product; addition of excess MeOH or methylthioglycolate allows disubstitution [66]. (Nitroarene)FeCp$^+$ complexes have also been used with carbon nucleophiles [87] and again reactivity similar to the chloroarene complexes is observed. Application of carbon nucleophile substitution chemistry for the synthesis of heterocycles such as **36** is described [88].

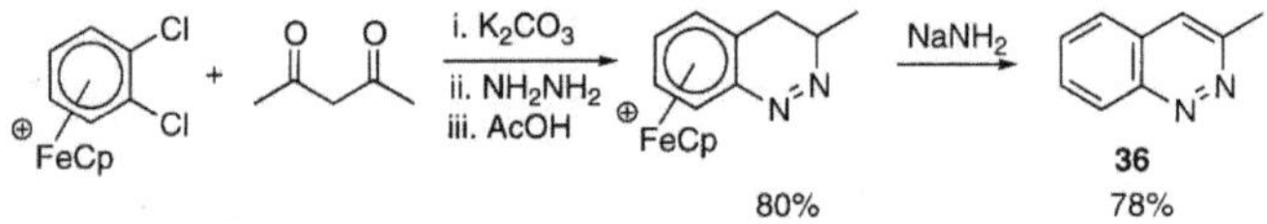

(20)

2.3 Halobenzene-RuCp$^+$ Complexes

RuCp$^+$ is very similar to FeCp$^+$ as an arene activating group, albeit a little less electrophilic. The advantages of Ru compared to Fe are mild complexation and demetallation conditions. Efficient substitution for chloride in (chloroarene)RuCp$^+$ by aryloxides in presence of base-sensitive functionality was demonstrated [62]. This initial success lead to the synthesis of a series of vancomycin-related biaryl ethers via intramolecular S_NAr cyclization in a preformed dipeptide derivative (e.g., 37) [89, 90].

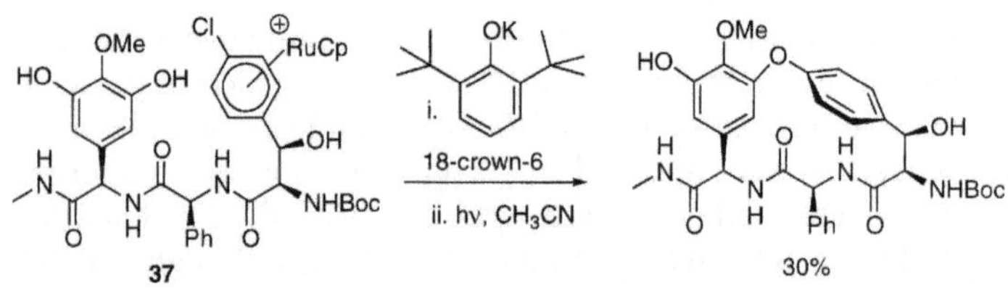

(21)

(Chloroindole)RuCp$^+$ as well as (nitroindole)RuCp$^+$ react with a variety of O, S, N, and C-nucleophiles to give substitution products [91]. One example of an enolate nucleophile effecting an intramolecular cyclization is shown in Eq. (22) [92].

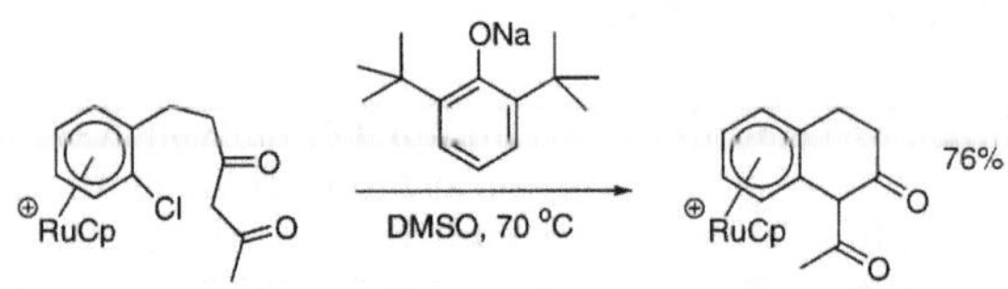

(22)

Soluble polymers are obtained in reaction of (*p*-dichlorobenzene)RuCp$^+$ with aryldioxides and disulfides [93]. Selective substitution is again possible with O and N nucleophiles in reactions with *m*- and *p*-(dichlorobenzene)RuCp$^+$ complexes [94]. Electron-rich arenes resulting from multiple nucleophilic substitutions are very good ligands for RuCp$^+$ and are therefore hard to remove by conventional methods. The requirement for a stoichiometric amount of Ru is somewhat costly; however, it was shown that it can be recycled after the reaction in the form of the tris-acetonitrile complex $(CH_3CN)_3RuCp^+$ [92].

2.4
Halobenzene-Mn(CO)$_3$$^+$

(Chlorobenzene)Mn(CO)$_3$$^+$ is easily prepared and undergoes direct substitution with alkoxy, phenoxy, thiolate, amine, and azide nucleophiles as well as aniline and selenide nucleophiles [2]. (Fluorobenzene)Mn(CO)$_3$$^+$ is known, but so reactive it is said to solvolyze rapidly during purification [97]. A promising application is the introduction of an asymmetric amine unit that allows differentiation of the two diastereotopic *meta* positions (in **39**) [96].

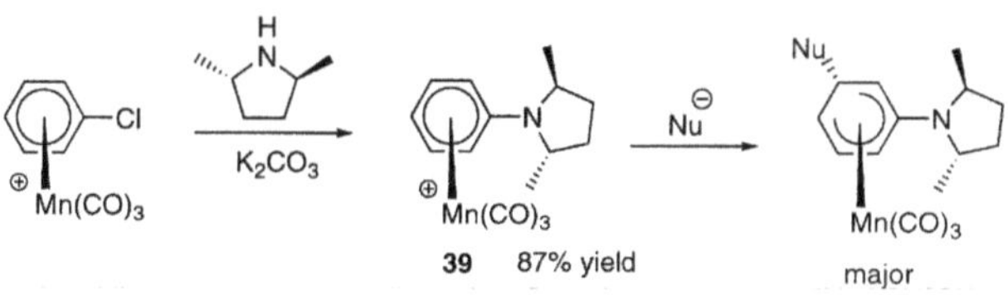

(23)

In several cases spontaneous loss of the metal unit was observed as a competing side-reaction during attempts to effect nucleophilic substitution [95, 98].

2.5
Halobenzene-RhCp(Et)$^{2+}$ Catalytic S$_N$Ar

The only known example of the catalytic nucleophilic aromatic substitution involves the use of Rh(III) complexes (e.g., **40**). The first demonstration was made in 1980 [99]. The metal center is so electrophilic that an internal alcohol nucleophile substitutes directly without an added base.

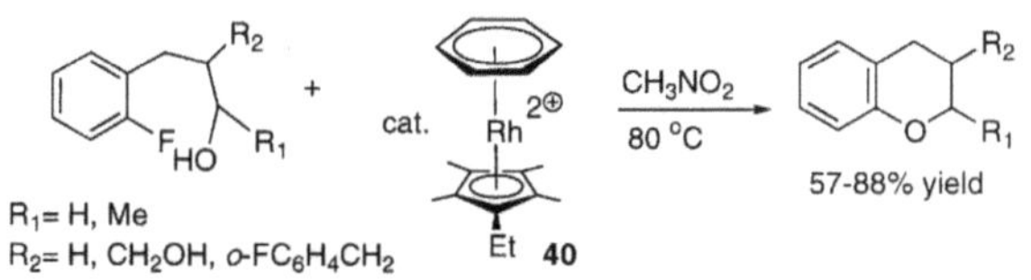

(24)

It soon became apparent, however, that the system is severely limited: attempts to cyclize the corresponding ketones (enol), amides, or amines failed [99]. It was also shown that a nitro substituent on the arene prevents the cyclization (possibly by disfavoring the complexation of this electron-poor arene to the metal) while an electron-donating methoxy substituent slows down the nucleophilic substitution reaction significantly. Strongly coordinating solvents like DMF or methanol prevent the reaction, probably by coordinating to the metal and thereby removing it from the catalytic cycle [99, 100]. Later investigations involved an intermolecular version of the reaction. Methanol was successfully used as a nucleophile on a variety of fluoroarene substrates; *p*-fluorotoluene gave 79% yield based on the arene and 980% yield based on the catalyst [100, 101].

3
Addition/Oxidation: Formal Nucleophilic Substitution for Hydrogen

3.1
Addition/Oxidation with Arene-$Cr(CO)_3$ Complexes

3.1.1
General Features

Once it is recognized that cyclohexadienyl anionic complexes of chromium (**41**) can be generated by addition of sufficiently reactive nucleophiles and that simple oxidizing techniques convert the anionic intermediates to free substituted arenes, a general substitution process becomes available which does not depend on a specific leaving group on the arene [2]. The process is general for carbanions derived from carbon acids with pK_a>22 or so; only one example of a heteroatom nucleophile is reported [102].

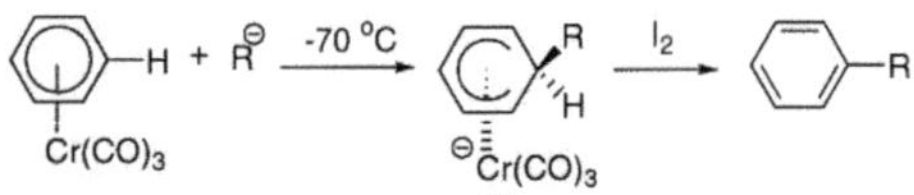

41

(25)

With anions such as ester enolate anions, reaction occurs within minutes at -78 °C and the intermediate cyclohexadienyl complex can be observed spectroscopically [103]. The intermediates are exceedingly air sensitive and are generally quenched directly, without purification. In one case, from the addition of 2-lithio-1,3-dithiane, the adduct has been crystallized and fully characterized by X-ray diffraction analysis [103]. Oxidation with excess iodine (at least 2.5 mol-

Table 1 Reactivity of carbanions (R-Li) toward benzene-$Cr(CO)_3$

Unreactive	Successful	Metallation
1. $LiCH(CO_2Me)2$	7. $LiCH_2CO_2$*t*-Bu	18. *n*-BuLi
2. $LiCH_2COtBu$	8. $LiCH_2CN$	19. MeLi
3. MeMgBr	9. KCH_2CO*t*-Bu	20. *s*-BuLi
4. *t*-BuMgBr	10. LiCH(CN)(OR)	
5. Me_2CuLi	11. $LiCH_2SPh$	
6. LiC(CN)(Ph)(OR)	12. 2-Li-1,3-dithiane	
	13. $LiCH{=}CH_2$	
	14. LiPh	
	15. LiC=CR	
	16. $LiCH_2CH{=}CH_2$	
	17. $LiC(Me)_3$	

eq of I_2) also proceeds at -78 °C and is complete within hours below room temperature. The products are the substituted arene, HI, Cr(III), and CO. For acid-sensitive products such as trialkylsilyl-substituted arenes, an excess of amine (conveniently, diisopropylamine) can buffer the mixture. Table 1 provides a representative sample of the carbanions which have been tested with η^6-(benzene)Cr(CO)$_3$. In addition to those nucleophiles, several dithiane derivatives have been tested as well [104, 105].

3.1.2
Regioselectivity

Regioselectivity has been the subject of numerous studies with a variety of mono- and polysubstituted arene ligands. The addition of synthetically-interesting carbanions to the typical arene ligand, especially those bearing electron-donating substituents, can be reversible at rates comparable to the rate of addition: it is not always obvious whether to assume kinetic or thermodynamic control [35, 106–109]. However, correlations have been suggested which allow prediction or rationalization of regioselectivity with a modest degree of confidence. With some significant exceptions as discussed below, the difference between the kinetic and thermodynamic selectivity has not been determined or is small.

With arenes bearing a single resonance donor substituent (NR_2, OMe, F), the addition is strongly preferred at the *meta* position, with small amounts of *ortho* substitution (0–10%) (Eq. 26) [110–112]. Representative examples are shown in Table 2.

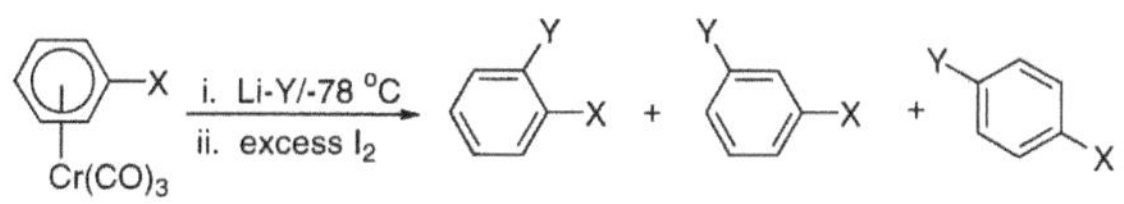

(26)

Table 2 Addition/oxidation with anisole, *N,N*-dimethyl aniline, and fluorobenzene Cr(CO)$_3$ complexes

Entry	X	LiY	Ratio (o:m:p)	(Combined yield)
1	OMe	$LiCH_2CN$	3:97:0	(38%)
2	OMe	$LiC(Me)_2CN$	3:97:0	(93%)
3	OMe	LiCH2CO$_2$*t*-Bu	6:94:0	(86%)
4	OMe	LiCH(Me)CO$_2$*t*-Bu	4:96:0	(93%)
5	OMe	LiC(Me)$_2$CO$_2$*t*-Bu	0:100:0	(76%)
6	OMe	LiC(CN)(OR)CH$_2$Ph[a]	0:100:0	(75%)
7	OMe	2-Lithio-1,3-dithiane	10:90:0	(35%)[b]
8	$N(Me)_2$	$LiC(Me)_2CN$	1:99:0	(92%)
9	F	$LiC(Me)_2CN$	2:98:0	(84%)

[a] The R group is CH(Me)(OEt)
[b] A major side reaction is *ortho* deprotonation of the anisole ligand

The *meta* acylation of the anisole ligand in **42**, using a carbonyl anion equivalent (**43**) as the nucleophile, illustrates the unique regioselectivity available with the $Cr(CO)_3$ activation, producing **44** [110].

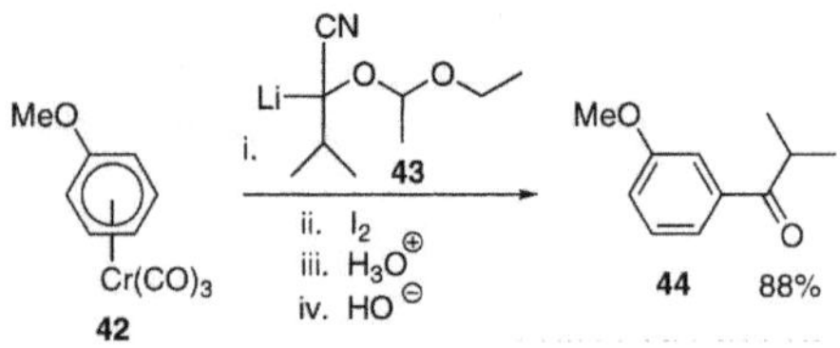

(27)

Selectivity is more complicated with a methyl or chloro substituent. Again, *meta* substitution is always significant, but *ortho* substitution can account for 50–70% of the mixture in some cases [2]. More reactive anions (1,3-dithianyl) and less substituted carbanions (e.g., *tert*-butyl lithioacetate) tend to favor *ortho* substitution. Representative examples are shown in Table 3. Entries 2–4 show that variation of reaction temperatures from -100 °C to 0 °C has no significant effect in that highly selective system. The added activating effect of the Cl substituent allows addition of the pinacolone enolate anion (entry 11), whereas no addition to the anisole nor toluene ligand is observed with the same anion.

Table 4 also includes results with *para* directing substituents, CF_3 and Me_3Si (entries 14 and 15). Another example of a *para*-directing group is phenylsulfonyl which activates the ring towards a variety of nucleophiles including allyl MgBr [113]. In general, electron withdrawing substituents such as acyl and cyano are

Table 3 Addition/oxidation with chlorobenzene and toluene $Cr(CO)_3$ complexes

Entry	X	LiY	Ratio (o:m:p)	(Combined yield)
1	Me	$LiCH_2CN$	35:63:2	(88%)
2	Me	$LiC(Me)_2CN$ (-100 °C/1.5 min)	2:96:2	(52%)
3	Me	$LiC(Me)_2CN$ (-78 °C/1.5 min)	1:97:2	(95%)
4	Me	$LiC(Me)_2CN$ (0 °C/20 min)	1:97:2	(86%)
5	Me	$LiCH_2CO_2$*t*-Bu	28:72:0	(89%)
6	Me	$LiC(Me)_2CO_2$*t*-Bu	3:97:0	(96%)
7	Me	2-Lithio-1,3-dithiane	52:46:2	(94%)
8	Cl	$LiCH_2CO_2$*t*-Bu	54:45:1	(98%)
9	Cl	$LiCH(Me)CO_2$*t*-Bu	53:46:1	(88%)
10	Cl	$LiC(Me)_2CO_2$*t*-BU	5:95:0	(84%)
11	Cl	$LiCH_2COC(Me)_3$	70:24:0	(87%)
12	Cl	$LiC(Me)_2CN$	10:89:1	(84%)
13	Cl	2-Lithio-1,3-dithiane	46:53:1	(56%)
14	CF_3	LiC(CN)(Me)(OR)[a]	0:30:70	(33%)
15	$SiMe_3$	$LiC(Me)_2CN$	0:2:98	(65%)

[a] The R group is -CH(Me)(OEt)

Table 4 Steric effects on regioselectivity

Entry	Complex	Anion	Product ratio (o:m:p)	Combined yield
1	X=Me	$LiCH_2CN$	35:63:2	88
2	X=*t*-Bu	$LiCH_2CN$	28:48:24	51
3	X=Me	LiC(OR)MeCN[a]	0:96:4	75
4	X=Et	LiC(OR)MeCN[a]	0:96:4	89
5	X=*i*-Pr	LiC(OR)MeCN[a]	0:80:20	88
6	X=*t*-Bu	LiC(OR)MeCN[a]	0:35:65	86
7	X=Me	$LiC(Me)_2CN$	1:97:2	95
8	X=*t*-Bu	$LiC(Me)_2CN$	0:55:45	78
9	X=CH(*t*-BU)$_2$	$LiC(Me)_2CN$	0:0:100	63
10	X=Me	$LiCH_2SPh$	52:46.2	96
11	X=*i*-Pr	$LiCH_2SPh$	47:46:7	86
12	X=*t*-Bu	$LiCH_2SPh$	45:32:23	88

[a]R=C(Me)OEt

also activated toward nucleophile addition to the substituent itself (C=O or CN) and this process competes with addition to the ring. Cyano-stabilized nucleophiles, on the other hand, give mixtures of *m*- and *p*-addition products [114].

Regioselectivity in *o*-disubstituted arenes is often high and useful. A series of examples is summarized in Fig. 2. The resonance donor substituent (OMe) appears to dominate the directing influences, favoring addition at the less hindered position *meta* to itself [2]. However, with two identical substituents, the addition is preferred in the adjacent position [2]. For example, with the complex of benzocyclobutene (**45**), six carbon nucleophiles were tested and each gave addition exclusively at C-3 [2, 115]. With the complex of indane (**46**), selective addition at C-3 was observed with dithianyl anions, but cyano-stabilized anions gave up to 20% of the isomeric product from addition at C-4. Substituents at the benzylic carbons in the indane ligand have a strong effect on selectivity, and can lead to the 1,2,4-substitution product [2].

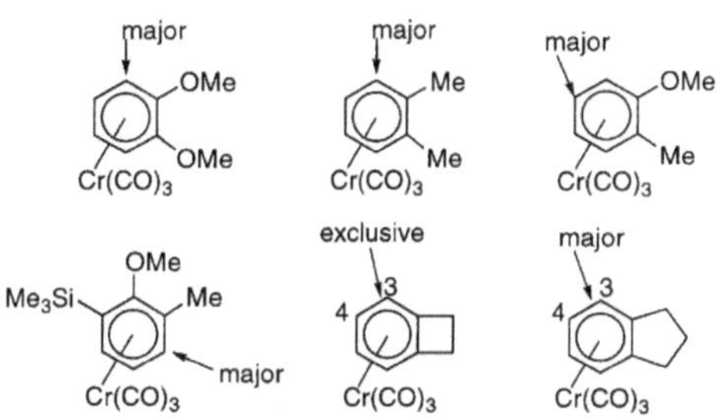

45 **46**

Fig. 2 Summary of regioselectivity of addition to substituted (Arene)Cr(CO)$_3$ complexes

Addition to *o*-alkyl aniline complexes has been the subject of a detailed study, and presents a clear case of different products from kinetic vs thermodynamic control [108]. Addition of 2-lithio-2-cyanopropane to *N*-methyl-(tetrahydroquinoline)-$Cr(CO)_3$ (**47**), with variation in time, temperature, and solvent (THF or THF with 4 mol-eq of HMPA added) gives product distributions ranging from 2:1 for C-5/C-3 (**48/49**) at -78 °C for one min to >96% addition at C-5 when the adducts (**50** and **51**) are allowed to equilibrate (8.6 h at -78 °C). The data show that equilibration is occurring within minutes at -78 °C, and the product distribution does not further change upon warming to 20 °C. The effect of added HMPA was also observed in halide substitution (addition/elimination) and clearly established during addition to simple arenes [35, 116]. With the less sterically demanding anion, $LiCH_2CN$, addition is incomplete after 22 h at -78 °C, and favored at C-3 (82:18 for C-3/C-5).

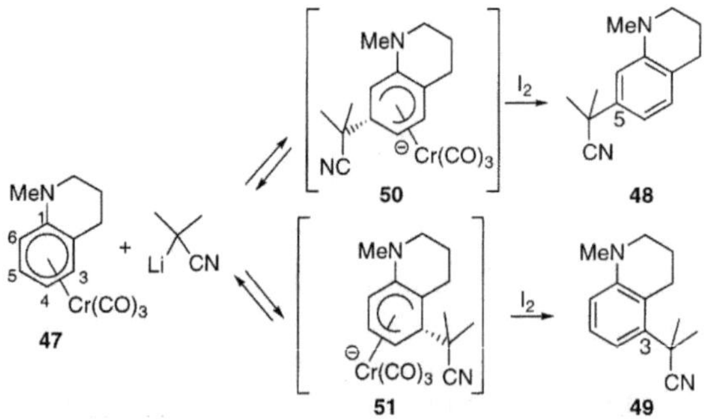

Scheme 4 Addition of 2-lithio-2-cyanopropane to N-methyl-(tetrahydroquinoline)$Cr(CO)_3$

The same study shows that, with the complex of *N,N*-dimethyl *o*-toluidine, the selectivity depends on time, temperature, and the nature of the anion [107]. Again equilibration occurs with the tertiary nitrile-stabilized anion, favoring the C-4 substitution product, while lithioacetonitrile favors addition at C-3. The 2-methyl-1,3-dithianyl anion gives precisely the same product mixture at -78 °C and at -30 °C; there is no evidence for equilibration with this anion.

Indole is a particularly interesting case, because the $Cr(CO)_3$ unit selectively activates the six-membered ring (**52**) [104, 117, 118], while in free indole the five-membered ring dominates the (electrophilic addition) reactivity. The selectivity in addition to the $Cr(CO)_3$ complexes of indole derivatives shows a preference for addition at C-4 (indole numbering) and C-7, with steric effects due to substituents at C-3 and N-1 as well as anion type influencing the selectivity [118].

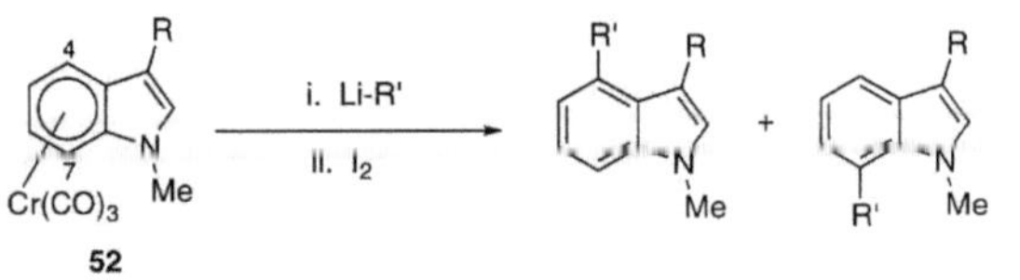

(28)

A. R' = $C(Me)_2CN$; R=H; Y=Me	99	:	1	(92%)
B. R' = (1,3-dithianyl); R=H ; Y=Me	14	:	86	(68%)
C. R' = $C(Me)_2CN$; R=CH_2SiMe_3	17	:	83	(82%)
D. R' = $C(Me)_2CN$; R=CH_2SiMe_3 ; Y=Si(*t*-Bu)$(Me)_2$	95	:	5	(78%)

Another example in which the regioselectivity of addition is different under kinetic vs thermodynamic control is the naphthalene series. In the addition of $LiC(Me)_2CN$ to naphthalene-$Cr(CO)_3$ (**53**), a mixture of products is observed from addition at C-α and C-β in the ratio 42:58 under conditions where equilibration is minimized (0.3 h, -65 °C, THF/HMPA). With the same reactants, but in THF and at 0 °C, the product is almost exclusively the α-substituted naphthalene (**54**) [119].

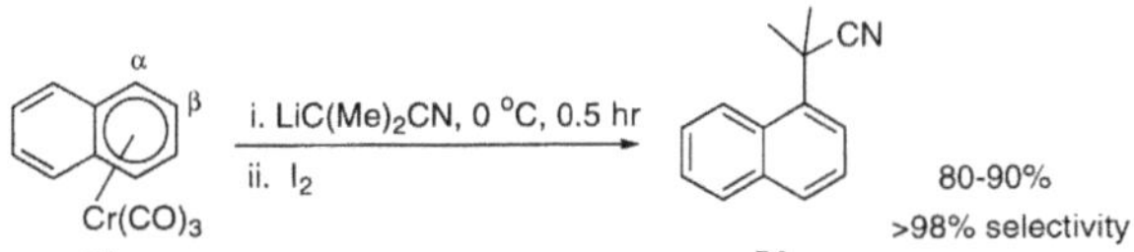

(29)

Using the standard procedures, 1,4-dimethoxynaphthalene is complexed at the less substituted ring with high selectivity to give (**55**) [35]. Again, under conditions of minimum equilibration of anion position, $LiC(Me)_2CN$ gave a mixture (after iodine oxidation) of the 1,4-dimethoxy-β-substituted and 1,4-dimethoxy-α-substituted products in the ratio 78:22. After equilibration, α-substitution (**56**) was essentially the only product found.

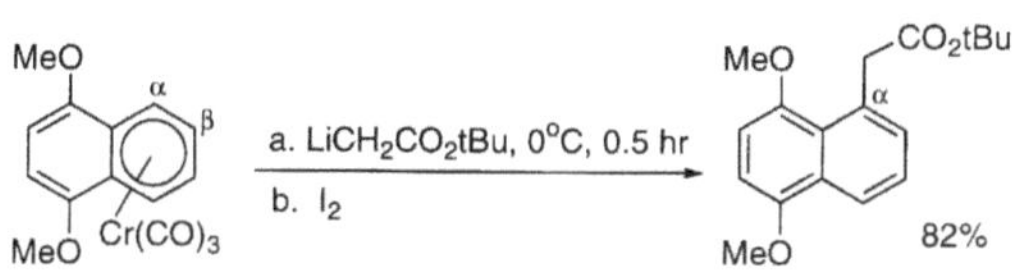

(30)

The 1,4-dimethoxynaphthalene ligand was used to probe for the parameters which influence the rate of equilibration. The change from pure THF to a mixture of 3:1 THF:HMPA slows the rate of equilibration by a factor of 50,000 [35, 106]. Replacement of Li with K in the carbanion nucleophile also slows equilibration, by a factor of 500. The equilibration is shown to be intermolecular (dissociation of the nucleophile) by elegant "crossover" experiments with doubly labeled substrates. With the 1,4-dimethoxynaphthalene ligand, cyano-stabilized anions (including cyanohydrin acetal anions) and ester enolates equilibrate even at low temperature and strongly favor addition at the α-position (C-5). The kinetic site of addition is also generally C-α. However, the 2-lithio-1,3-dithiane anion and phenyl lithium do not equilibrate over the temperature range -78 °C to 0 °C. The sulfur-stabilized anions favor addition at C-α while phenyl lithium gives a mixture favoring C-β.

The steric effect on regioselectivity shows clearly in the series in Eq. (31), Table 4. Comparing similar anion type, except for size, entries 1, 3, and 7 show that *ortho* substitution is very significant with a primary carbanion but essentially absent with a tertiary cyano-stabilized anion. It is striking that as the size of the alkyl substituent on the arene increases, not only is *ortho* substitution disfavored, but *meta* is as well: compare entries 3–6 and compare 7–9. With the very large $CH(t\text{-}Bu)_2$ group (entry 9), only *para* substitution is observed. Regioselectivity is also dependent on the electronic nature of the nucleophile. Most remarkably, addition of the primary sulfur-stabilized anion shows nearly equal amounts of *ortho* and *meta* substitution with the toluene ligand, but, as the size of the arene substituent increases, the *para* substitution product increases at the expense of the meta product (entries 10–12). The dependence of selectivity on anion type and ring substituent is understood as a change in the balance of charge vs orbital control [2].

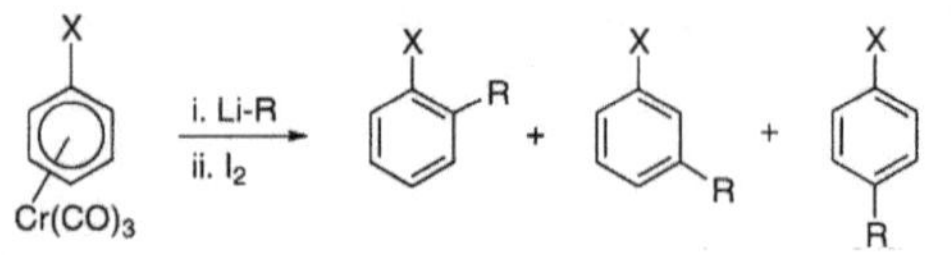

(31)

Imine-type substituents on the arene ligand can also favor ortho addition of organolithium reagents, presumably via coordination of the lithium with the side chain imine [120]. The *ortho* selectivity was particularly high with the less stabilized reagents such as MeLi, *t*-BuLi, PhLi, vinyl-Li and $LiCH_2S(O)Ar$. Figure 3 displays the results with two complexes, and the suggested picture for the selectivity. The products, after an addition/hydride abstraction protocol using an asymmetric ligand and trityl cation, are planar chiral complexes and can be produced with up to 95% enantioselectivity [121].

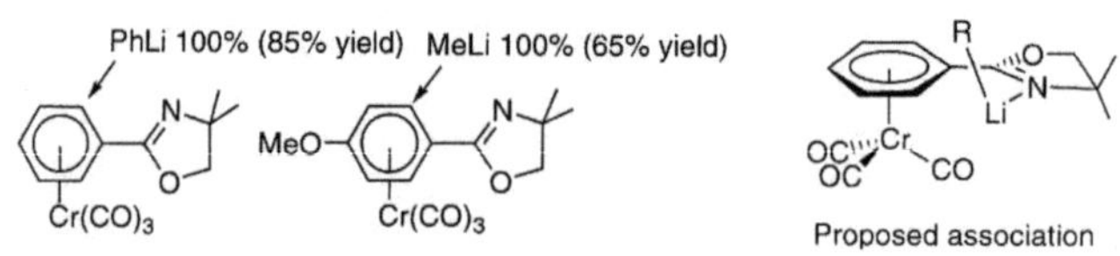

Fig. 3 Ortho directing effects of an oxazoline substituent

3.1.3
Intramolecular Addition/Oxidation

Intramolecular addition/oxidation with reactive carbanions is generally successful; most of the examples involve cyano-stabilized carbanions. Formation of a six-membered fused ring (**57**) is efficient, but five-membered fused ring formation is not [2]. In that case, the only addition/oxidation product is a cyclic dimer, the [3.3]-metacyclophane, (**58**).

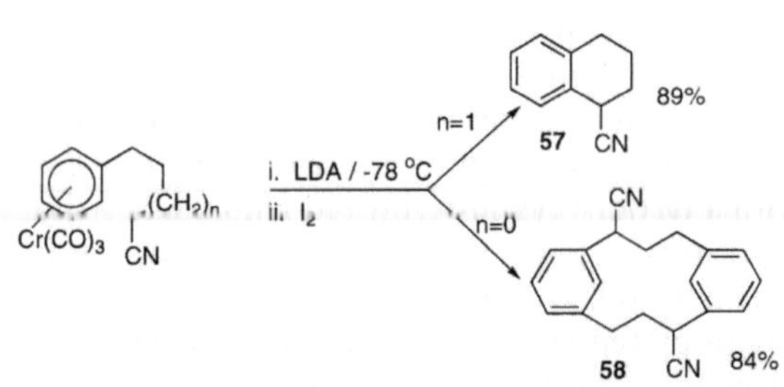

Scheme 5 Intramolecular cyclyzation of cyano-stabilized carbanion

Reversibility again is apparent with the higher homologue (**59**). At low temperature (0.5 h/-78 °C), a mixture of cyclohexadienyl anionic intermediates is formed with the spiro ring isomer (**60**) preferred by a factor of almost 3:1. An alternative quenching procedure, using trifluoroacetic acid (see below) retains the spiro ring and produces the spiro[5.5]cyclohexadiene product, (**61**). If the initial adduct is allowed to warm to 0 °C or above, essentially complete rearrangement to the fused ring adduct (**62**) occurs, and oxidative quenching gives the seven-membered ring (**63**) in good yield, [2]. Efficient fusion of the 6-membered ring was observed in formation of perhydroisoquinoline derivatives [122].

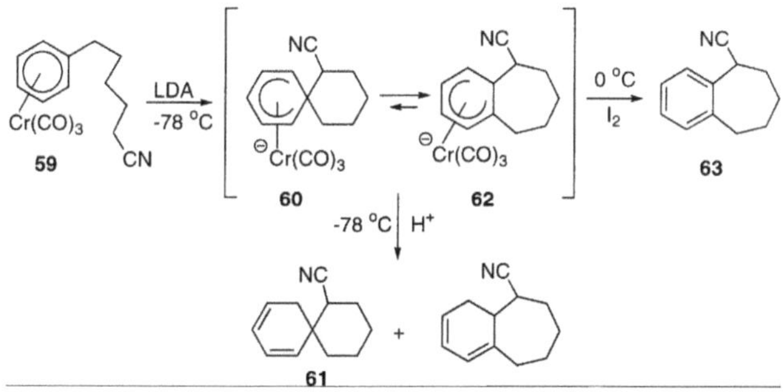

Scheme 6 Intramolecular cyclyzation of cyano-stabilized carbanion (evidence of reversibility)

3.2
Addition/Oxidation with (Arene)FeCp Cation Complexes

It has been recognized for many years that nucleophiles will add to η^6-benzene-FeCp$^+$ complexes. Reactive nucleophiles such as cyanide, hydride [123], methyl-, and phenyl lithium [2] add to unactivated (arene)FeCp$^+$ complexes, whereas less reactive nucleophiles such as phenyl acetylide, lithium salts of acetonitrile, nitromethane and ketones [84, 124] require an electron-withdrawing group to be present on the arene ring of the complex. On the other hand, cyanide or an anion derived from propionitrile added to several unactivated arene complexes [67, 76, 123–126].

(Dichlorobenzene)FeCp$^+$ reacted with a series of enol silyl ethers (e.g., **64**) activated by fluoride [127].

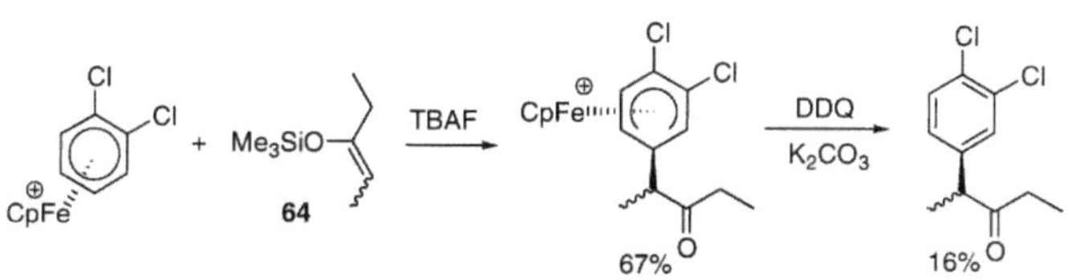

(32)

The regioselectivity of hydride addition has been determined for a number of monosubstituted arene ligands [2]. Chloro and carbomethoxy substituents are *ortho*-directing, while methoxy is a *meta* director, and a methyl group leads to similar amounts of addition at the *o*, *m*, and *p* positions. In general, electron withdrawing groups (nitro, halogen, benzoyl, cyano, and sulfonyl) favor *ortho* addition, as exemplified for the addition of the acetone enolate in Eq. (33) [2].

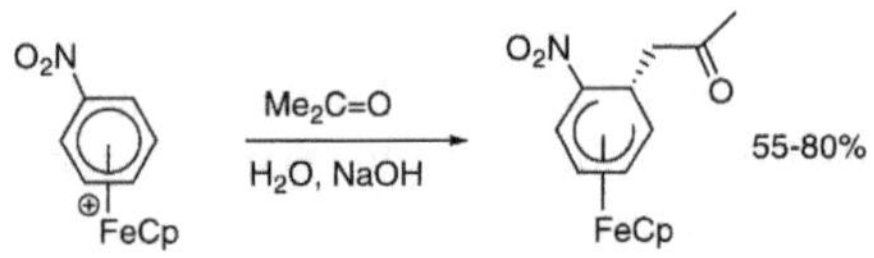

(33)

In most cases the intermediate neutral η^5-(6-substituted)-cyclohexadienyl-FeCp complexes can be isolated and characterized. No efficient and general procedure for demetallation exists, however, and this severely limits the applicability of this chemistry.

3.3
Addition/Oxidation with Cationic (Arene)RuCp Complexes

Analogous chemistry with Ru complexes is much less developed. Hydride and phenyllithium add to (arene)$RuCp^+$ complexes in good yields [128]. Inter- [127] and intramolecular [129] enolate additions are also successful. The resulting η^5 species can be rearomatized using DDQ in good yield; demetallation was a competing process (15% yield of the free arene), Eq. (34) [127].

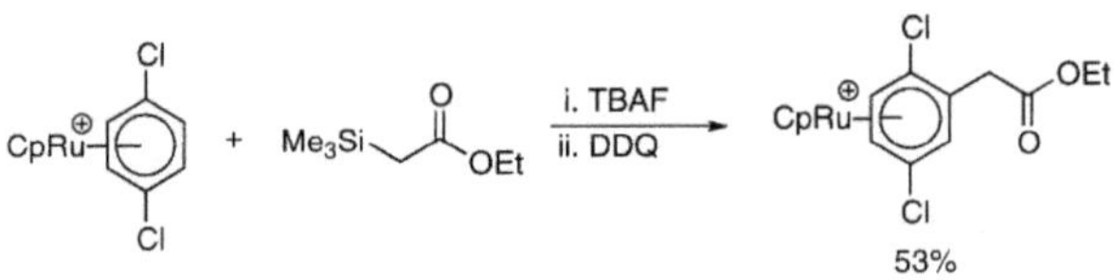

(34)

3.4
Addition/Oxidation with Arene-$Mn(CO)_3$ Cation Complexes

(Arene)$Mn(CO)_3{}^+$ complexes are significantly more electrophilic than (arene)$FeCp^+$ or (arene)$Cr(CO)_3$ analogs. Sensitive functionality readily tolerate the mild complexation and demetallation conditions. A list of successful nucleophiles includes stabilized and unstabilized organolithiums [2, 130], Grignard reagents [131], a variety of hydride reagents [132] and cyanide [133, 134]. Neutral η^5-intermediates (e.g., **65**) can be isolated and purified. Hydride abstraction from the intermediate is possible after it rearranges via hydride migration to expose an *exo* hydrogen [135]. Oxidation with Jones reagent or DDQ in the presence of a strong sigma donor solvent such as acetonitrile provides the substituted arene and Mn(I) in a recyclable form [136].

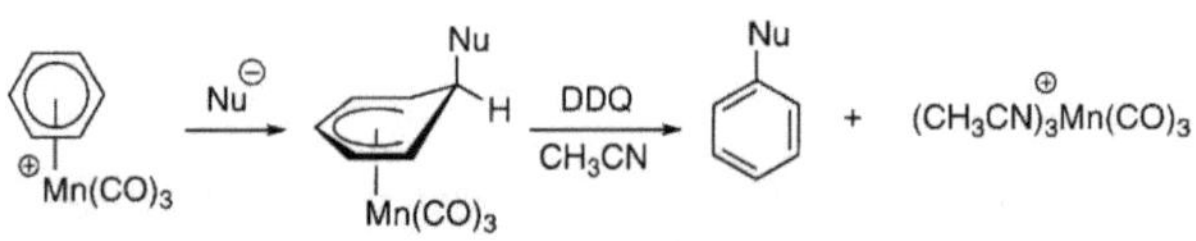

65

(35)

The regioselectivity is similar to the (arene)$Cr(CO)_3$ case. Strong electron donors such as OMe and NMe_2, as well as NHC(O)R direct addition to the *meta* position, whereas Cl (electron-withdrawing) favors the *ortho* position [2]. Alkyl substituents typically yield mixtures of regioisomers. The steric effects become important for the (*tert*-butylbenzene)$Mn(CO)_3^+$ complex which yields large amounts of the *p*-product [2]. Ketone enolates are sufficiently reactive to add to the arene ring [137]. A chiral auxiliary on the nucleophile allows diastereoselective phenylation Eq. (36) [138].

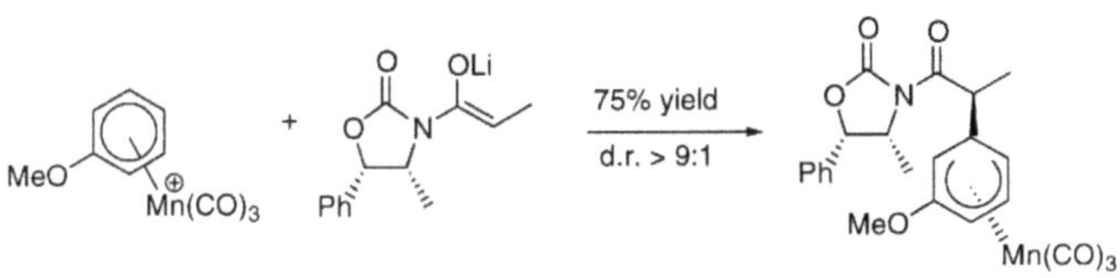

(36)

Other chiral nucleophiles that have been employed include Schollkopf's (**66**) and Williams's (**67**) chiral glycine enolate equivalent [139, 140].

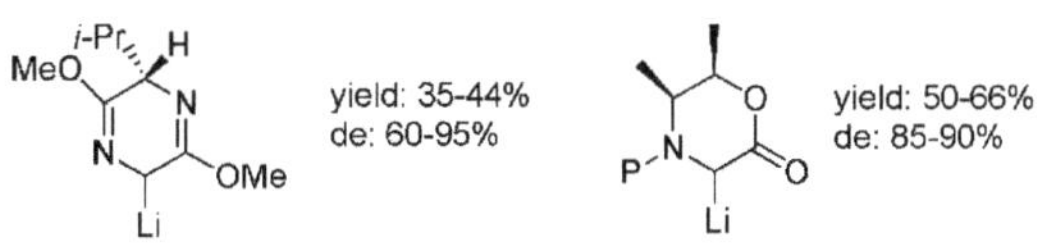

66 **67**

Fig. 4 Chiral nucleophiles added to (Arene)$Mn(CO)_3^+$ complexes

A chiral auxiliary can also be positioned on the arene ring itself [96, 141, 142]. Addition of nucleophiles then proceeds regioselectively away from the metal and on the side not blocked by the methyl group of this auxiliary. Reversal of the selectivity for certain nucleophiles is not well understood.

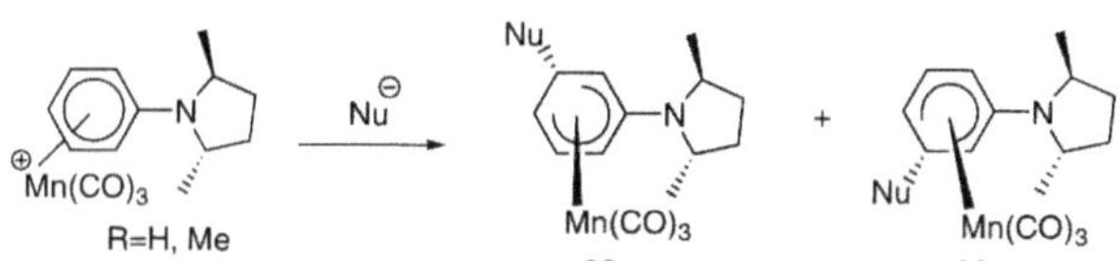

68 **69**

(37)

Nu	combined yield (%)	ratio **68:69**
PhLi	52	2:1
PhMgBr	60	19:1
CH_2=CHMgBr	50	2:1
CH_2=$CHCH_2MgBr$	65	4:1
MeLi	60	1:3

A wide variety of nucleophile-stabilizing groups has been found compatible. They include imine, cyano, sulfonyl, nitro, and alkoxycarbonyl [143, 144] as well as more esoteric substituents such as diazo [145], phosphine [146], phosphate [147], metal carbene [148], and $Cr(CO)_3$ [149, 150]. Malonate, on the other hand, adds but the resulting carbon-arene bond is too weak and is cleaved under demetallation conditions to regenerate the starting material [151].

Polyaromatic and heteroaromatic arenes can be coordinated to the $Mn(CO)_3^+$ moiety and attacked by a variety of nucleophiles [2]. An interesting example is the attack on the (thiophene)$Mn(CO)_3^+$ complex [152, 153]. Hydride and cyanide add to the carbon adjacent to the S [148] whereas organomagnesium and cuprate reagents add to the sulfur itself.

Substituted arene complexes are typically attacked at the unsubstituted position. For example, the tetralin complex undergoes addition to the carbon *ortho* to the ring [131]. On the other hand, the biphenylene complex is attacked at the ring junction, possibly due to the relief of the angle strain (Eq. 38) [154].

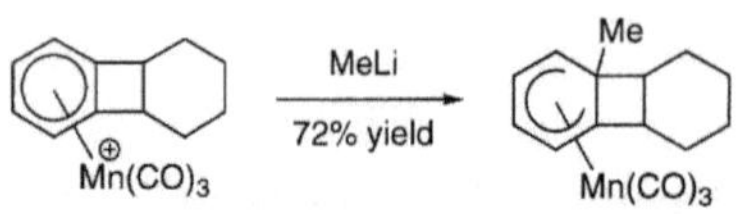

(38)

When amines were used as nucleophiles, attack at the CO ligand occurred [155]. Phosphines add to the arene ring in a reversible fashion, although in the presence of light, irreversible substitution for CO occurs [156].

References

1. Gill US, Moriarty RM, Ku YY, Butler IR (1991) J Organomet Chem 417:313
2. Semmelhack MF (1995) Transition metal arene complexes: nucleophilic addition. In: Abel EW, Stone FGA, Wilkinson G (eds) Comprehensive organometallic chemistry II. Pergamon, Oxford New York Tokyo, Sect 9.1
3. Rose-Munch F, Gagliardini V, Renard C, Rose E (1998) Coord Chem Rev 178/180:249
4. Pike RD, Sweigart DA (1999) Coord Chem Rev 187:183
5. Balas L, Latxague L, Grelier S, Morel Y, Hamdani M, Ardoin N, Astruc D (1990) Bull Soc Chim Fr 127:401
6. Rose-Munch F, Rose E (1999) Curr Org Chem 3:445
7. Sutherland RG, Iqbal M, Piorko A (1986) J Organomet Chem 302:307
8. Astruc D (1991) Top Curr Chem 60:47
9. Abd-El-Aziz AS, Bernardin S (2000) Coord Chem Rev 203:219
10. Moriarty RM, Gill US, Ku YY (1988) J Organomet Chem 350:157
11. Pigge FC, Coniglio JJ (2001) Curr Org Chem 5:757
12. Sun S, Dullaghan CA, Sweigart DA (1996) J Chem Soc Dalton Trans 4493
13. Knipe AC, McGuinness SJ, Watts WE (1981) J Chem Soc Perkins Trans II 193
14. Bunnett JF, Hermann H (1971) J Org Chem 36:4081
15. Maiorana S, Baldoli C, Licandro E, Casiraghi L, de Magistris E, Paio A, Provera S, Seneci P (2000) Tetrahedron Lett 41:7271
16. Semmelhack MF, Hilt G, Colley JH (1998) Tetrahedron Lett 39:7683
17. Gibson SE, Hales NJ, Peplow MA (1999) Tetrahedron Lett 40:1417
18. Whiting MC (1966) US Patent 3 225 071
19. Hamilton J, Mahaffy CAL (1986) Synth React Inorg Met-Org Chem 16:1363
20. Davies SG, Hume WE (1995) J Chem Soc Chem Commun 251
21. Semmelhack MF, Schmalz H-G (1996) Tetrahedron Lett 37:3089
22. Dickens MJ, Gilday JP, Mowlem TJ, Widdowson DA (1991) Tetrahedron 47:8621
23. Hamilton J, Mahaffy CAL (1986) Synth React Inorg Met-Org Chem 16:61
24. Baldoli C, Del Buttero P, Licandro E, Maiorana S (1988) Synthesis 344
25. Kirschke K, Deutsch J, Niclas H-J (1996) Phosphorus 117:293

26. Alemagna A, Cremonesi P, Del Buttero P, Licandro E, Maiorana S (1983) J Org Chem 48:3114
27. Perez M, Potier P, Halazy S (1996) Tetrahedron Lett 37:8487
28. Maiorana S, Baldoli C, Del Buttero P, Di Ciolo M, Papagni A (1998) Synthesis 735
29. Baldoli C, Del Buttero P, Maiorana S (1992) Tetrahedron Lett 33:4049
30. Nelson SG, Hilfiker MA (1999) Org Lett 1:1379
31. Andersson Y, Langstrom B (1994) J Chem Soc Perkin Trans I 1395
32. Del Buttero P, Baldoli C, Molteni G, Pilati T (2000) Tetrahedron: Asymmetry 11:1927
33. Baldoli C, Del Buttero P, Maiorana S, Papagni A (1985) J Chem Soc Chem Commun 1181
34. Litvak VV, Goryunov LI, Shteingarts VD (1986) Zh Org Khim 22:157
35. Kündig EP, Desobry V, Simmons DP, Wenger E (1989) J Am Chem Soc 111:1804
36. Hong F-E, Sun S-Y, Wang S-L, Liao F-L (1998) Polyhedron 17:1643
37. Qian C, Guo J, Sun J, Chen J, Zheng P (1997) Inorg Chem 36:1286
38. Deutsch J, Bartoszek M, Schneider M, Zabel V, Niclas H-J (1997) J Prakt Chem 339:171
39. Deutsch J, Bartoszek M, Schneider M, Zabel V, Niclas H-J (1997) J Prakt Chem 339:261
40. Rose-Munch F, Aniss K, Rose E (1990) J Organomet Chem 385:C1
41. Rose-Munch F, Aniss K, Rose E, Vaisserman J (1991) J Organomet Chem 415:223
42. Calafat SV, Durantini EN, Silber JJ, Chiacchiera SM (1999) Organometallics 18:2727
43. Chaari M, Jenhi A, Lavergne J-P, Viallefont P (1991) Tetrahedron 47:4619
44. Seebach D, Boes M, Naef R, Schweizer WB (1983) J Am Chem Soc 105:5390
45. Kamikawa K, Uemura M (1996) Tetrahedron Lett 37:6359
46. Rose E, Rose-Munch F (2002) Eur J Inorg Chem 1269
47. Rose-Munch F, Chavignon R, Tranchier J-P, Gagliardini V, Rose E (2000) Inorg Chim Acta 300/302:693
48. Boutonnet J-C, Rose-Munch F, Rose E, Semra A (1987) Bull Soc Chim Fr 640
49. Schmalz H-G, Schellhaas K (1995) Tetrahedron Lett 36:5511
50. Hong F-E, Lo S-C, Liou M-W, Chou L-F, Lin C-C (1996) J Organomet Chem 516:123
51. Rose-Munch F, Rose E, Semra A, Jeannin Y, Robert F (1988) J Organomet Chem 353:53
52. Rose-Munch F, Khourzom R, Djukic J-P, Perrotey A, Rose E, Brocard J (1994) J Organomet Chem 467:195
53. Persson M, Hacksell U, Csoregh I (1991) J Chem Soc Perkin Trans I 1453
54. Djukic J-P, Rose-Munch F, Rose E, Simon F, Dromzee Y (1995) Organometallics 14:2027
55. Djukic J-P, Rose-Munch F, Rose E (1993) J Am Chem Soc 115:6434
56. Djukic JP, Geysermans P, Rose-Munch F, Rose E (1991) Tetrahedron Lett 32:6703
57. Djukic J-P, Rose-Munch F, Rose E (1991) J Chem Soc Chem Commun 1634
58. Abd-El-Aziz AS, Schriemer DC, de Denus CR (1994) Organometallics 13:374
59. Abd-El-Aziz AS, de Denus CR, Todd EK, Bernarni SA (2000) Macromolecules 33:5000
60. Storm JP, Andersson C-M (2000) J Org Chem 65:5264
61. Pearson AJ, Gelormini AM (1994) J Org Chem 59:4561
62. (a) Pearson AJ, Park JG, Yang SH, Chuang Y-H (1989) J Chem Soc Chem Commun 1363; (b) Pearson AJ, Zhang P, Lee K (1996) J Org Chem 61:6581; (c) Pearson AJ, Zhang P, Bignan G (1997) J Org Chem 62:4536
63. Abd-El-Aziz AS, Lei Y, de Denus CR (1995) Polyhedron 14:1585
64. Pearson AJ, Gelormini AM (1995) J Org Chem 60:281
65. Abd-El-Aziz AS, Epp KM, Lei Y, Kotowich S (1995) J Chem Res (S) 182
66. Moriarty RM, Gill US, Ku YK (1988) Polyhedron 7:2685
67. Piorko A, Abd-El-Aziz AS, Lee CC, Sutherland RG (1989) J Chem Soc Perkin Trans I 469
68. Pearson AJ, Gelormini AM, Fox MA, Watkins D (1996) J Org Chem 61:1297
69. Roberts RMG, Johnsen E (1997) J Organomet Chem 544:197
70. Abd-El-Aziz AS, Epp KM, de Denus CR, Fisher-Smith G (1994) Organometallics 13:2299
71. Cambie RC, Rutledge PS, Stevenson RJ, Woodgate PD (1994) J Organomet Chem 471:133

72. Brown RA, Fernando SIS, Roberts RMG (1994) J Chem Soc Perkin Trans I 197
73. Sutherland RG, Chang PC, Lee CC (1982) J Organomet Chem 234:197
74. Manzur C, Baeza E, Millan L, Fuentealba M, Hamon P, Hamon J-R, Boys D, Carrillo D (2000) J Organomet Chem 608:126
75. Pearson AJ (1997) Tetrahedron Lett 38:5123
76. Lee CC, Zhang CH, Abd-El-Aziz AS, Piorko A, Sutherland RG (1989) J Organomet Chem 364:217
77. Ruhland T, Bang KS, Andersen K (2002) J Org Chem 67:5257
78. Sutherland RG, Piorko A, Gill US, Lee CC (1982) J Heterocyclic Chem 19:801
79. Holden MS, Cole KA (1992) Synth Commun 22:2579
80. Abd-El-Aziz AS, Piorko A, Lee CC, Sutherland RG (1989) Can J Chem 67:1618
81. Lee CC, Abd-El-Aziz AS, Chowdry RL, Piorko A, Sutherland RG (1986) Synth React Inorg Met-Org Chem 16:541
82. Dabirmanesh Q, Fernando SIS, Roberts RMG (1995) J Chem Soc Perkin Trans I 743
83. Abd-El-Aziz AS, Piorko A, Baranski AS, Sutherland RG (1989) Synth Commun 19:1865
84. Sutherland RG, Chowdhury RL, Piorko A, Lee CC (1986) Can J Chem 64:2031
85. Abd-El-Aziz AS, de Denus CR, Hutton HM (1995) Can J Chem 73:289
86. Gill US, Moriarty RM (1986) Synth React Inorg Met-Org Chem 16:485
87. Abd-El-Aziz AS, Boraie W, Al-Salem N, Sadek SA, Epp KM (1997) J Chem Soc Perkin Trans I 1469
88. Sutherland RG, Abd-El-Aziz AS, Piorko A, Gill US, Lee CC (1988) J Heterocyclic Chem 25:1107
89. West CW, Rich DH (1999) Org Lett 1:1819
90. Pearson AJ, Heo J-N (2000) Tetrahedron Lett 41:5991
91. Moriarty RM, Ku Y-Y, Gill US (1988) Organometallics 7:660
92. Pigge FC, Fang S (2001) Tetrahedron Lett 42:17
93. Dembek AA, Fagan PJ, Marsi M (1993) Macromolecules 26:2992
94. Pearson AJ, Hwang J-J (2001) Tetrahedron 57:1489
95. Pearson AJ, Shin H (1992) Tetrahedron Lett 48:7527
96. Pearson AJ, Zhu PY, Youngs WJ, Bradshaw JD, McConville DB (1993) J Am Chem Soc 115:10376
97. Bhasin KK, Balkeen WG, Pauson PL (1981) J Organomet Chem 204:C25
98. Lee SS, Lee J-S, Chung YK (1993) Organometallics 12:4640
99. Houghton RP, Voyle M, Price R (1984) J Chem Soc Perkin Trans 925
100. Litvak VV, Goryunov LI, Shteingarts VD (1984) Izv SOAN SSSR 132
101. Goryunov LI, Litvak VV, Shteingarts VD (1988) Zhurn Org Khimii 24:401
103. Semmelhack MF, Hall HT, Farina R, Yoshifuji M, Clark G, Bargar T, Hirotsu K, Clardy J (1979) J Am Chem Soc 101:3535
102. Keller K, Times-Marshall K, Behar S, Richards K (1989) Tetrahedron Lett 30:3373
104. Semmelhack MF, Garcia JL, Cortes D, Farina R, Hong R, Carpenter BK (1983) Organometallics 2:467
105. Cambie RC, Clark GR, Gallagher SR, Rutledge PS, Stone MJ, Woodgate PD (1988) J Organomet Chem 342:315
106. Kündig EP, Cunningham AF, Paglia P, Simmons DP, Bernardinelli G (1990) Helv Chim Acta 73:386
107. Ohlsson B, Ullenius C (1988) J Organomet Chem 350:35
108. Ohlsson B, Ullenius C, Jagner S, Grivet C, Wenger E, Kündig EP (1989) J Organomet Chem 365:243
109. Semmelhack MF, Clard G (1977) J Am Chem Soc 99:1675
110. Semmelhack MF, Clark GR, Farina R, Saeman M (1979) J Am Chem Soc 101:217
111. Fernando SIS, Roberts RMG (1994) J Organomet Chem 474:133
112. Cambie RC, Rutledge PS, Stevenson RJ, Woodgate PD (1994) J Organomet Chem 471:149
113. Narasaka K, Sakurai H, Liu C (1994) Bull Chem Soc Jpn 67:1156

114. Camps F, Coll J, Moreto JM, Pages L, Roget J (1990) J Chem Res (S) 236
115. Kündig EP, Grivet C, Wenger E, Bernardinelli G, Williams AF (1991) Helv Chim Acta 74:2009
116. Kündig EP, Desbory V, Simmons DP (1983) J Am Chem Soc 105:6962
117. Kozikowski AP, Isobe K (1978) J Chem Soc Chem Commun 1076
118. Semmelhack MF, Wulff W, Garcia JL (1982) J Organomet Chem 240:C5
119. Kündig EP, Inage M, Bernardinelli G (1991) Organometallics 10:2921
120. Fretzen A, Ripa A, Liu R, Bernardinelli G, Kündig EP (1993) Organometallics 12:3724
121. Fretzen A, Kündig EP (1997) Helv Chim Acta 80:2023
122. Yeh M-CP, Chuang C-N, Yiu C-H (1997) Tetrahedron Lett 38:7387
123. Abd-El-Aziz AS, Armstrong DA, Bernardin S, Hutton HM (1996) Can J Chem 74:2073
124. Sutherland RG, Chowdhury RL, Piorko A, Lee CC (1987) J Org Chem 52:4618
125. Guennec N, Moinet C (1995) J Organomet Chem 503:171
126. Abd-El-Aziz AS, Boraie W, Gavel NT (1999) Organometallics 18:1562
127. Cambie RC, Coulson SA, Mackay LG, Janssen SJ, Rutledge PS, Woodgate PD (1991) J Organomet Chem 409:385
128. Huckett SC, Angelici RJ (1988) Organometallics 7:1491
129. Pigge FC, Fang S, Rath NP (1999) Org Lett 1:1851
130. Krow GR, Miles WH, Smiley PM, Lester WS, Kim YJ (1992) J Org Chem 57:4040
131. (a) Lee T-Y, Lee SS, Chung YK, Lee SW; (b) Winkhaus G, Pratt L, Wilkinson G (1961) J Chem Soc 3807
132. Woo K, Carpenter GB, Sweigart DA (1994) Inorg Chim Acta 220:297
133. Walker PJC, Mawby RJ (1973) J Chem Soc Dalton Trans 622
134. Evans DJ, Kane-Maguire LAP (1981) J Organomet Chem 215:C27
135. Munro GAM, Pauson PL (1979) Z Anorg Allg Chem 458:211
136. Miles WH, Smiley PM, Brinkman HR (1989) J Chem Soc Chem Commun 1897
137. Pearson AJ, Richards IC (1983) J Organomet Chem 258:C41
138. Miles WH, Brinkman HR (1992) Tetrahedron Lett 33:589
139. Pearson AJ, Bruhn PR, Gouzoules F, Lee S-H (1989) J Chem Soc Chem Commun 659
140. Lee S-H, Nam S-W (1998) Bull Korean Chem Soc 19:613
141. Pearson AJ, Milletti MC, Zhu PY (1995) J Chem Soc Chem Commun 995
142. Pearson AJ, Gontcharov AV, Zhu PY (1997) Tetrahedron 53:3849
143. Balssa F, Gagliardini V, Le Corre-Susanne C, Rose-Munch F, Rose E, Vaisserman J (1997) Bull Soc Chim Fr 134:537
144. Gagliardini V, Balssa F, Rose-Munch F, Rose E, Susanne C, Dromzee Y (1996) J Organomet Chem 519:281
145. Reau R, Reed RW, Dahan F, Bertrand G (1993) Organometallics 12:1501
146. Lee S-G, Chung YK, Yoon T-S, Shin W (1993) Organometallics 12:2873
147. Bae HK, Jung IN, Chung YK (1986) J Organomet Chem 317:C1
148. Rose-Munch F, Le Corre-Susanne C, Balssa F, Rose E, Vaisserman J, Licandro E, Papagni A, Maiorana S, Meng W-D, Stephenson GR (1997) J Organomet Chem 545/546:9
149. Renard C, Valentic R, Rose-Munch F, Rose E, Vaisserman J (1998) Organometallics 17:1587
150. Rose E, Le Corre-Susanne C, Rose-Munch F, Renard C, Gagliardini V, Teldji F, Vaissermann J (1999) Eur J Chem 421
151. Walker PJC, Mawby RJ (1973) Inorg Chim Acta 7:621
152. Chen J, Young VG, Angelici RJ (1996) Organometallics 15:325
153. Lee SS, Lee T-Y, Choi DS, Lee JS, Chung YK (1997) Organometallics 16:1749
154. Dullaghan CA, Carpenter GB, Sweigart DA (1997) Chem Eur J 3:75
155. Angelici RJ, Blacik LJ (1972) Inorg Chem 11:1754
156. Brown DA, Glass WK, Kreddan KM (1991) J Organomet Chem 413:233

Topics Organomet Chem (2004) 7: 71–94
DOI 10.1007/b94492

Dearomatization via η^6-Arene Complexes

E. Peter Kündig · Andrew Pape

Department of Organic Chemistry, University of Geneva, 30 Quai Ernest Ansermet, 1211 Geneva 4, Switzerland
E-mail: peter.kundig@chiorg.unige.ch

Abstract The chapter presents a detailed overview of the dearomatization of arenes by temporary complexation to the electrophilic complex fragments $Cr(CO)_3$ and $Mn(CO)_3^+$. The activation of the η^6-coordinated arene in the complex enables nucleophilic addition. In the case of the fragment $Cr(CO)_3$, the resulting anionic cyclohexadienyl complexes can be reacted with electrophiles (H^+, RX) to yield, after decomplexation, *trans*-disubstituted cyclohexadienes. With H^+, rapid isomerization results in mixtures of dienes. With anisol, this can be directed to yield synthetically useful 5-substituted cyclohexenones. With C-electrophiles (primary and secondary allyl-, benzyl-, and alkyl-iodides and -bromides), the overall reaction sequence is a *trans*-addition of a C-nucleophile and a C-electrophile across an arene double bond. With alkyl electrophiles, a CO insertion step yields ketone products. Highly asymmetric reactions have been developed by using either chiral directing groups on the arene, chiral nucleophiles, chiral Cr bound ligands, or planar chiral arene complexes. Reactions with the $Mn(CO)_3^+$ activating group enlarges the range of nucleophiles in the initial addition step. As this results in a neutral cyclohexadienyl complex, a reactivation by ligand exchange is often necessary at this stage in order to follow up with a second nucleophilic addition.

Keywords Arene-chromium complexes · Arene-manganese complexes · Dearomatization · Cyclohexadiene · Organo lithium reagents · Asymmetric arene transformations

1 Introduction

Coordination of an arene to an electrophilic transition metal complex fragment renders the arene susceptible to nucleophilic addition. In the preceding chapter the scope of nucleophiles, questions of regioselectivity and reversibility, and aromatic substitution via this methodology were discussed. In the present chapter we will focus on the transformation of arenes into functionalized alicyclic molecules via the same cyclohexadienyl intermediates.

Arenes and aromatic heterocycles are widely available. They are readily derivatized through electrophilic and nucleophilic substitution, or via ortho-lithiation followed by reaction with electrophiles (see also Chap. 3). Routes to differentially substituted aromatic products are thus well established. This is not the case for substitutive dearomatization reactions, primarily because this requires disruption of the aromatic π system. Benzene and its derivatives are attractive starting materials because they have the potential to provide a rapid entry into complex alicyclic synthetic building blocks containing unmasked functionality, new carbon-carbon bonds, and new stereogenic centers [1].

There are a number of efficient synthetic methods available that enable arenes to be transformed into alicyclic molecules. Most established is the Birch reduction and newer variants of this reaction which achieve dearomatization via single electron transfer [2–4]. The photocycloaddition of arenes to alkenes has received attention [5, 6], and microbial oxidation of benzene derivatives with *Pseudomonas putida* delivers synthetically important *cis*-cyclohexadienediols [7, 8]. Closer to the topic of this chapter are methods which rely on nucleophilic addition to an electron deficient arene [9]. We note conjugate addition of organolithium reagents to bulky naphthylesters [10], and extensive elegant chemistry with naphthyloxazolines [11]. Both approaches are limited to fused aromatics and to pyridine because these molecules are easier to dearomatize than benzene and its derivatives. However, more recently dearomatization of benzaldehydes and acetophenone via conjugate addition of carbanions in the presence of ATPH has been realized [12]. Other new developments are conjugate addition to arenes bearing a Fischer carbene appendage [13], and Clayden has shown that this reaction is also possible with bulky *N*-benzoyl amides [14].

Arene dearomatization reactions can also be induced by temporary complexation of its π-system to a transition metal complex. Specifically this review will cover the synthesis of alicyclic molecules from arene metal complexes of $Cr(CO)_3$ and $Mn(CO)_3^+$, and preliminary results with $Mo(CO)_3$. New synthetic methodology will be the prime subject and particular attention is focused on asymmetric methodologies. The development of dearomatization methodology via π-arene complexes of these complex fragments is at different stages of development. Whilst the fundamental reactions have been established, new reactions and in particular asymmetric methodologies continue to be found and applications in this area flourish. The present review aims to present the scope and limitations of the methodologies currently available.

2 Chromium Mediated Dearomatization Reactions

The range of nucleophiles that add to (arene)$Cr(CO)_3$ complexes to give anionic cyclohexadienyl complexes has been detailed in the previous chapter. For successful dearomatization, the forward reaction, leading ultimately to a cyclohexadiene needs to be much faster than the back reaction that regenerates the starting complex (Scheme 1).

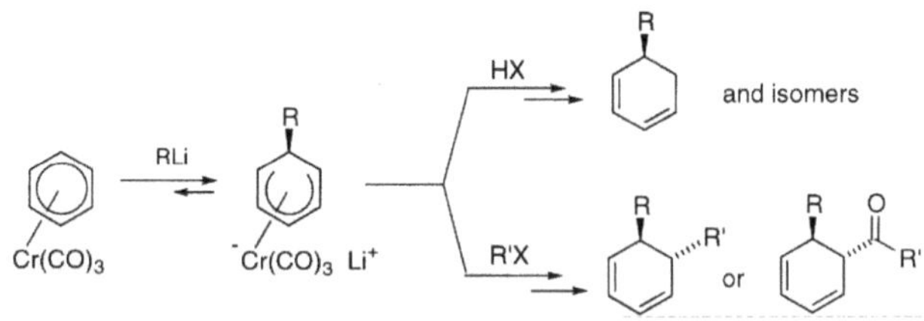

Scheme 1

Reversibility, even at low temperatures, has been shown to be fast for stabilized carbanions (e.g., nitrile stabilized carbanions, ester enolates) whereas (most) sulfur stabilized carbanions and simple organo lithium compounds add irreversibly. Nevertheless protonation is more rapid than anion dissociation even for the first category of anions mentioned and nucleophile addition/protonation reactions allows efficient conversion to a dearomatized product.

2.1 Nucleophile Addition/Protonation Reactions

Reaction of the anionic cyclohexadienyl $Cr(CO)_3$, obtained by addition of a nitrile stabilized carbanion to [Cr(benzene)$(CO)_3$], with MeI, regenerates the starting complex. However, treatment of the same intermediate with a strong acid at low temperature affords a mixture of isomeric cyclohexadienes. With time, the reaction tends to converge to the most stable diene (Scheme 2) [15–18]. It has also been reported that protonation under a CO atmosphere allows recycling of $Cr(CO)_6$ [19].

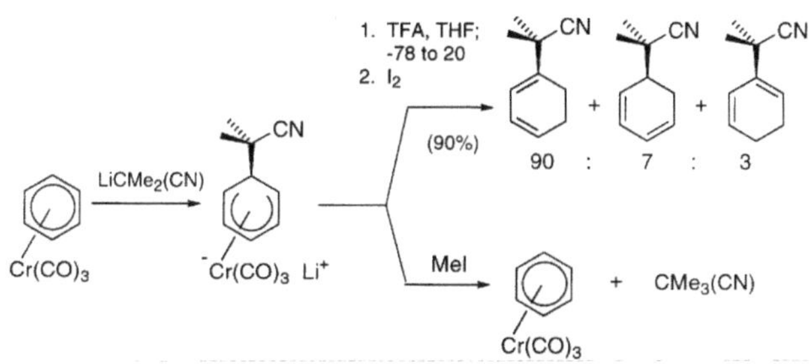

Scheme 2

The mechanism of the double bond isomerization has been probed in the analogous [Cr(benzene)$(PF_3)_3$] complex. The interconversion of isomeric diene complexes occurs via both 1,5-hydride migrations and 1,2-hydride exchange processes in the agostic complex intermediate. These processes are rapid and only cooling to -130 °C stops the isomerization (Scheme 3) [20].

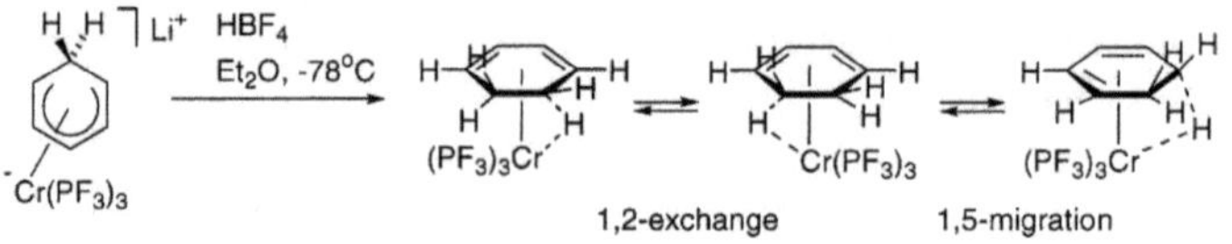

Scheme 3

Isomerization is also demonstrated in the reactions with [Cr(anisole)$(CO)_3$], as shown in Scheme 4. Arguably the most useful nucleophile addition/protonation sequence, it ultimately permits the controlled synthesis of substituted cyclohexenones [18, 21].

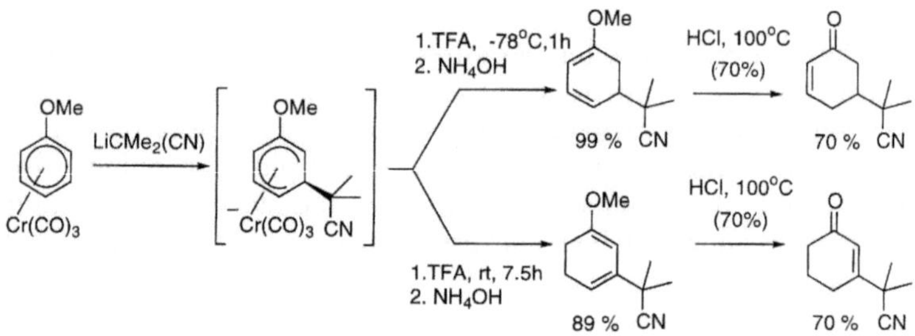

Scheme 4

An asymmetric variant makes use of a planar chiral complex. This arene to substituted cyclohexenone conversion has been elegantly used in a synthesis of (+)-ptilocaulin. (Scheme 5; see also Chap. 8) [22]. The sequence is unusual in that after nucleophilic addition, acid free $ClSiMe_3$ is added. When a strong acid

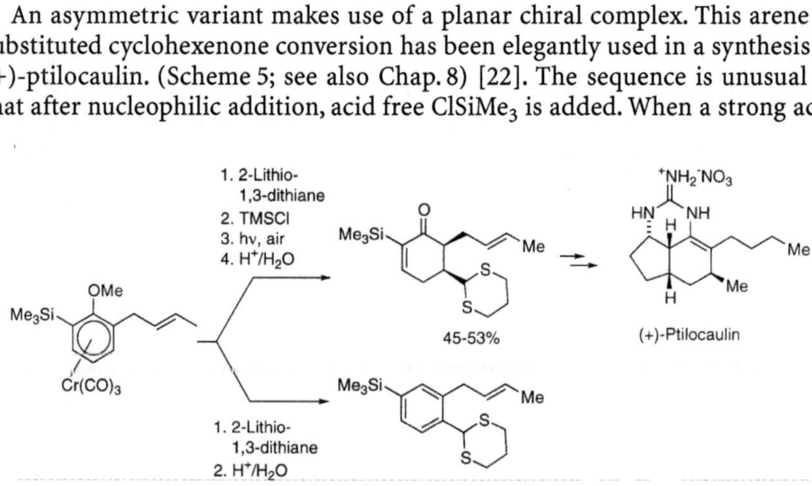

Scheme 5

(TFA) is used instead, an aromatic product is formed rather than the requisite diene. This presumably occurs via *tele*-substitution of the OMe group (see preceeding chapter).

Another route of access to non-racemic cyclohexenone products is via an arene bound chiral auxiliary. This has received attention from two research groups. The starting complexes are accessible by an efficient $Cr(CO)_3$ mediated nucleophilic substitution.

Semmelhack's results in this area are centered on menthol derived chiral auxiliaries [23]. Product yields are good but diastereoselectivity in these reactions are modest (<48% ee). The example in Scheme 6 shows that higher enantiomeric excesses are obtained at elevated temperatures (0 °C as opposed to -78 °C) and this could be indicative of a change in the reaction from kinetic control at low temperature to thermodynamic control at the higher temperature. This is in keeping with the ready reversibility of nucleophilic addition of nitrile stabilized carbanions at temperatures above -70 °C [24, 25].

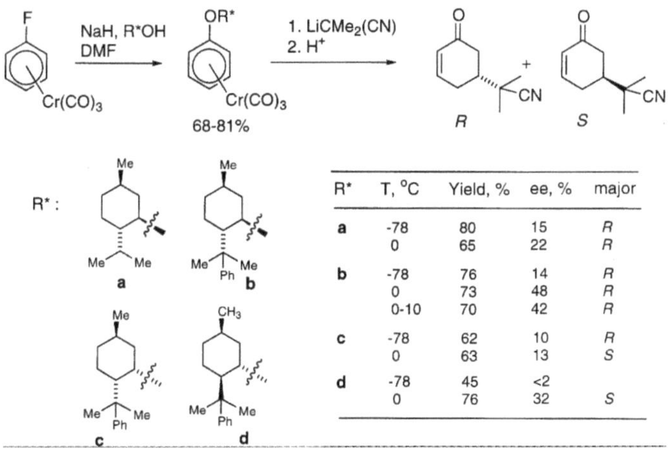

R*	T, °C	Yield, %	ee, %	major
a	-78	80	15	R
	0	65	22	R
b	-78	76	14	R
	0	73	48	R
	0-10	70	42	R
c	-78	62	10	R
	0	63	13	S
d	-78	45	<2	
	0	76	32	S

Scheme 6

Pearson has taken a similar approach with the same target molecules but using terpenoid derived chiral auxiliaries. Higher product diastereoselectivities result as shown in Scheme 7 [26, 27]. A *para* Me group considerably enhances induction. In contrast to Semmelhack's results, diastereoselectivity varies little with temperature and, interestingly, regioselectivity changes from *meta* to *ortho* on warm-up when a *para* alkyl group is present.

Methyldithiane lithium adds to [$Cr(CO)_3$(1,4-dimethoxynaphthalene)] irreversibly at the β-position. Protonation/decomplexation in this case has been carried out with cerium ammonium nitrate in THF/H_2O (9:1) to give a daunomycinone precursor (Scheme 8) [28].

The extension of the nucleophile addition/protonation sequence to an intramolecular reaction is shown in Scheme 9 and elegantly illustrates the potential for the rapid access to elaborate organic compounds with high selectivity.

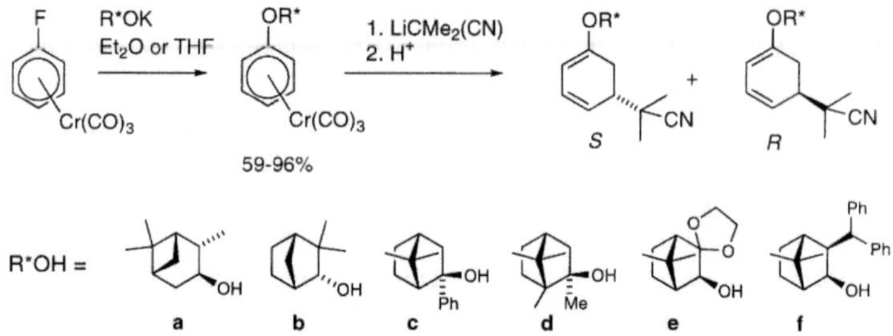

Auxiliary	R	$R:S$	Yield	Auxiliary	R	$R:S$	Yield
a	H	70 : 30	80	d	H	36 : 64	89
	Me	73 : 27	90		Me	10 : 90	81
b	H	78 : 22	67	e	H	12 : 88	86
	Me	88 : 12	76		Me	6 : 94	65
c	H	80 : 20	95	f	Me	8 : 92	76
	Me	96 : 4	63				

Scheme 7

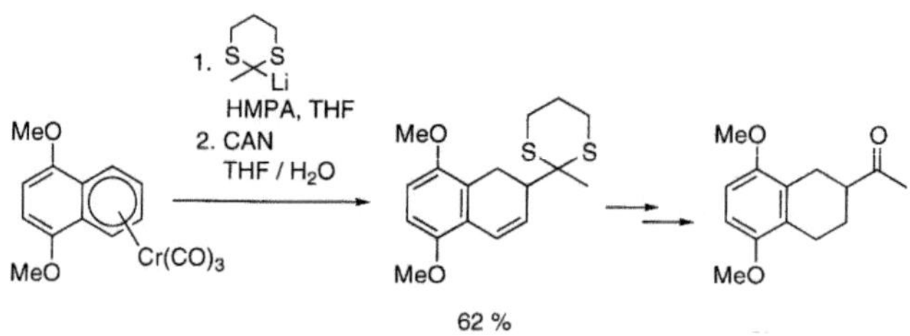

Scheme 8

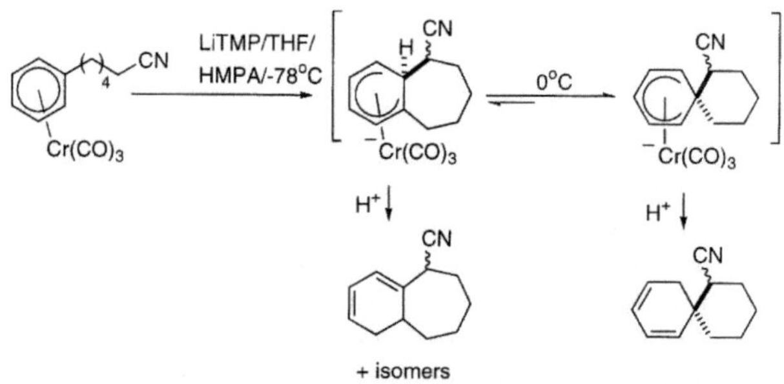

Time, h	Temp, °C	fused product	spiro-product	Yield, %
0.5	-78	72	28	72
0.25	0	37	63	79
4.0	0	25	75	81
24	0	3	97	70

Scheme 9

The products are the fused bi-cyclic ring system as a mixture of diene isomers, and the spirocyclic system [29]. The product ratio is dependent on the length of the chain and on the reaction conditions. Kinetic conditions in the example shown favor the fused bi-cyclic product, whereas thermodynamic conditions deliver the spirocyclic system. This approach was subsequently applied in a rapid synthesis of acorenone and acorenone B (see Chap. 8) [30].

Ipso versus vicinal nucleophilic addition was also probed in (arene)$Cr(CO)_3$ complexes made from Fischer carbene complexes. Products of higher complexity are accessible by a sequential benzannulation/nucleophilic addition reaction/oxidation sequence [31].

Schmalz has reported on the intramolecular radical cyclization of (arene)$Cr(CO)_3$ complexes (Scheme 10). Whilst most of the products from these reactions were transformed into aromatics, dearomatized products can potentially be accessed via this new and appealing approach [32–34].

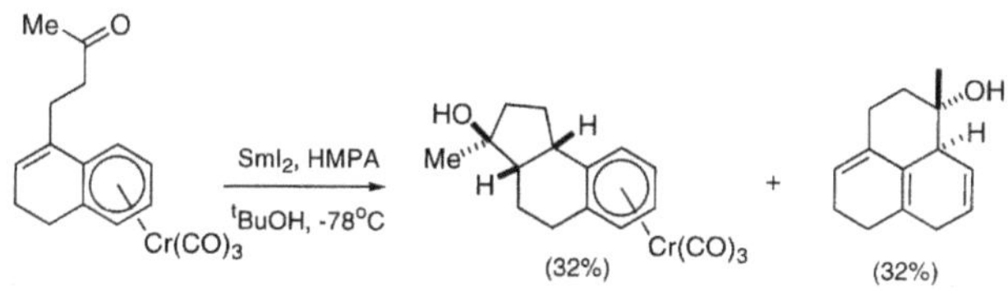

Scheme 10

2.2 Sequential C-Nucleophile/C-Electrophile Addition Reactions

The sequential *trans*-addition of a carbon nucleophile and a carbon electrophile across an arene double bond in (arene)$Cr(CO)_3$ was first reported in 1983 [35]. Since then this methodology has undergone extensive development, with recent efforts mainly directed towards enantioenriched products [36]. Anionic (cyclohexadienyl)$Cr(CO)_3$ complexes are very soft nucleophiles and this places restrictions on the electrophiles that can be used in this sequence. Specifically these reactions are successful when carbanion dissociation from the intermediate anionic cyclohexadienyl complex is slow compared to the reaction with the carbon electrophile. The sequential addition is usually carried out as a one-pot reaction and the proposed reaction sequence is that shown in Scheme 11. In contrast to the nucleophile addition/protonation sequence, products form with excellent 1,2-regioselectivity. It is likely that this is due to an irreversible transfer of the acyl, allyl, or propargyl group to one of the two termini of the cyclohexadienyl ligand.

With chromium, intermediates other than the anionic cyclohexadienyl complex have not been isolated, but three X-ray structures of the products of interception of anionic cyclohexadienyl chromium complexes with $ClSnPh_3$ provide indirect evidence for the proposed mechanism [37–39]. Electrophile selectivity was established as shown in Scheme 12.

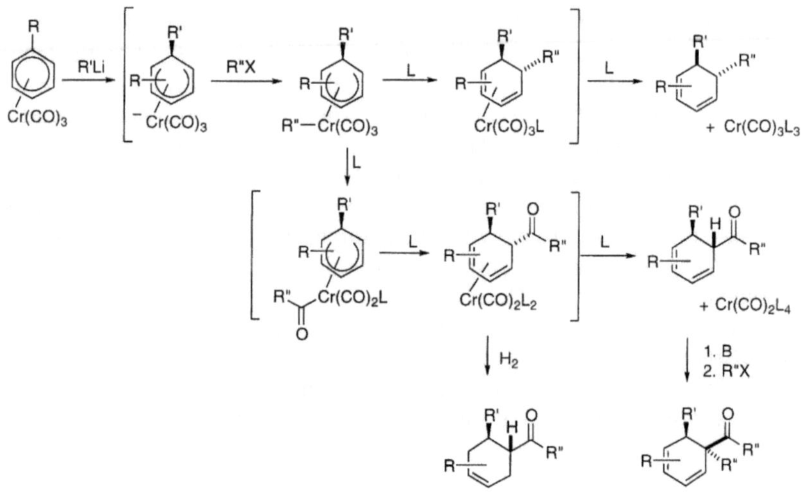

Scheme 11

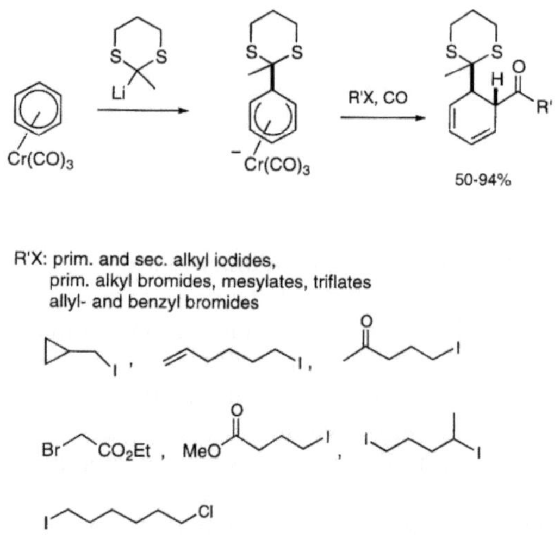

Scheme 12

Reactive or soft electrophiles work best in this reaction with leaving group preferences following the sequence OTf≈OMs≈I>Br>>Cl and yields are often better when polar co-solvents like HMPA or DMPU are added [37, 40]. Conservation of the three-membered ring in the reaction with cyclopropylmethyl iodide, and lack of cyclization in reactions with 6-iodo-hex-1-ene, argue for an S_N2 mechanism rather than an electron transfer reaction. Ketones, esters, alkenes, primary alkyl chlorides, and secondary alkyl iodides are conserved when present in the same molecule as a primary alkyl iodide. Substituted allyl bromides react to give the product without allylic rearrangement. Migratory CO insertion depends very much on the nature of the electrophile and the presence or absence of substituents on the arene ring. In the case of benzene, all groups R″ (Scheme 11), with the exception of propargyl, undergo carbonylation prior to reductive elimination and reactions are cleaner and yields higher when carried out under CO (up to 4 bar) or in the presence of an added ligand (PPh_3, $P(OPh)_3$, CH_3CN, NEt_3). When CO is used, $Cr(CO)_6$ can be partly recovered [40]. Carbonylation follows the expected order based on migratory aptitude of R″ to an adjacent CO group: ethyl>methyl>benzyl, allyl>>propargyl.

For [Cr(benzene)$(CO)_3$], the nucleophile addition/acylation sequence has been extended to include an in situ hydrogenation of the Cr coordinated diene product. This affords predominantly the cyclohexene resulting from 1,4-hydrogen addition. This procedure is based on an analogy between the postulated (η^4-diene)Cr(0) intermediate (Scheme 11) and that advanced for the '$Cr(CO)_3$' catalyzed hydrogenation of conjugated dienes (see Chap. 9). Thus adding either acetonitrile or benzonitrile with the electrophile (to make up for the consummation of one carbonyl ligand) and placing the reaction under 5 bar of hydrogen provides the expected substituted cyclohexene product as the major isomer (Scheme 13) [41, 42]. The synthetic potential of this reaction sequence is evident but it has not yet been explored.

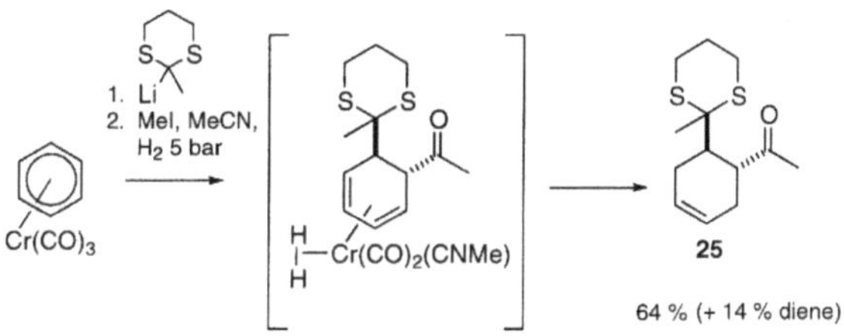

Scheme 13

In the reaction with anisol-, phenyl oxazoline- or benzaldehyde imine-complexes, carbonylation takes place with alkyl groups just as in [Cr(benzene)$(CO)_3$]. In contrast, with allyl bromide, allyl/cyclohexadienyl C-C bond formation takes precedent over CO insertion and the allyl ketone is but a side product formed in less then 15%. With propargyl bromide, direct reductive elimination is the sole pathway. In benzaldehyde-imine and phenyl oxazoline complexes, RLi reagents are principally directed to the *ortho* position (see Chap. 4). Presumably this involves coordination of the RLi reagent to the nitrogen of either the imine or the oxazoline [43–45]. This regioselectivity is also observed with hydrazones [46]. Reductive elimination is also highly regioselective and occurs at the unsubstituted terminus of the cyclohexadienyl ligand.

With acyl products, enolate formation allows the addition of a third carbon unit with complete diastereospecificity. This is particularly facile if an oxazoline or imine substituent is present [43, 45]. Remarkably, protonation in the imine intermediate takes place at the (harder) nitrogen center whereas alkylation is at the (softer) carbon adjacent to the carbonyl function (Schemes 14, 15 and 16) [45].

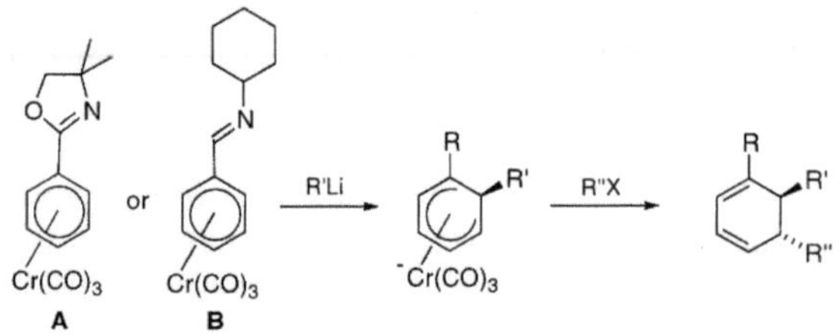

Complex	R'Li	R''X	Yield, %
A	MeLi	allylBr	68
B	MeLi	allylBr	57[a]
A		$CH_2CH(Me)CH_2Br$	61
A		$PhCH_2Br$	52
A		$HC{\equiv}CCH_2Br$	71
A		$Me_3SiC{\equiv}CCH_2Br$	72[b]
B		$Me_3SiC{\equiv}CCH_2Br$	76[a]
A	n-BuLi	allylBr	63
A		$HC{\equiv}CCH_2Br$	87[b]
A	PhLi	allylBr	67
A		$HC{\equiv}CCH_2Br$	77[b]
A	vinylLi	allylBr	54
A		$HC{\equiv}CCH_2Br$	82[b]
A		$Me_3SiC{\equiv}CCH_2Br$	88[b]
A	$NCCH_2Li$	allylBr	48[b]

[a] reaction proceeds with hydrolysis of the imine to give the aldehyde.
[b] use of HMPA as co-solvant

Scheme 14

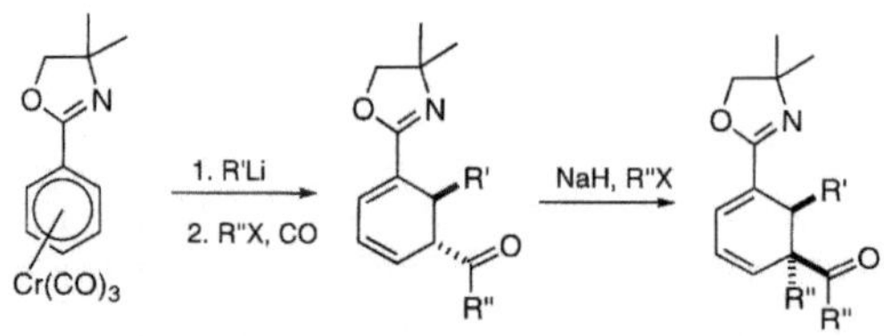

R'Li	R''X	Yield,%
MeLi	MeI	52
	EtI	62

Scheme 15

Asymmetric versions of these dearomatization reactions include both enantioselective and diastereoselective approaches. Taking a cue from Tomioka, Shindo, and Koga's work [47–49], chiral amine and ether ligands were used in combination with organo lithium reagents and toluene as solvent in order to strengthen ligand/RLi interaction. Cyclohexadiene products were formed highly

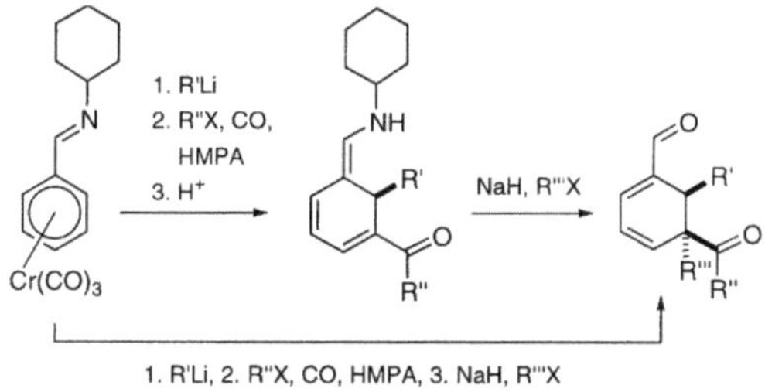

R'Li	R"X	R'''X	Yield, %
MeLi	MeI	MeI	90
PhLi	MeI	MeI	80
	MeI	MeI	56
MeLi	MeI	MeI	63[a]
MeLi	MeI	EtI	43[b]
MeLi	MeI	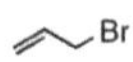	58[c]
MeLi	MeI	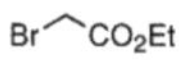	37[d]
MeLi	MeI	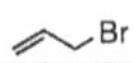	36[e]

Note: when R"=R''' the one-pot reaction sequence was followed.
Yield of enamine when R"= Me: 73%; Et: 45%.
Yield for enamine to dienal: [a] 86%; [b] 59%; [c] 79%;
[d] 50% (KOH used as base); [e] 79%.

Scheme 16

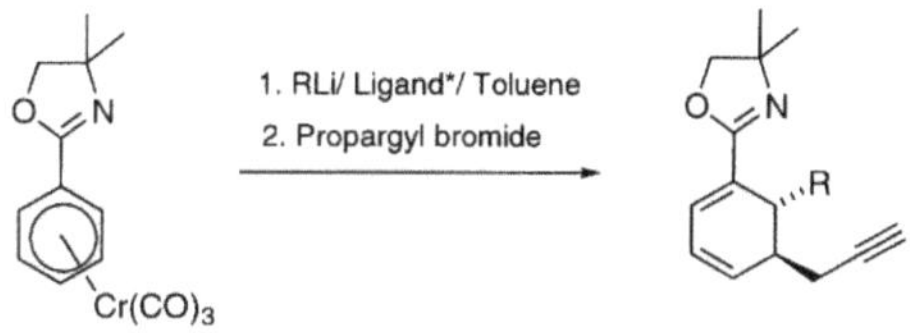

Ligand*	RLi	Yield,%	ee, %	Ligand*	RLi	Yield,%	ee, %
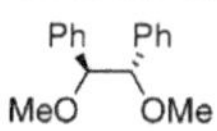	PhLi	66	93	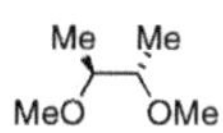	PhLi	72	81
	VinylLi	53	87		VinylLi	85	50
	MeLi	51	87		MeLi	60	47
	n-BuLi	67	65		n-BuLi	68	45
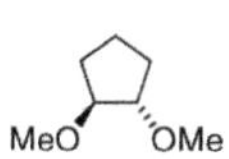	PhLi	66	81	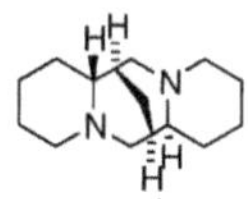	PhLi	72	54
	VinylLi	60	61		VinylLi	87	34
	MeLi	50	84		MeLi	70	47
	n-BuLi	75	61		n-BuLi	65	36

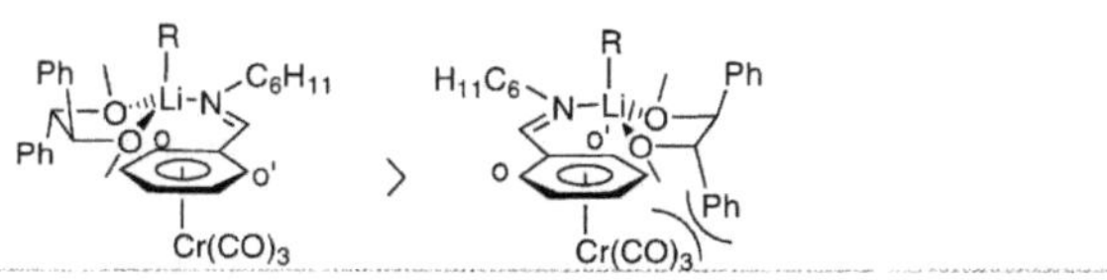

Scheme 17

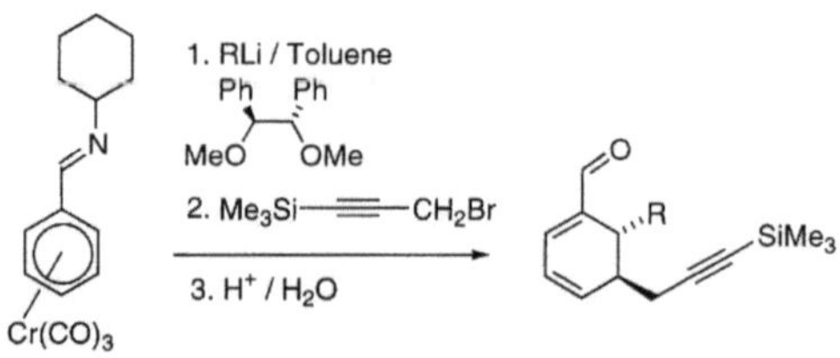

RLi	Yield,%	ee, %
PhLi	64	93
VinylLi	77	75
MeLi	62	91

Scheme 18

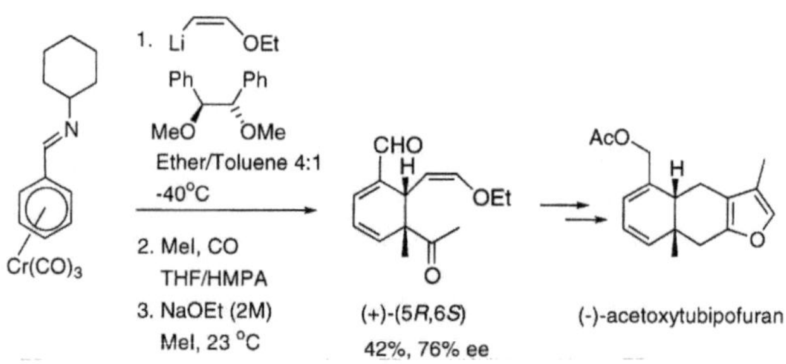

Scheme 19

enantioenriched and the best results were reported with the readily available enantiopure dimethoxy diphenyl ethane (Schemes 17 and 18).

Scheme 19 shows the enantioselective dearomatization of the imine complex as a key-step in the synthesis of (-)-acetoxytubipofuran [50].

The *ortho* and *meta* position of arene complexes bearing a chiral substituent are diastereotopic. Selective *meta* addition to chiral aryl ethers has been mentioned earlier in this review (Schemes 6 and 7). In the context of diastereoselective *ortho* additions, this approach is highly successful with oxazoline or hydrazone substituents (Schemes 20 and 21) and leads to products of high synthetic potential [51–53]. Asymmetric induction in the oxazolines derived from l-valinol and l-*tert*-butylglycinol is thought to stem from the co-ordination controlled nucleophile addition (Scheme 20). Steric interactions between the bulky oxazoline R group and the nucleophile are avoided in the favored *endo* conformation. Whilst there is no hard mechanistic evidence for this proposal, an X-ray structure of the starting complex (R=*i*-Pr) shows that the R group in the oxazoline is pseudo-equatorial and hence does not interact with the $Cr(CO)_3$ group in the *endo*-conformation [51,52]. The model shown (Scheme 20) is in accord with the observed diastereoselectivity, with the finding that better induction is achieved with R=*t*-Bu than with R=*i*-Pr, and with the subsequent reports of the same mode of induction in chiral oxazoline mediated diastereoselective *ortho* deprotonations of ferrocenes [54–56].

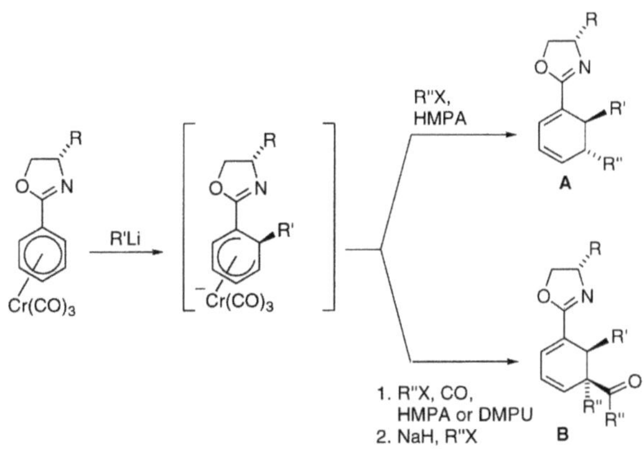

R = ^{i}Pr

R'Li	R"X	Yield, %	d.r.
MeLi	$CH_2{=}CHCH_2Br$	61[a]	96:4
BuLi	$CH_2{=}CHCH_2Br$	54[a]	95.5:4.5
PhLi	$CH_2{=}CHCH_2Br$	60[a]	81:19
$CH_2{=}CHLi$	$CH_2{=}CHCH_2Br$	48[a]	95.5:4.5
$CH_2{=}CHLi$	$Me_3Si{-}{\equiv}{-}CH_2Br$	86[a]	85:15
MeLi	$Me_3Si{-}{\equiv}{-}CH_2Br$	77[a]	93:7
MeLi	$PhCH_2Br$	51[a]	95:5
MeLi	MeI	54[b]	95:5

R = ^{t}Bu

R'Li	R"X	Yield, %	d.r.
MeLi	$CH_2{=}CHCH_2Br$	69[a]	>99:1
BuLi	$CH_2{=}CHCH_2Br$	62[a]	>99:1
PhLi	$CH_2{=}CHCH_2Br$	51[a]	95.5:4.5
MeLi	MeI	58[b]	>99:1

[a] product **A**. [b] product **B**

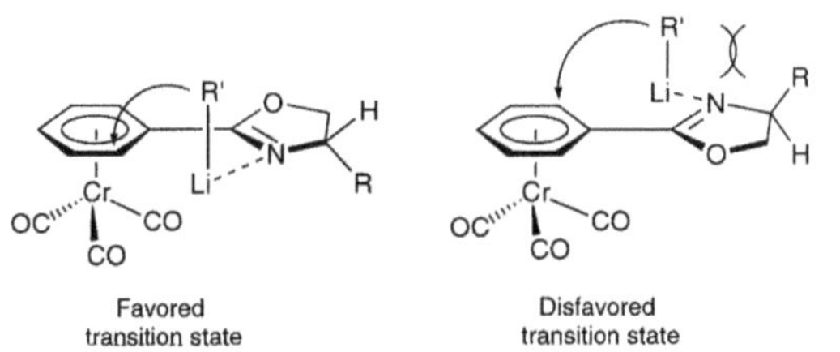

Scheme 20

Preliminary results of reactions with the chiral benzaldehyde imine complex depicted in Scheme 22 indicate moderate enantioselectivity. The *S*,*S*-product is obtained predominantly from the *S*-starting complex (78% yield, 58% ee) in THF. Interestingly, the use of toluene as solvent favors the opposite enantiomer (50% yield, 38% ee) [52].

The complex incorporating the chiral d-valinol derived imine auxiliary has been used in the synthesis of (+)-acetoxytubipofuran. This auxiliary delivered the dearomatized tetrasubstituted cyclohexadiene intermediate as a single enantiomer (Scheme 23) [50].

A third mode of asymmetric induction is offered via planar chiral complexes. Diastereoselective lithiation and enantioselective lithiation provide efficient

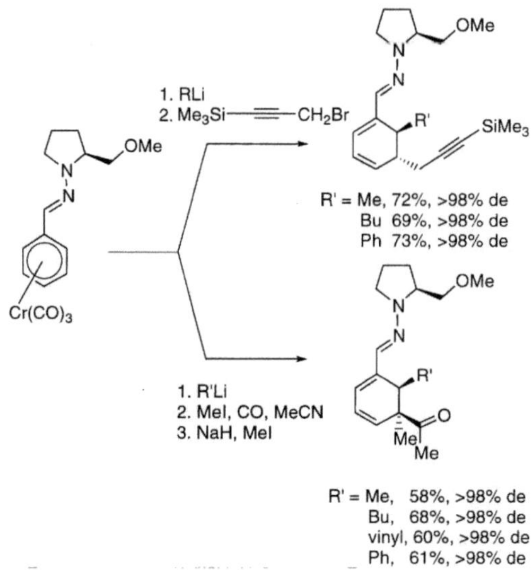

Scheme 21

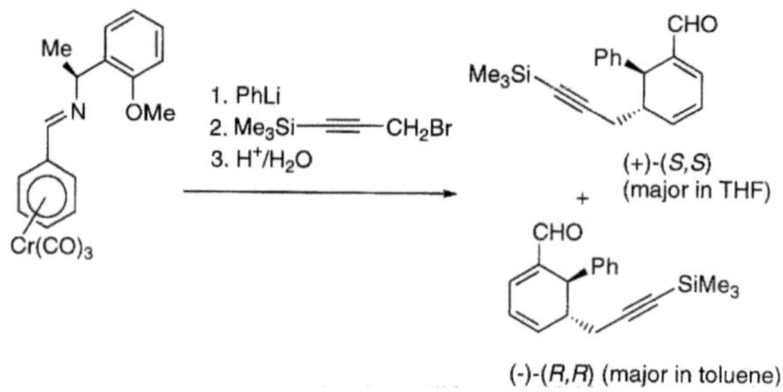

Scheme 22

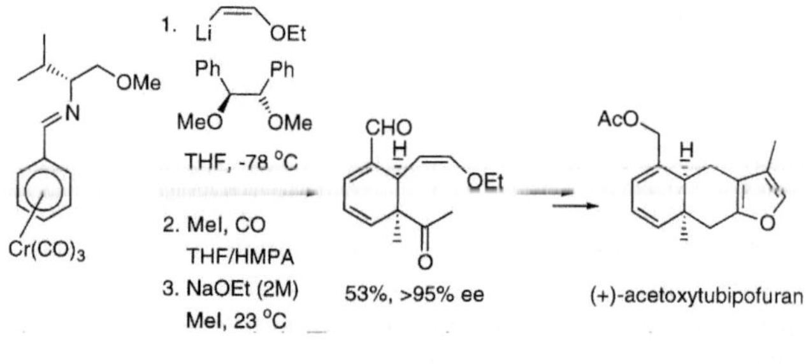

Scheme 23

routes to highly enantioenriched *o*-substituted benzaldehyde complexes and to *o*-substituted anisol complexes (see Chap. 4). The former are also accessible via enantioselective *o*-nucleophilic addition/hydride abstraction [57].

The *o*-$SiMe_3$ substituted anisole complex is the starting material in the synthesis of the fused alicyclic ring system shown in Scheme 24. Propargyl lithium addition is followed by reaction with allyl bromide, both reactions occurring with complete regio- and diastereoselectivity. The cyclohexadiene product was converted to the enone and the subsequent Pauson-Khand cyclization afforded the expected tricyclic product containing a 6,6-*trans* fused ring system in 90% ee and 100% de. This corresponds to the quantitative transfer of chiral information and illustrates the strength of this approach [58].

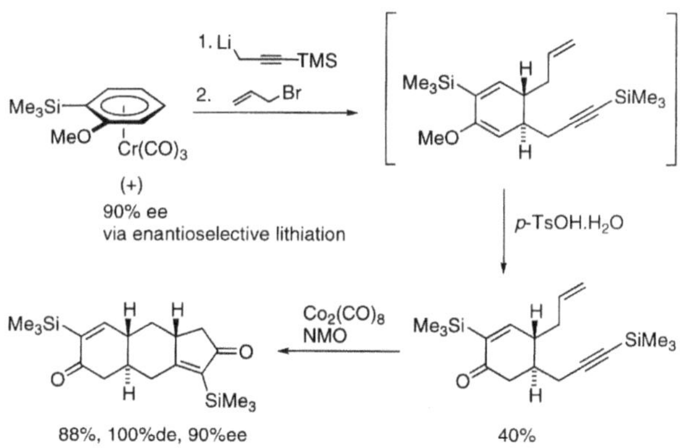

Scheme 24

Yet another approach is the use of a chiral ligand at chromium (Scheme 25). In contrast to the methods described above, asymmetric induction here does not occur in the nucleophilic addition step, but in either the CO insertion step or the reductive elimination step, or both. Work in this area is not at an advanced stage and as yet, enantioselectivities are modest and the requirement for the use of a chiral phosphorous ligand in a stoichiometric reaction is unlikely to become a method of synthetic utility. Mechanistically, the approach is highly interesting as it poses the question of the influence of the chiral ligand L^* on the migratory CO insertion step and on the reductive elimination—both being diastereomeric processes [38, 59].

The sequential nucleophile/electrophile addition can also be applied to the dearomatization of naphthalene and derivatives. Treatment of [$Cr(CO)_3$(1,4-dimethoxynaphthalene)] with 2-methyl-2-litihiodithiane affords a single regioisomeric dihydronaphthalene (Scheme 24) [35, 40]. In the second example in Scheme 26, an α-nitrile anion is used. HMPA is essential in this case to favor alkylation of the metal as opposed to anion dissociation [40].

[$Cr(CO)_3$(1-Methoxynaphthalene)], prepared in 91% yield as a single regioisomer, served as the starting complex in a route to the AB ring system of Ak-

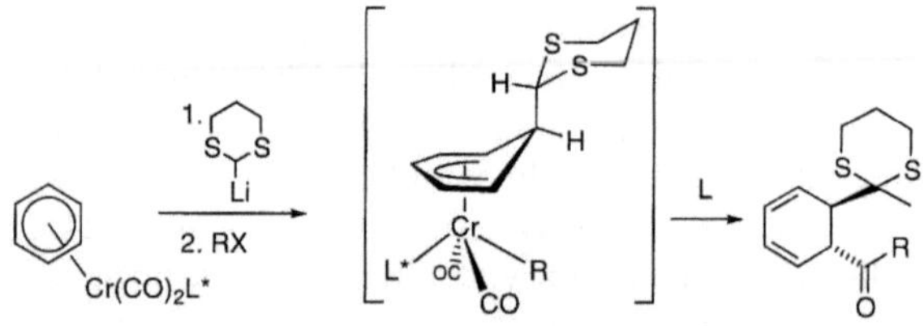

L*	R'X	L	Yield, %	ee, % (config.)
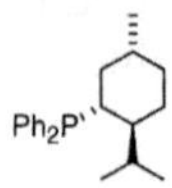	MeI	CO	60	8(SS)
	EtI	CO	44	19(SS)
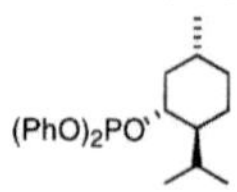	MeI	CO	55	14(RR)
	MeI	$P(OPh)_3$	91	16(RR)
	EtI	CO	67	28(RR)
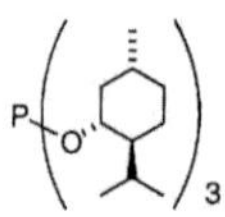	MeI	CO	54	40(SS)
	MeI	$P(OPh)_3$	70	31(SS)
	EtI	$P(OPh)_3$	57	21(SS)
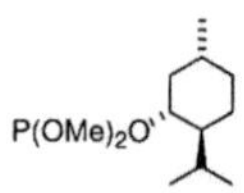	EtI	CO	82	26(RR)
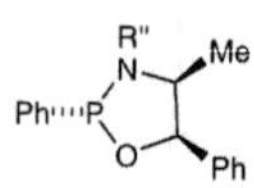	MeI	CO	71 (R"=Me)	41(SS)
			60 (R"=Cy)	56(SS)
			50 (R"=^{i}Pr)	69(SS)

Scheme 25

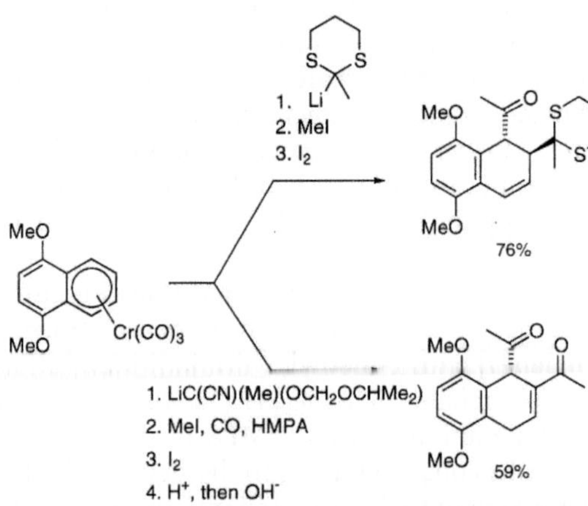

Scheme 26

lavinone. Nucleophilic attack on under kinetic conditions is non-selective and all four regioisomers are formed. However, on warming the reaction mixture to -10 °C, equilibration of the less stable anionic β-addition products to the more stable α-addition products takes place, with addition to C(5) being favored over addition to C(8) for steric reasons. Treatment with a co-solvent to suppress anion dissociation, followed by alkylation/migratory CO insertion and reductive elimination, gives a mixture of two isomers from which the key 1,5,6-substituted intermediate can be isolated by a single crystallization (Scheme 27) [60].

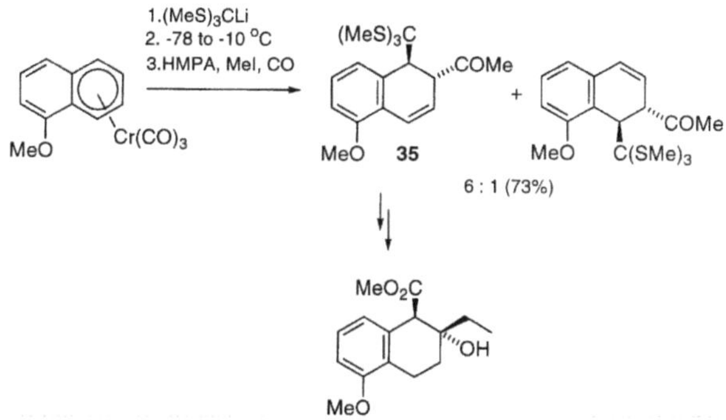

Scheme 27

3 Molybdenum Mediated Dearomatization Reactions

[(Benzene)Mo(CO)$_3$] is readily accessible from Mo(CO)$_6$ [61–65] (see also Chap. 2) but, surprisingly, analogous reactions to the ones listed above for the Cr(CO)$_3$ complex have not been reported prior to 2002. The arene-M bond is stronger in [Mo(benzene)(CO)$_3$] than in the analogous Cr complex (Chap. 2) but it is far more labile. This results in a more difficult handling of the Mo compounds and has retarded their use in synthesis. Very recently, the first examples of nucleophile/electrophile addition reactions with [Mo(benzene)(CO)$_3$] appeared in the literature [66]. Besides demonstrating the viability of using π-arene molybdenum complexes in this reaction sequence, the results show that intermediates that have eluded characterization in the Cr-mediated reaction sequence can be isolated and that reactions of Cr and Mo complexes can have different selectivities. The addition sequence detailed in Scheme 28 also attests to the decreased tendency to carbonylation of the Mo-bound allyl group when compared to the analogous chromium mediated transformation.

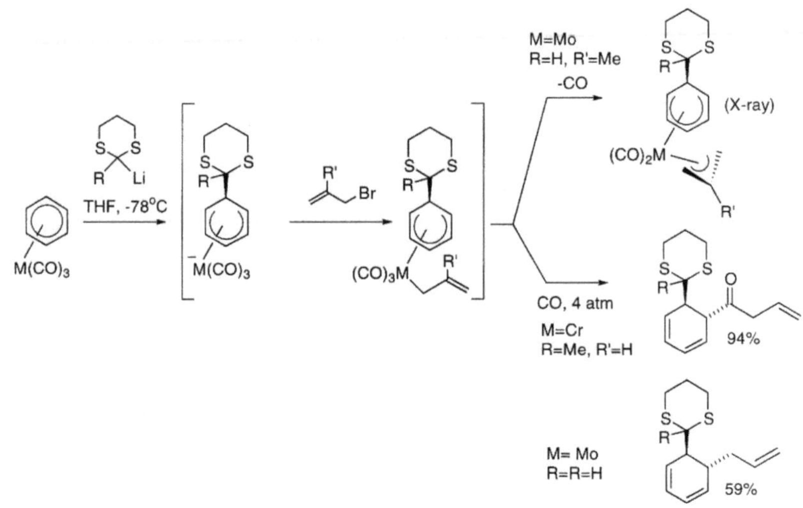

Scheme 28

4 Manganese Mediated Dearomatization Reactions

Arene tricarbonyl manganese complexes are open to attack by a broad range of nucleophiles because of their increased electrophilic character when compared to arene chromium tricarbonyl complexes [67]. The reactivity of the resulting $Mn(CO)_3$(cyclohexadienyl) complexes differs markedly from the $Cr(CO)_3$ analogues. The low nucleophilicity of the neutral $Mn(CO)_3$(cyclohexadienyl) complexes is insufficient for reactions with C-electrophiles. Protonation under oxidative conditions (stoichiometric amount of Jones reagent—CrO_3/H_2SO_4/acetone) generally leads to rearomatization. Where the dearomatized product is isolated, a complex mixture of diene isomers is obtained, separation of which has proved problematic. Routes to dearomatized products via $Mn(arene)(CO)_3^+$ complexes, therefore, rely upon the methods of the addition of two nucleophiles (Scheme 29). Depending on the sequence used, the products are either *cis*- or

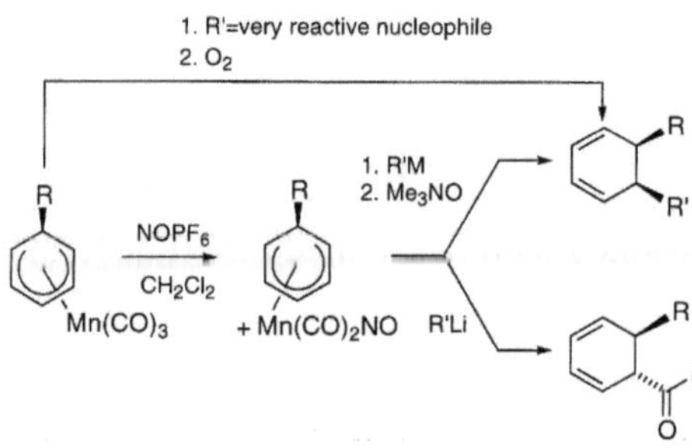

Scheme 29

trans-di-substituted cyclohexadienes when two C-nucleophiles are added or mono-substituted cyclohexadienes when the first nucleophile added is a hydride.

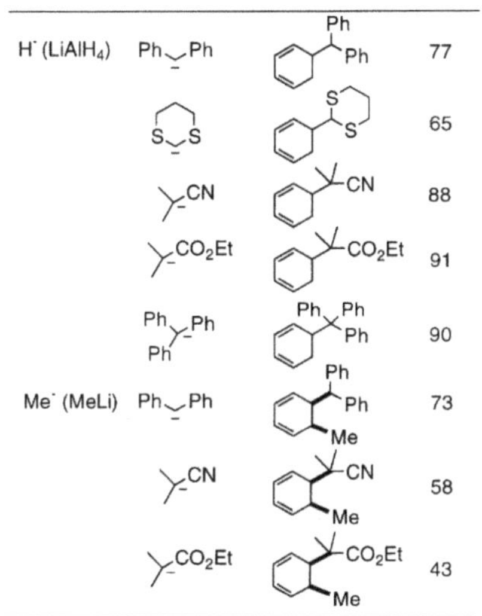

X	R	L	Nucleophile	Yield %	Product
H	Me	CO	$NaCH(CO_2Et)(COMe)$	67	A
			$NaCH(CO_2Et)CN$	59	A
			$NaCH(CO_2Me)(SO_2Ph)$	67	A
			$NaSCH_2CH_2OH$	60	A
H	Me	$P(OPh)_3$	$NaCH(CO_2Et)(COMe)$	79	A
			$NaCH(CO_2Et)CN$	82	A
			$NaCH(CO_2Me)(SO_2Ph)$	83	A
			$NaSCH_2CH_2OH$	83	A
OMe	Me	CO	$NaCH(CO_2Et)(COMe)$	18	C
			$NaCH(CO_2Et)CN$	27	B
			$NaCH(CO_2Me)(SO_2Ph)$	28	B
			$NaSCH_2CH_2OH$	50	B
OMe	Me	$P(OPh)_3$	$NaCH(CO_2Et)(COMe)$	62	-
			$NaCH(CO_2Et)CN$	80	-
			$NaCH(CO_2Me)(SO_2Ph)$	82	-
H	Ph	CO	$NaCH(CO_2Et)(COMe)$	66	A
			$NaCH(CO_2Et)CN$	59	A
			$NaCH(CO_2Me)(SO_2Ph)$	62	A
			$NaSCH_2CH_2OH$	87	A
			$NaBH_4$	85	-
			$NaCH(CN)_2$	-	-
			NaCp	-	
H	Ph	PMe_3	$NaBH_4$	85	A
			$NaCH(CN)_2$	94	-
			NaCp	62	-
			$NaCH(CO_2Me)_2$	80	-

Scheme 30

The first and more direct method involves treatment of [Mn(benzene)$(CO)_3$] sequentially with two nucleophiles. The reaction appears to be limited to the use of either H^- (from $LiAlH_4$) or MeLi as first nucleophile. The second nucleophile again adds *exo* to the manganese unit at a cyclohexadienyl terminus, and this affords a *cis*-substituted cyclohexadiene (Scheme 30) [68].

The second method available for double nucleophilic addition to an arene manganese tricarbonyl complex involves 'reactivation' of the cyclohexadienyl complex by substitution of a CO ligand by NO^+. This cationic cyclohexadienyl complex can now react with less reactive nucleophiles [69, 70]. The cationic cyclohexadienyl $Mn(CO)_2NO^+$ complex is in fact more electrophilic than the original Mn(arene)$(CO)_3{}^+$ complex. Grignard reagents apparently react by single-electron transfer and this does not give the desired diene. This problem has been circumvented by further replacement of a CO ligand with PMe_3, although it can be argued that this renders the procedure rather awkward [71, 72]. Phosphorus and nitrogen nucleophiles have been shown to add, but perhaps the most useful conversions are those resulting from either hydride or carbon nucleophile addition [73–76]. Hydride and stabilized enolates add directly to the unsaturated ligand. Hydride adds *endo*, whilst stabilized carbon nucleophiles add *exo* to the manganese unit (Scheme 31) [77–78]. A number of carbon nucleophiles have been used in the second nucleophilic addition, although the first is restricted to methyl and phenyl Grignard reagents.

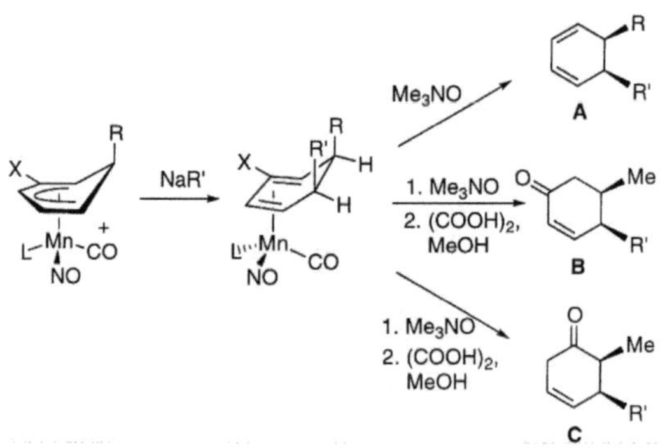

Scheme 31

A different pathway is followed in the reaction of the cationic (cyclohexadienyl)$Mn(CO)_2NO$ complex with phenyl and methyl lithium reagents. Reaction now occurs at a CO ligand, and this is followed by rapid reductive elimination of the resulting cyclohexadienyl/acyl moieties and decomplexation to yield a *trans*-disubstituted 1,3-cyclohexadiene [72]32.

Two approaches have been studied to access non-racemic products: a chiral nucleophile and a chiral auxiliary. In an approach based on Pearson's earlier work with dienyliron and dienemolybdenum complexes [79], Miles used a chiral enolate derived from *N*-acyloxazolidinones in a reaction with a prochiral Mn(arene)$(CO)_3{}^+$ complex to give a 3.5:1 mixture of chiral η^5-dienylmanganese

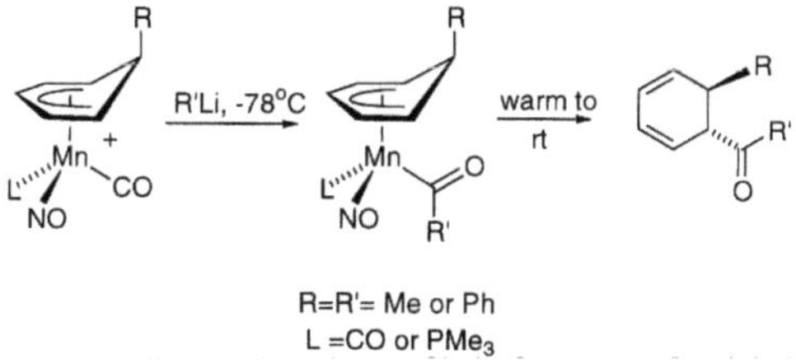

Scheme 32

complexes. Following diastereoisomeric enrichment by fractional crystallization, this intermediate was advanced to a highly enantiomerically enriched ketal product, an intermediate in the total synthesis of the juvenile hormone (+)-juvabione (Scheme 33) [80].

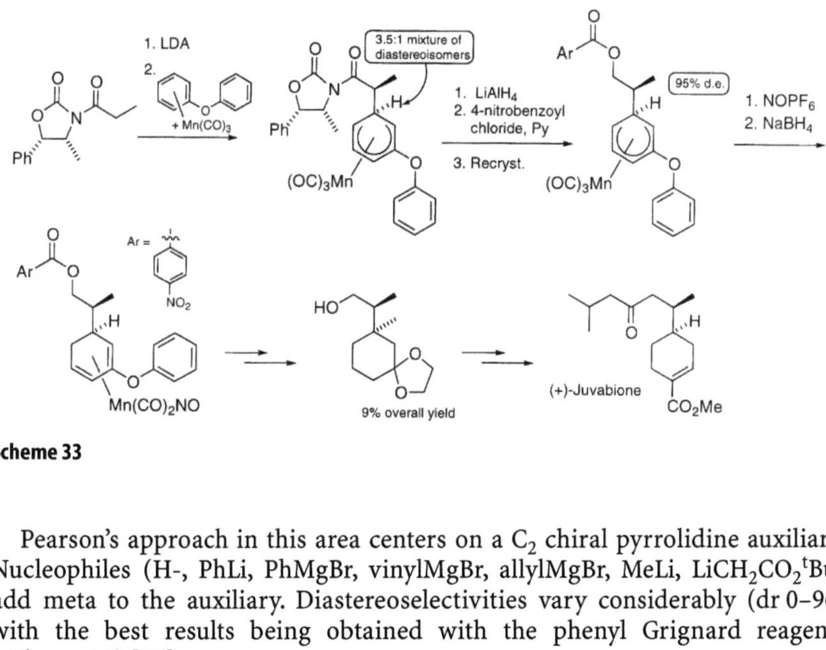

Scheme 33

Pearson's approach in this area centers on a C_2 chiral pyrrolidine auxiliary. Nucleophiles (H-, PhLi, PhMgBr, vinylMgBr, allylMgBr, MeLi, $LiCH_2CO_2{}^tBu$) add meta to the auxiliary. Diastereoselectivities vary considerably (dr 0–96) with the best results being obtained with the phenyl Grignard reagents (Scheme 34) [81].

Scheme 34

Manipulation through to the dearomatized products has however proved impossible with this system. In order to address this problem, Pearson has recently described (albeit racemic) methodology which enables the nucleophilic addition to arenemanganese complexes (bearing an amino linkage) and transformation to substituted cyclohexenones (Scheme 35) [82].

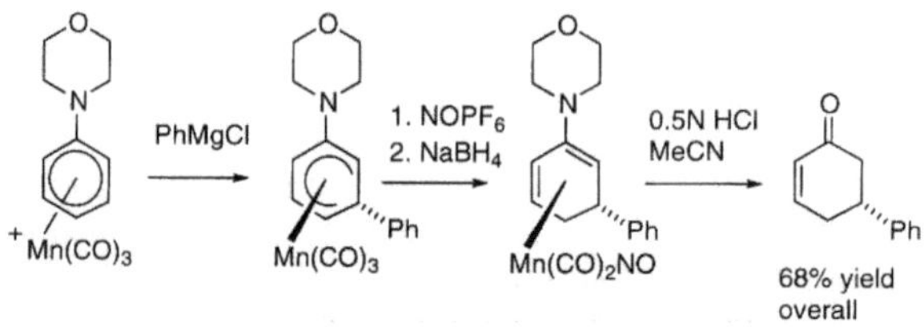

Scheme 35

Conclusion

Transition metal-mediated dearomatization using electrophilic chromium, molybdenum and manganese complex fragments provide rapid access to compounds which otherwise require long and tedious manipulations. Efficient use of the described methodology can lead to the synthesis of complex organic products with high regio-, chemo-, stereo-, and often enantioselectivity. The chemistry covering intermolecular manipulations is extensive and there is a broad understanding of the respective limitations for application of each approach. Research within the asymmetric arena is still at the development stage, although in the chromium mediated chemistry numerous methods which deliver asymmetric induction have been designed and adopted. There is no doubt that development in this area will be pursued vigorously and, with efficient methodology established, applications in the synthesis of organic compounds of high complexity will follow.

References

1. Bach T (1996) Angew Chem Int Edn Eng 35:729
2. Rabideau PW, Marcinow Z (1992) Org React 42:1
3. Guo Z, Schultz AG (2001) J Org Chem 66:2154
4. Donohoe TJ, House D (2002) J Org Chem 67:5015
5. Wender PA, Siggel L, Nuss JM (1991) In: Paquette LA, Trost BM, Fleming I (eds) Comprehensive organic synthesis, vol 5. Pergamon, Oxford, p 645
6. Cornelisse J (1993) Chem Rev 93:615
7. Ley SV (1990) Pure Appl Chem 62:2031
8. Hudlicky T (1994) Pure Appl Chem 66:2067
9. Paradisi C (1991) In: Semmelhack MF, Trost BM, Fleming I (eds) Comprehensive organic synthesis, vol 4. Pergamon, Oxford, p 423
10. Tomioka K, Shindo M, Koga K (1993) Tetrahedron Lett 34:681
11. Gant TG, Meyers AI (1994) Tetrahedron 50:2297

12. Maruoka K, Ito M, Yamamoto H (1995) J Am Chem Soc 117:9091
13. Barluenga J, Trabanco AA, Florez J, Garcia-Granda S, Martin E (1996) J Am Chem Soc 118:13099
14. Clayden J, Forricher YJY, Lam HK (2002) Eur J Org Chem 3558
15. Semmelhack MF, Hall JHT, Yoshifuji M (1976) J Am Chem Soc 98:6387
16. Semmelhack MF, Hall HT, Yoshifuji M, Clark G (1975) J Am Chem Soc 97:1247
17. Semmelhack MF, Hall JHT, Farina R, Yoshifuji M, Clark G, Bargar T, Hirotsu K, Clardy J (1979) J Am Chem Soc 101:3535
18. Semmelhack MF, Clark GR, Garcia JL, Harrison JJ, Thebtaranonth Y, Wulff W, Yamashita A (1981) Tetrahedron 37:3957
19. Boutonnet J-C, Levisalles J, Normant JM, Rose E (1983) J Organomet Chem 255:C21
20. Kündig EP, Amurrio D, Bernardinelli G, Chowdhury R (1993) Organometallics 12:4275
21. Semmelhack MF, Harrison JJ, Thebtaranonth Y (1979) J Org Chem 44:3275
22. Schellhaas K, Schmalz H-G, Bats JW (1998) Chem Eur J 4:57
23. Semmelhack MF, Schmalz H-G (1996) Tetrahedron Lett 37:3089
24. Kündig EP, Desobry V, Simmons DP, Wenger E (1989) J Am Chem Soc 111:1804
25. Ohlsson B, Ullenius C (1988) J Organomet Chem 350:35
26. Pearson AJ, Gontcharov AV, Woodgate PD (1996) Tetrahedron Lett 37:3087
27. Pearson AJ, Gontcharov AV (1998) J Org Chem 63:152
28. Kündig EP, Desobry V, Simmons DP (1983) J Am Chem Soc 105:6962
29. Semmelhack MF, Thebtaranonth Y, Keller L (1977) J Am Chem Soc 99:959
30. Semmelhack MF, Yamashita A (1980) J Am Chem Soc 102:5924
31. Chamberlain S, Wulff WD (1992) J Am Chem Soc 114:10667
32. Schmalz H-G, Siegel S, Bats JW (1995) Angew Chem Int Edn Eng 34:2383
33. Schmalz H-G, Schellhaas K (1995) Tetrahedron Lett 36:5511
34. Hoffmann O, Schmalz H-G (1998) Synlett 1426
35. Kündig EP, Simmons DP (1983) J Chem Soc Chem Commun 1320
36. Pape AR, Kaliappan KP, Kündig EP (2000) Chem Rev 100:2917
37. Kündig EP, Cunningham JAF, Paglia P, Simmons DP, Bernardinelli G (1990) Helv Chim Acta 73:386
38. Bernardinelli G, Cunningham JA, Dupré C, Kündig EP, Stussi D, Weber J (1992) Chimia 46:126
39. Djukic JP, Rose-Munch F, Rose E, Dromzee Y (1993) J Am Chem Soc 115:6434
40. Kündig EP (1985) Pure Appl Chem 57:1855
41. Bernardinelli G, Cunningham JA, Dupré C, Kündig EP, Stussi D, Weber J (1992) Chimia 46:126
42. Kündig EP, Quattropani A, Inage M, Ripa A, Dupré C, Cunningham JAF, Bourdin B (1996) Pure Appl Chem 68:97
43. Kündig EP, Bernardinelli G, Liu R, Ripa A (1991) J Am Chem Soc 113:9676
44. Kündig EP, Ripa A, Liu R, Amurrio D, Bernardinelli G (1993) Organometallics 12:3724
45. Kündig EP, Ripa A, Liu R, Bernardinelli G (1994) J Org Chem 59:4773
46. Fretzen A, Ripa A, Liu R, Bernardinelli G, Kundig EP (1998) Chem Eur J 4:251
47. Tomioka K, Shindo M, Koga K (1989) J Am Chem Soc 111:8266
48. Tomioka K, Shindo M, Koga K (1990) Tetrahedron Lett 31:1739
49. Tomioka K, Shindo M, Koga K (1993) Tetrahedron Lett 34:681
50. Kundig EP, Cannas R, Laxmisha M, Liu R, Tchertchian S (2003) J Am Chem Soc 125:5642
51. Kündig EP, Ripa A, Bernardinelli G (1992) Angew Chem Int Edn Eng 31:1071
52. Bernardinelli G, Gillet S, Kündig EP, Liu R, Ripa A, Saudan L (2001) Synthesis: 2040
53. Kündig EP, Amurrio D, Anderson G, Beruben D, Khan K, Ripa A, Ronggang L (1997) Pure Appl Chem 69:543
54. Sammakia T, Latham HA, Schaad DR (1995) J Org Chem 60:10
55. Nishibayashi Y, Uemura S (1995) Synlett 79
56. Zhang W, Adachi Y, Hirao T, Ikeda I (1996) Tetrahedron Asymmetry 7:451

57. Fretzen A, Kundig EP (1997) Helv Chim Acta 80:2023
58. Quattropani A, Anderson G, Bernardinelli G, Kündig EP (1997) J Am Chem Soc 119:4773
59. Kündig EP, Quattropani A, Inage M, Ripa A, Dupré C, Cunningham JAF, Bourdin B (1996) Pure Appl Chem 68:97
60. Kündig EP, Inage M, Bernardinelli G (1991) Organometallics 10:2921
61. Fischer EO, Oefele KZ (1958) Naturforsch B 13:458
62. Strohmeier W (1961) Chem Ber 94:3337
63. Nesmeyanov AN, Kaganovich VS, Krivykh VV, Rybinskaya I (1975) J Organomet Chem 102:185
64. Magomedov GKI (1990) J Organomet Chem 385:113
65. Hudecek M, Toma S (1990) J Organomet Chem 393:115
66. Kündig EP, Fabritius CH, Grossheimann G, Robvieux F, Romanens R, Bernardinelli G (2002) Angew Chem Int Edn Eng 41:4577
67. Kane-Maguire LAP, Honig ED, Sweigart DA (1984) Chem Rev 84:525
68. Roell BC, McDaniel KF, Vaughan WS, Macy TS (1993) Organometallics 12:224
69. Ashford PK, Baker PK, Connelly NG, Kelly RC, Woodley VA (1982) J Chem Soc Dalton Trans 477
70. Chung YK, Choi HS, Sweigart DA, Connelly NG (1982) J Am Chem Soc 104:4245
71. Pike RD, Ryan WJ, Carpenter GB, Sweigart DA (1989) J Am Chem Soc 111:8535
72. Pike RD, Sweigart DA (1990) Synlett 565
73. Chung YK, Choi HS, Sweigart DA, Connelly NG (1982) J Am Chem Soc 104:4245
74. Pike RD, Ryan WJ, Carpenter GB, Sweigart DA (1989) J Am Chem Soc 111:8535
75. Chung YK, Honig ED, Robinson WT, Sweigart DA, Connelly NG, Ittel SD (1983) Organometallics 2:1479
76. Ittel SD, Whitney JF, Chung YK, Williard PG, Sweigart DA (1988) Organometallics 7:1323
77. Chung YK, Sweigart DA, Connelly NG, Sheridan JB (1985) J Am Chem Soc 107:2388
78. Lee T-Y, Kang YK, Chung YK, Pike RD, Sweigart DA (1993) Inorg Chim Acta 214:125
79. Pearson AJ, Khetani VD, Roden BA (1989) J Org Chem 54:5141
80. Miles WH, Brinkman HR (1992) Tetrahedron Lett 33:589
81. Pearson AJ, Gontcharov AV, Zhu PY (1997) Tetrahedron 53:3849
82. Pearson AJ, Vickerman RJ (1998) Tetrahedron Lett 39:5931

Topics Organomet Chem (2004) 7: 95–127
DOI 10.1007/b94492

The Dearomatization of Arenes by Dihapto-Coordination

W. Dean Harman

Department of Chemistry, University of Virginia, P.O. 400319 Charlottesville, Virginia 22904-4319, USA
E-mail: wdh5z@virginia.edu

Abstract Progress in the development of a dearomatization methodology for arenes based on their dihapto-coordination is reviewed. The fragments $\{Os(NH_3)_5\}^{2+}$, {TpRe(CO)(L)} (where L=1-methylimidazole, pyridine, PMe_3, *tert*-butylisonitrile, NH_3) and {TpMo(NO)(MeIm)} form stable η^2-coordinate complexes with a wide variety of arenes. The act of coordination greatly reduces the aromatic character of these ligands and, as a consequence, activates them toward various organic reactions. In particular, the addition of carbon-based electrophiles to arenes is notably enhanced relative to this reaction for free aromatic molecules. The resulting arenium intermediates are stabilized by metal back-bonding to the point that they may be isolated and subsequently subjected to a variety of carbon-based nucleophiles. The one-electron oxidation of the metal removes the organic ligand. The overall vicinal-difunctionalization of two ring carbons may be accomplished with excellent stereo- and regiocontrol.

Keywords Dearomatization · Osmium · Rhenium · Molybdenum · Arene

List of Abbreviations

9-BBN	9-Borabicyclo[3.3.1]nonane
CAN	Ceric ammonium nitrate
DDQ	2,3-Dichloro-5,6-dicyanobenzoquinone
DIEA	Diisopropylethylamine (Hünig's base)
DMAc	*N,N*-Dimethylacetamide
DME	1,2-Dimethoxyethane
MMTP	1-Methoxy-2-methyl-1-trimethylsiloxy-2-propene
MVK	Methyl vinyl ketone
MeIm	*N*-Methylimidazole
NMM	*N*-Methylmaleimide
OTf	Triflate
TBS	*tert*-Butyldimethylsilyl
TMB	1,2,3,4-Tetramethylbenzene
Tp	Hydridotris(pyrazolyl)borate

1 Introduction

For decades, the dearomatization of arenes has been recognized as a chemical transformation of fundamental importance. It provides the connection between this robust and abundant source of hydrocarbons and the alicyclic frameworks common to many biologically active products. Thus, dearomatization methods have become powerful tools for organic synthesis [1,2]. The chemical properties of arenes are influenced by their coordination to a transition metal [3]. For example, complexes such as (η^6-arene)$Cr(CO)_3$ and its cationic analogs (e.g., $FeCp^+$, $RuCp^+$, $Mn(CO)_3{}^+$) are susceptible to nucleophilic substitution [4] or addition [5], ultimately leading to the formation of substituted arenes or cyclohexadienes, respectively. During the past three decades, the application of η^6-arene complexes to organic synthesis has been widely explored [3], as is described elsewhere in this book. In this chapter, an alternative approach to the dearomatization of arenes is described in which the aromatic system may be activated toward organic transformations by η^2-coordination [6]. Here, the metal-arene bond is stabilized primarily by the interaction of filled metal dπ orbitals with the π^* system of the arene, an interaction having two important consequences for the activation of the aromatic ring. Through π backbonding, the aromatic π system becomes more electron-rich, similar to what is observed for organic arenes bearing electron-donating groups. Additionally, structural data for complexes containing η^2-coordinated aromatic rings show significant distortions in the bond lengths of the ring, consistent with a localization of π electron density. Together, these effects activate η^2-bound aromatic systems toward electrophilic rather than nucleophilic addition (Fig. 1) [6]. This is similar in concept to the reactions of the $Mn(CO)_3$(η^4-benzene)- which also undergo reactions with electrophiles [7].

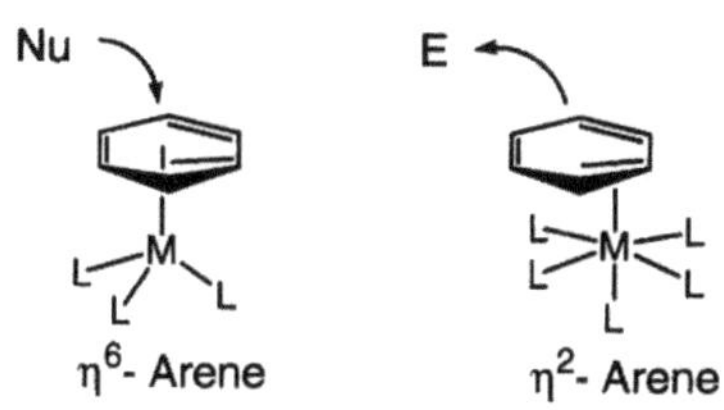

Fig. 1 A comparison of general reactivities for η^6-arene and η^2-arene complexes

Of the handful of transition metal systems that are known to form stable η^2 complexes with aromatic molecules [8–14], only d^6 octahedral metal complexes have been shown to enhance the reactivity of the aromatic ligand toward electrophiles to date [15]. For nearly a decade, despite our best efforts, this mode of arene activation was known only for the pentaammineosmium(II) system. However, in the past few years a new generation of dearomatization agents have been developed based on a careful matching of the d^5/d^6 reduction potential of rhenium(I) [16] and molybdenum(0) complexes [17] with that of pentaammineosmium(II).

This chapter will initially cover several aspects of dihapto-coordination of aromatic molecules, including the scope of the dearomatization agent and the aromatic substrate. The primary focus of this work, however, will be the fundamental organic reactions of these complexes with electrophiles and the subsequent reactions of those products. Several applications of this methodology will also be illustrated. Owing largely to its earlier discovery, the majority of the organic transformations reviewed will be with osmium(II), however, recent arene transformations promoted with rhenium(I) and molybdenum(0) will also be discussed, with an emphasis on differences in reactivity compared to those of osmium.

2 Binding Characteristics

2.1 Survey of π Bases

Table 1 lists a number of examples of transition metals forming stable, isolable complexes with arenes at ambient temperature. Given that the 1,2-double bond of naphthalene is considerably more localized than that of benzene, complexes of the former have greater kinetic and thermodynamic stability relative to complexes of the parent arene. Note that while examples are shown in Table 1 for transition metals in many of the columns and rows of the periodic table, only a handful, all formally octahedral d^6, have been demonstrated to active the arene toward electrophiles. The primary interaction responsible for the stability of η^2-arene complexes is the π backbonding interaction. Indeed DFT calculations with pentaammineosmium(II) indicate that despite its dicationic charge, there is a net transfer of electron density to the arene ligand [18]. Thus, it is essential that the metal be highly electron-rich. Of note, metal fragments bearing a Cp ligand are more susceptible to oxidative addition of the metal to the C-H bond compared to those with a pseudo-octahedral geometry. [19] In addition, the Cp ligand can compete with the arene as the site of electrophilic addition.

2.2 Survey of Aromatic Ligands

Pentaammineosmium(II) complexes have been prepared with a diverse array of η^2-bound aromatic ligands (e.g., **2–8** in Fig. 2) including benzenes [20, 21], anisoles [22, 23], anilines [24], phenols [25, 26], naphthalenes, and other polyaromatic hydrocarbons [27–29], with yields typically >90%. Where convenient, the ligand is used in approximately a tenfold excess to minimize the formation of binuclear impurities. However, if the aromatic substrate is precious, as little as 1.1 equivalents may be used, provided the reaction is carried out in water or methanol [23]. Carrying out the complexation procedure without a large excess of ligand should be avoided, as these conditions increase the possibility of forming stable binuclear arene complexes (see Fig. 2, **1**) [20].

Table 1 Stability of η^2-arene complexes in solution

Metal Fragment	d^n	Stability of Naphthalene Complex t=1/2 (T, °C)	Stability of Benzene Complex t=1/2	Known Example of Electrophilic Additon	Reference
$[Os(NH_3)_5]^{2+}$	6	4 h (100) acetone[a]	5.5 h (20) acetone[a]	yes	*6*
TpRe(CO)(MeIm)	6	1.3 h (100) acetone[a]	1.6 h (23) acetone[a] > 1 month (20) in benzene	yes	*16*
TpRe(CO)(py)	6	0.7 h (100) acetone[a]	not isolated	yes	*16*
$TpRe(CO)(PMe_3)$	6	5.9 (23) acetone[a]	not isolated	yes	*16*
TpRe(CO)(*t*-BuNC)	6	12.1 (23) acetone[a]	not isolated	yes	*16*
TpMo(NO)(MeIm)	6	236 (20) acetone[a]	not isolated	yes	*17*
Cp*Ru(NO)	8	0.4 h(75) benzene	~8 h (20) in benzene	no	*10*
Cp*Rh(CO)	8	18 h (51) benzene[b]	not stable[b]	no	*8*
$Ni(L_2)$ (L_2 = diphosphine)	10	stable at 40°C in THF		no	*13*

a) arene displaced by acetone b) in equilibrium with aryl hydride

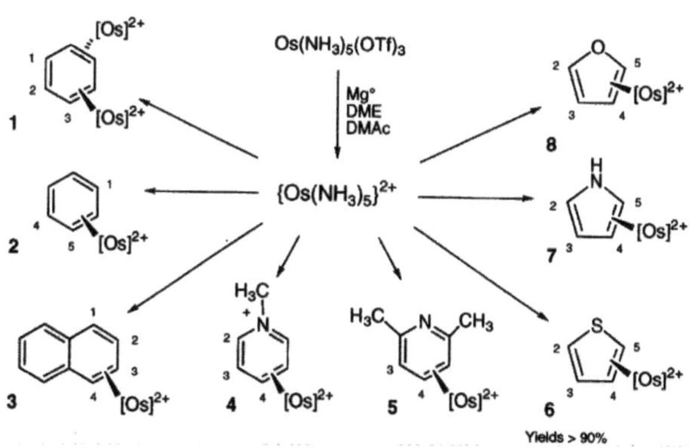

Fig. 2 A survey of pentaammineosmium(II) complexes with dihapto-bound aromatic ligands

The TpRe(CO) and TpMo(NO) systems are less extensively investigated than the pentaammineosmium system, so the range of known arene complexes is not as broad. All of these systems (Table 1) form stable complexes with naphthalene with substitution half-lives in acetone ranging from hours to months at 20 °C. In addition, benzene, anisole, and *N*-ethylaniline complexes have been reported for TpRe(CO)(MeIm) [16], as of the date of this writing. Once formed, these arene complexes are substitution-inert at ambient temperature, resisting ligand displacement even by strong π acids (e.g., CO, electron-deficient olefins) and good σ donors (e.g., primary amines). Substitution half-lives for aromatic complexes are largely independent of the incoming ligand, consistent with a dissociative substitution mechanism [30]. The substitution half-live for the complex {TpRe(CO)(MeIm)(benzene)} is 1.6 h at 23 °C while the pentaammineosmium(II) benzene complex half life is almost four times this period.

2.2.1
Binding Selectivities

The systems in Table 1 that form stable complexes with arenes are highly reducing (e.g., E° for d^5/d^6: $[Os(NH_3)_5(X)]^{2+/+}$, ~-0.75 V; $TpRe(CO)(MeIm)(X)^{0/-1}$, ~-1.4 V; $TpMo(NO)(MeIm)(X)^{0/-1}$, -1.3 V vs NHE, where X=halide or triflate) and show a strong preference for π-acceptor ligands. For example, when the $\{Os(NH_3)_5\}^{2+}$ system (abbreviated herein as $[Os]^{2+}$) is presented with the unhindered amine aniline, the η^2-arene complex successfully competes with nitrogen coordination [31]. Other organic π-acceptor ligands that form stable complexes with one or more of these metal centers and potentially compete with complexation of aromatic molecules include nitriles [32], aldehydes and ketones [33, 34], esters [35], alkenes [34], and alkynes [36]. Conversely, amides, ethers, alcohols, water, and most amines (protonated) generally do not interfere with complexation.

As a rule, the π base will select a binding site where it causes the minimum disruption to the π system of the organic molecule. Jones et al. [8] have demonstrated this principle for the system $Rh^{I}Cp^{*}PMe_3$ as has Krüger et al. for $Ni^0(PR_3)_2$ (R=cyclohexyl) [37]. Figure 3 further illustrates this point by showing a survey of pentaammineosmium(II) complexes (**9–16**) as their dominant linkage isomers (favored by >20:1 in all cases). For arenes bearing a single substituent (C1), the metal preferentially binds at C5 and C6, allowing linear conjugation of the substituent and unbound portion of the aromatic ring. This is true for *both* electron-rich (e.g., anisole, aniline) *and* electron-deficient (e.g., benzophenone) arenes, although the effect is most pronounced in the former [22]. For the anisole and aniline complexes, the 5,6-η^2 isomer is approximately 3–5 kcal/mol more stable than its 4,5-η^2 form based on a comparison between isomerization rate data for these substituted benzene complexes and those of their parent [38].

When a vinyl group is conjugated to an aromatic ring, coordination is thermodynamically favored outside the aromatic system. In contrast, when a pendant acetyl group is present, the thermodynamic isomer often involves coordination of the aromatic system. For polycyclic aromatic systems, the metal favors the coordination site that minimizes the loss of aromatic stabilization. Thus,

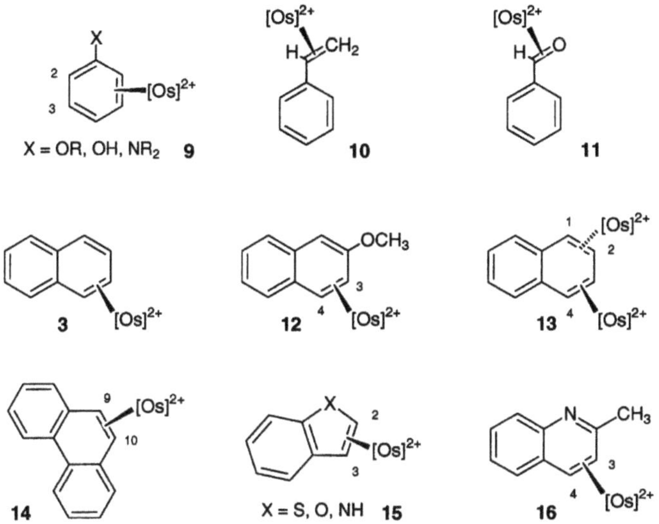

Fig. 3 Pentaammineosmium(II) complexes of aromatic ligands showing the most stable linkage isomer (in all cases, the species shown is favored by >3 kcal/mol over other possible linkage isomers)

naphthalene is bound preferentially by π bases at C1 and C2, while phenanthrene is bound at C9 and C10 [29]. Finally, when two π bases bind naphthalene, the thermodynamically favored isomer has these metals bound to a common ring (i.e., 1,2-η^2 and 3,4-η^2) in order to minimize disruption to the π system [8, 10, 39, 40].

2.2.2
Coordination Diastereomers with TpRe(CO) and TpMo(NO) Systems

X-ray data for the rhenium and molybdenum systems show that the coordinated C=C bond of an aromatic ligand always lies perpendicular to the Re-CO or Mo-NO bond axis. In this alignment, the arene ligand undergoes a backbonding interaction with a filled d orbital that is orthogonal to the π^* orbitals of the diatomic π acid. Adaptation of the quadrant analysis scheme used by Gladysz [41] shows that when the auxiliary ligand, L, is small (e.g., L=MeIm in Fig. 4), Quadrant D is the least sterically congested. Both the coordination site and face of the arene will be selected such that the largest feature of that ligand projects into quadrant D, away from the pyrazolyl rings of Tp. In contrast, when L is large (e.g., PMe_3), quadrant A has the lowest steric profile and will best accommodate any steric bulk of the aromatic ligand. The aromatic ligand can readily adjust its binding site such that the thermodynamically favored position may be reached (see below). Quadrant C is the most sterically congested. The steric strain in this quadrant between a methyl group attached to a bound alkene carbon and the pyrazole ring trans to the CO in the system {TpRe(CO)(MeIm)} has been estimated at 15 kcal/mol [34].

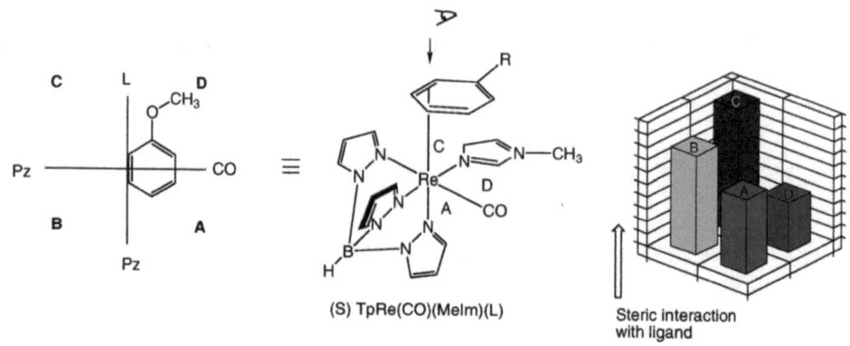

Fig. 4 The asymmetric environment of TpRe(CO)(L)(arene) complexes and the assignment of quadrants

3 Organic Reactions of Arene and Arenium Complexes

3.1 Reactions Promoted by Osmium(II)

3.1.1 *Preparation of Osmium(II) η^2-Arene Complexes*

The precursor to η^2-complexes of pentaammineosmium(II), $[Os(NH_3)_5(OTf)](OTf)_2$ (OTf = trifluoromethanesulfonate), is commercially available from Aldrich Chemical and can be prepared in three steps from OsO_4 (92% overall yield) [42]. Thermally stable complexes of benzenes, naphthalenes, anisoles, anilines, and phenols can be prepared with yields >90% by reducing this osmium(III) salt with an excess of the aromatic compound. Typically, the reductions are performed under a dinitrogen atmosphere with either Mg^0 in a DMA/DME mixture or Zn/Hg in methanol. Isolation of the η^2-complexes is carried out by removing the excess reductant and precipitating the osmium salt in a solution of diethyl ether/methylene chloride. Excess ligand can be recovered or recycled from the resulting filtrate. Dihapto-coordinated arene complexes of $[Os]^{2+}$ are stable in solution under an inert atmosphere, and olefin complexes that result from dearomatization are stable to oxygen.

3.1.2 *Addition Reactions with Heteroatom-Substituted Benzenes*

3.1.2.1 *Hydrogenation Reactions*

The product of hydrogenation for the anisole complex of pentaammineosmium(II) depends on the reaction medium [43]. In the presence of dry methanol, bound anisole of **17** is converted cleanly to 3-methoxycyclohexene (i.e., **18**), where the methoxy group is oriented toward the metal in the final complex. However, if the methanol is not carefully dried, water intercepts the dihydroanisole intermediate producing the cyclohexenone complex (**19**, Fig. 5).

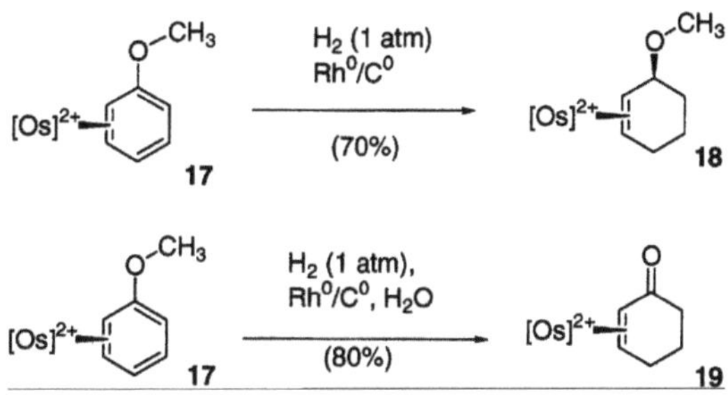

Fig. 5 The hydrogenation of anisole bound to pentaammineosmium(II)

3.1.2.2
Electrophilic Addition Reactions to Arenes

Most osmium complexes of phenols [26, 44], anilines [24, 45], and anisoles [23, 46, 47] undergo electrophilic addition with a high regiochemical preference for para addition. While electrophilic additions to phenol complexes are typically carried out in the presence of an amine base catalyst, the other two classes generally require a mild Lewis or Brønsted acid to promote the reaction. The primary advantage of the less activated arenes is that the 4*H*-arenium species resulting from electrophilic addition are more reactive toward nucleophilic addition reactions (see below).

a. Phenols

When an acetonitrile solution of the phenol complex is treated with one equivalent of MVK and pyridine (Fig. 6), a 4-alkylated 4*H*-phenol complex (**21**) is obtained that is a product of conjugate addition based on NOE data. Addition occurs to the face of the phenol ring opposite to that of metal coordination. This complex shows no signs of decomposition even after standing in acidic CH_3CN solution for 24 h. However, upon exposure to a moderate base (e.g., Hünig's

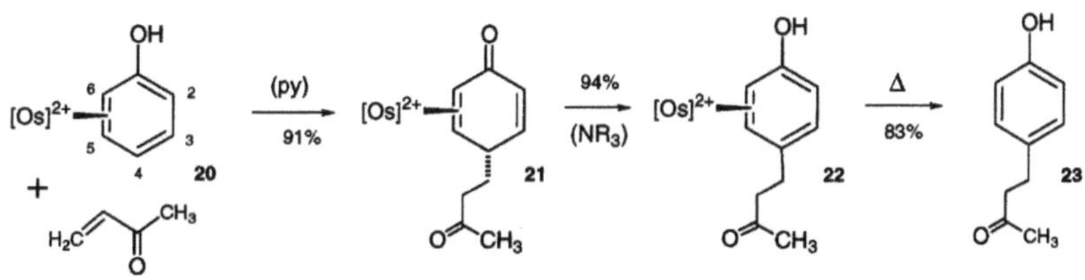

Fig. 6 The osmium(II)-promoted *para* alkylation of phenol with methyl vinyl ketone (MVK)

base) rearomatization occurs, yielding the 4-(2-oxobutyl)phenol complex (**22**). Heating this complex releases the raspberry ketone product (**23**) in high yield [44]. An interesting feature of this reaction sequence is that, in contrast to the 4*H*-phenol complexes generated from isomerization of the phenol ligands, the dienone complex **21** resists isomerization to the arene in acidic solution. This stability is a direct result of the C4 substituent that sterically interferes with potential bases.

The η^2-phenol complex undergoes conjugate addition at C4 with a variety of Michael acceptors (Fig. 7), including those with β substituents [44]. In most cases, the addition reaction is accomplished with an amine base as catalyst (see **25**). Less reactive electrophiles, such as methyl acrylate or acrylonitrile, require a Lewis acid co-catalyst (e.g., **24**). An example of the versatility of this reaction is shown in Fig. 7, where the aromatic steroid β-estradiol (**26**) is complexed (**27**) and subsequently alkylated exclusively at C10 (i.e., *para*) at -40 °C. Since the osmium preferentially binds the α face of the steroid in **27**, conjugate addition occurs from the β face, yielding the stereochemistry found in testosterone [26]. The overall yield of this transformation *after* decomplexation of the dienone product **29** is 69%.

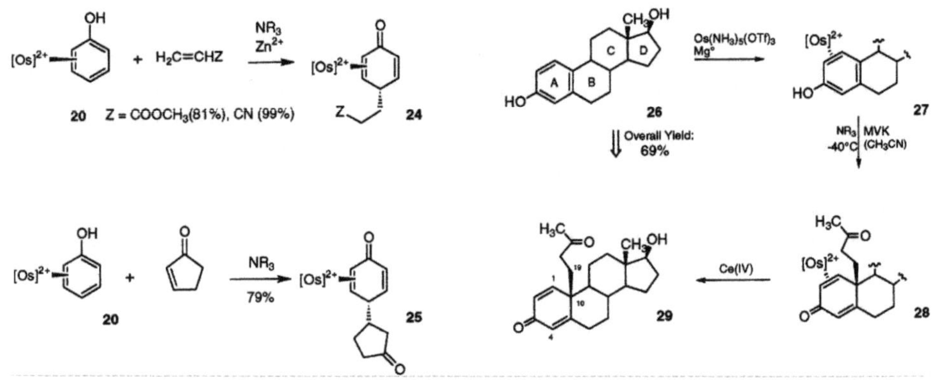

Fig. 7 Conjugate addition reactions with the η^2-phenol complex and various Michael acceptors

Although C4 addition occurs with phenol complexes even for cases where C4 is substituted, in many cases, *ortho* addition is *thermodynamically* favored. In this scenario, the regiochemistry can be effectively controlled by adjusting reaction variables such as temperature, time, and catalyst [44]. Under basic conditions, the active form of the phenol complex is the phenoxide species, which can undergo reversible Michael reactions at C4 and C2, provided that the resulting enolate is not protonated. For instance, the addition of MVK to the osmium complexes of *para*-cresol (**31**) or estradiol (**27**, Fig. 8) occurs at C4 to give the 4*H*-phenol product (**28**, **32**) at -40 °C with an amine base. However, if the reaction is carried out at 20 °C or is run in the presence of a Zn^{2+} co-catalyst, the initially formed enolate may undergo retroaddition, and ultimately, the reaction yields the orthoalkylated product (**30**, **33**; see Fig. 8). Electrophilic addition at the *ortho* position of η^2-arene complexes has been observed primarily in phenol complexes to date, although one example with 2-methylanisole has been reported [23]. In general, for anisole and N-substituted anilines, the heteroatom substituent is oriented away from metal coordination (5,6-η^2) and, as a consequence, it blocks the uncoordinated *ortho* carbon (C2) from undergoing reaction with most electrophiles.

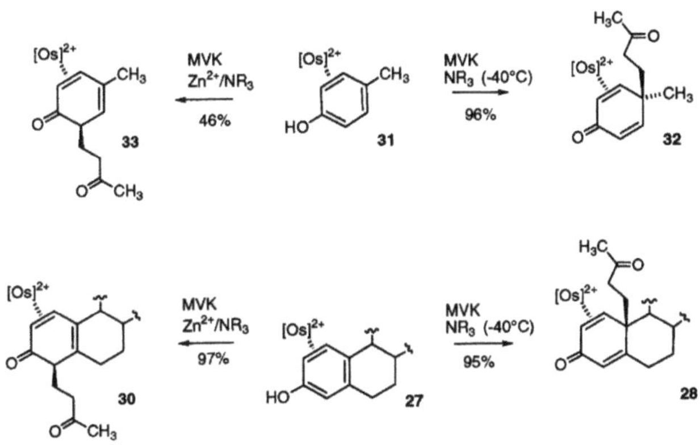

Fig. 8 Regiocontrol of the electrophilic addition reaction of MVK and phenol

Another interesting example of *ortho*-addition to phenol complexes is its reaction with aldehydes. In the presence of a weak base, the aldehyde undergoes an aldol condensation at C2 to give a rare example of a stable *o*-quinone methide complex. In Fig. 9 citronellal (**34**) is combined with phenol complex **20** to generate the quinone methide **35**. Upon oxidative decomplexation a hetero-Diels-Alder reaction occurs to form the benzochromene core **37** in 52% yield [48].

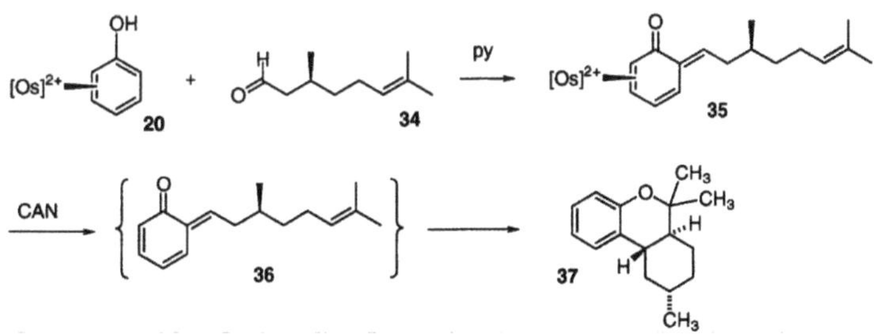

Fig. 9 Formation of an *o*-quinone methide complex, decomplexation, and subsequent cycloaddition of the free quinone methide

b. Anilines

Since deprotonation of the heteroatom with a moderate base is difficult for anilines and impossible for anisoles, the electrophilic addition reactions for osmium(II) complexes of these ligands must be activated by Lewis acids. The additives TBSOTf, $BF_3{\cdot}OEt_2$, $Zn(OTf)_2$, and $Sn(OTf)_2$ among others are tolerated by the metal and may be used to catalyze electrophilic addition reactions with η^2-anilines. Examples of useful electrophiles include Michael acceptors and acetals, as well as acylating agents. Several of these reactions are summarized in Table 2 [45]. Note that in some cases, the preferred reaction conditions involve water as a solvent.

Table 2 Electrophilic addition reactions of η^2-aniline complexes

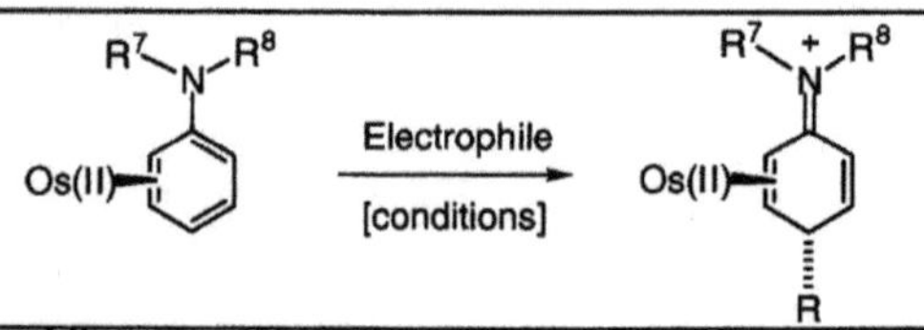

Electrophile (Rxn Cond.)[a]	R^7	R^8	R	isomers [b]	Yield(%)[c]
acrolein (D)	H	Et	$-CH_2CH_2CHO$	-	81
methyl acrylate (B)	H	Et	$-CH_2CH_2CO_2Me$	-	86
acrylonitrile (A)	H	Et	$-CH_2CH_2CN$	-	95
2-cyclopenten-1-one (A, B,C)	H	Et	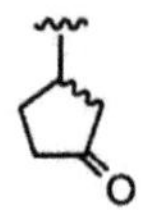	4:1 (C)	77 (C)
benzaldehyde dimethylacetal[a] (A)	H	Et	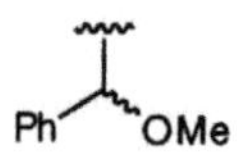	1:1	53
2,2-dimethoxy-propane (A)	H	Et	$-C(CH_3)_2OMe$	-	82
2-cyclohexen-1-one (A)	Me	Me	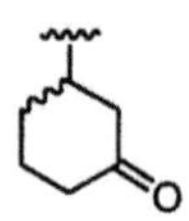	6:1	80
dimethyl fumarate (A)	Me	Me	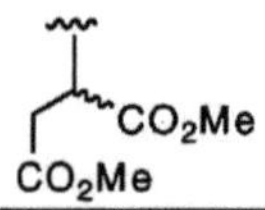	2:1	63

[a] Conditions: **A:** 1) TBSOTf, CH_3CN, 20°C; 2) H_2O. **B:** Same as **A**, but at -40°C. **C:** 1) HOTf (1 equiv), 2) CH_3OH, 20°C. **D:** Same as **C**, but using water as the solvent. [b] Isomer ratio for the stereogenic center located in R. This ratio is based on the relative heights of ^{13}C NMR resonances and is thus approximate.
[c] unoptimized yields.

For the parent aniline complex a complication arises in that the nitrogen-bound isomer is thermodynamically competitive with the ring-bound isomer. To avoid this complication, the nitrogen is first protected with a TBS group, then complexed. Upon deprotection, this material (**38**) cleanly gives the 4*H*-anilinium species (**39**) shown in Fig. 10. This compound resists rearomatization and is prepared free of any nitrogen-bound species. Finally, the 4*H*-anilinium species may be treated with a carbon electrophile to generate 4-alkylated 4*H*-anilinium species (**40–42**) in high yield [45].

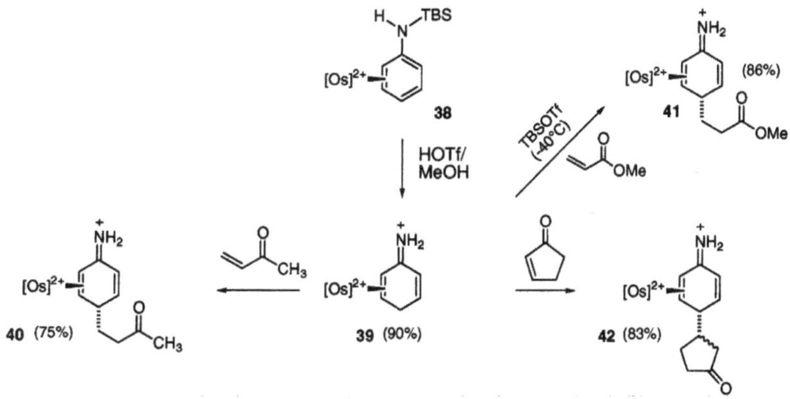

Fig. 10 Alkylations of aniline using a silicon protecting group to prevent nitrogen complexation

c. Anisoles

In contrast to the 4*H*-anilinium complex **39**, the 4*H*-anisolium analog undergoes decomposition in the presence of acid at 20 °C, yielding unidentified osmium(III) salts and approximately 0.5 equivalents of both anisole and benzene. However, at -40 °C these 4*H*-anisolium species are stable. Reactions of the anisole complex with carbon electrophiles are therefore carried out in the presence of a Lewis or Brønsted acid at low temperature [23, 46]. Table 3 summarizes a series of reactions in which the anisole complex has been alkylated and then deprotonated to give the corresponding 4-substituted anisole complex. Suitable electrophiles for this reaction include Michael acceptors, acetals, and nitrilium salts. As with other aromatic systems, the organic ligand is obtained by heating a solution of the complex in acetonitrile or by exposing a solution to a one-electron oxidant (e.g., Ag^+, DDQ, Ce(IV)).

Several deviations to the general reactivity pattern of anisoles arise when substituents are present in the anisole ring. A heteroatom substituent at C3 is in conjugation with the ketonium system resulting from conjugate addition at C4 and a donor group at this carbon greatly facilitates electrophilic addition and stabilizes the 4*H*-anisolium species to the point that the corresponding complex may be isolated [23, 46]. A methyl substituent at C2 has the opposite effect because the methyl group interferes with the methoxy group lying in the

Table 3 Electrophilic addition reactions with η^2-anisole complexes

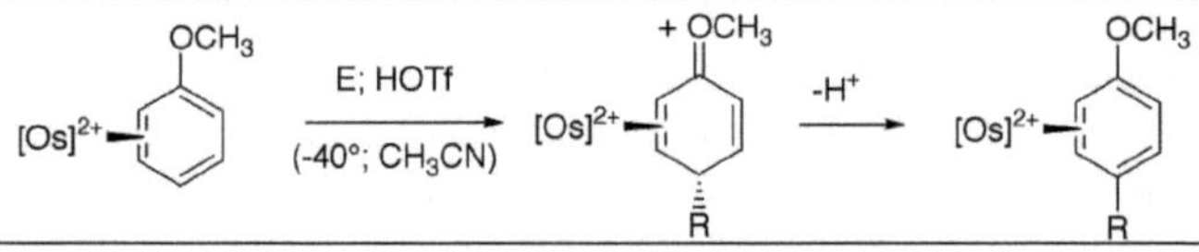

E	R	Yield(%)[a]
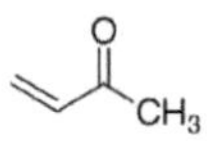	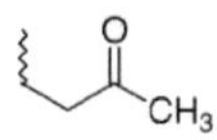	95
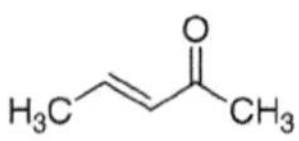	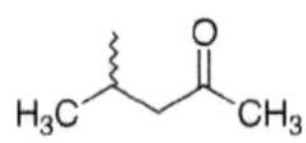	87(81% de)
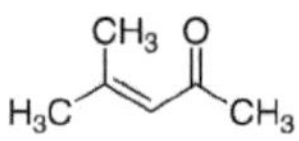	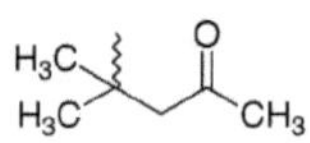	> 90[b]
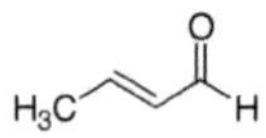	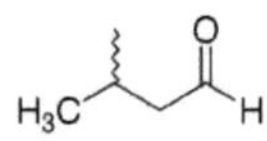	95 (> 90% de)
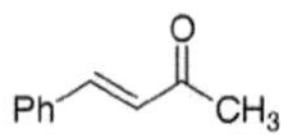	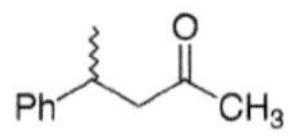	> 90(>90% de)[b]
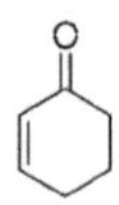	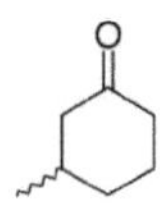	95 (0% de)
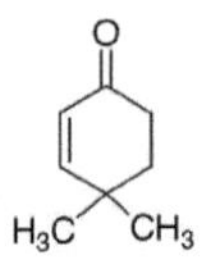	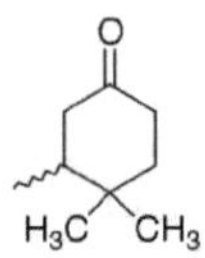	91(90%de)
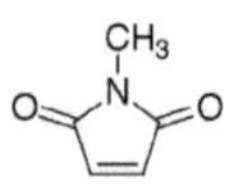	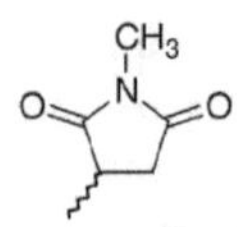	80(>90%de)
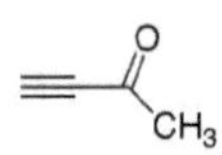	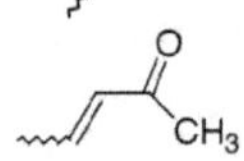	98
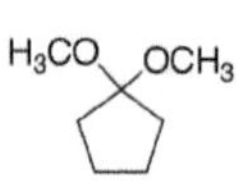	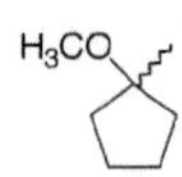	88
$CH(OEt)_3$	CHO	86
$CH_3C{\equiv}\overset{+}{N}CH_3$	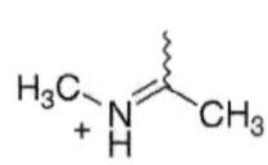	88

a Represents isolated yield; b Represents yield in solution (NMR)

arene plane (Fig. 11, 43). Not only does this retard electrophilic addition at the ring, but it forces the metal to the less congested 4,5-η^2 position (Fig. 11), where the only reasonable site left for electrophilic addition is C6. Thus, 2,6-disubstituted anisoles may be prepared (e.g., 45). In some but not all cases, a methyl group at C4 will redirect alkylation to C2. An illustrative comparison is shown in Fig. 12 for the *para*-methylanisole complex (46). Addition of methylnitrilium triflate results in an ortho electrophilic substitution to form **47** while the softer electrophile MVK adds to the para position to form the 4*H*-anisolium complex **49**.

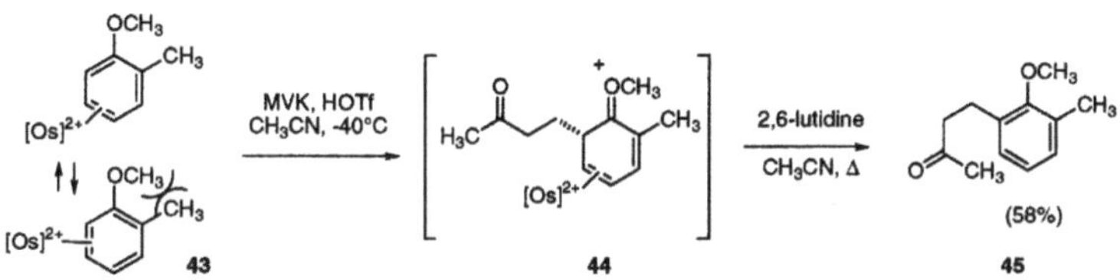

Fig. 11 Coordination and the C6-selective alkylation of 2-methylanisole

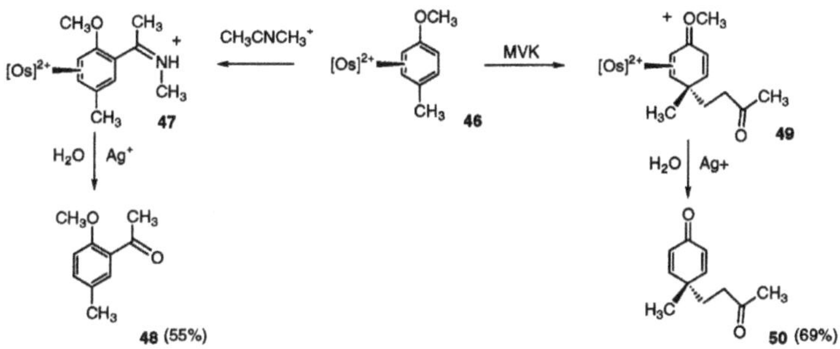

Fig. 12 Differing regiochemistry in the reaction of the 4-methylanisole complexes with carbon electrophiles

3.1.2.3
Intermolecular Nucleophilic Addition Reactions of 4H-Arenium Complexes

An important feature of this *dearomatization* methodology lies in the ability of the metal to stabilize, through metal-to-ligand backbonding, the dienone, anilinium, and anisolium intermediates derived from phenol, aniline, and anisole, respectively. This stabilization sets the stage for subsequent nucleophilic addition to the *meta* position, thereby preventing rearomatization of the ring upon decomplexation of the metal.

The strong backbonding interaction of the pentaammineosmium(II) system with the 2,5-cyclohexadienone ligands formed from phenols significantly reduces the susceptibility of these ligands toward nucleophilic attack. Thus, complexes of 4*H*-phenol (i.e. 2,5-cyclohexadienone) shown in Fig. 7 do not undergo reaction with sodium borohydride at 20 °C at any of the ring carbons [44], nor do they react with non-basic carbon nucleophiles, such as silyl enol ethers or silyl ketene acetals. Under basic conditions, retro-Michael addition or deprotonation at C4 occurs to return the aromatic precursors or electrophilic substitution products, respectively [44]. 4*H*-Anilinium complexes, such as those shown in Table 2, are readily reduced by borohydride reagents (e.g., Bu_4NBH_3CN), but difficulties often arise in stopping the reduction at the dienamine stage. The action of Li(9-BBNH) at -40 °C on the 4*H*-anilinium **51** followed by protonation generates the eneiminium species (**52**) in 60% yield (Fig. 13) [45]. Like their phenol-based analogs, the η^2-4*H*-aniliniums resist reaction with mild carbon nucleophiles such as silyl ketene acetals and enol ethers, and attempts to carry out intra- or intermolecular addition of basic carbon nucleophiles (e.g., enolates) often results in retroaddition of the electrophile or deprotonation at C4 [45].

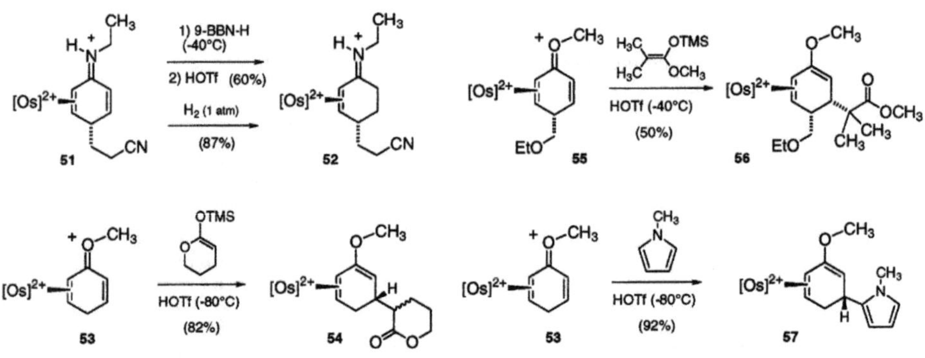

Fig. 13 C3 nucleophilic addition to 4*H*-anisolium and 4*H*-anilinium complexes

Even though an η^2-bound anisole is much less nucleophilic than an η^2-aniline or η^2-phenoxide, a 4*H*-anisolium complex is highly electrophilic, and a wide range of reactions for these systems have been uncovered. 4*H*-anisolium complexes readily undergo reaction at C3 with mild carbon electrophiles, and numerous examples of both intra- and intermolecular addition are known. In every case, the nucleophilic addition occurs *anti* to the metal. Three examples of intermolecular nucleophilic addition to 4*H*-anisolium complexes (**53, 55**) are shown in Fig. 13. Like their nitrogen analogs, 4*H*-anisolium complexes readily react with mild hydride sources, but stopping the reduction at the methoxydiene stage (i.e., **59** in Fig. 14) is often difficult, as any excess of acid produces an oxonium species (**60**) that is easily attacked by the hydride. However, useful compounds may be obtained from the stereospecific reduction of 4*H*-anisoliums to allyl ethers (**61**; see Fig. 14). Treatment of these materials with acid causes the elimination of alcohol, and the resulting π-allyl species (**62**) are in turn useful precursors to 3,6-disubstituted cyclohexenes (e.g., **63**) [46].

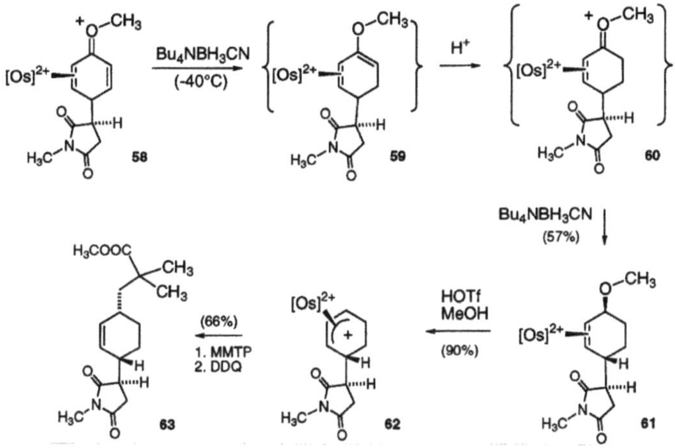

Fig. 14 Reduction of 4*H*-anisolium ligand to cyclohexadiene and the subsequent protonation to form a pentaammineosmium π-allyl species

3.1.2.4
Cyclization Reactions

a. [4+2] Cycloaddition Reactions

In the presence of $BF_3 \cdot OEt_2$, anisole complexes undergo Michael addition at C4 to generate an oxonium-boron-enolate (**65**, Fig. 15). Where the Michael acceptor is *N*-methylmaleimide, this intermediate either protonates to give a 4*H*-anisolium species (**69**) or undergoes ring closure at C1 to provide what is formally a cycloaddition product of the arene and olefin (**66**) [47]. If the pentaammineosmium(II) cycloaddition product is allowed to stand in solution, loss of ammonia occurs to generate the η^4-coordinated tetraammineosmium analog (**68**).

When the anisole ligand is substituted at C4, Michael addition still occurs *para* to the methoxy group provided that the electrophile is sterically unhindered. When 3-butyn-2-one is used as the Michael acceptor and care is taken to eliminate all proton sources, the initially formed boron enolate (**70**) closes to form an η^2-barrelene complex (**71**; see Fig. 15), which eliminates ammonia to give the tetraammine analog (**72**) [46].

b. [4+2] Michael-Michael Ring Closures

The 4*H*-anisolium reaction product of MVK and 4-methylanisole (**73**) offers the possibility of an intramolecular nucleophilic addition at C3 (Fig. 16). When the 4*H*-anisolium is allowed to stand in acidic methanol, ring closure of the purported enol occurs, resulting in a *cis*-decalin ring system (**74**; see Fig. 16) [46]. When this reaction sequence is repeated using ethyl vinyl ketone as the Michael acceptor, the methyl analog is obtained as a 6:1 ratio of diastereomers. The resulting enonium system may be either deprotonated at C2 to give the methoxydiene complex (**75**) or hydrolyzed and oxidized to provide the organic decalin (**76**) [46].

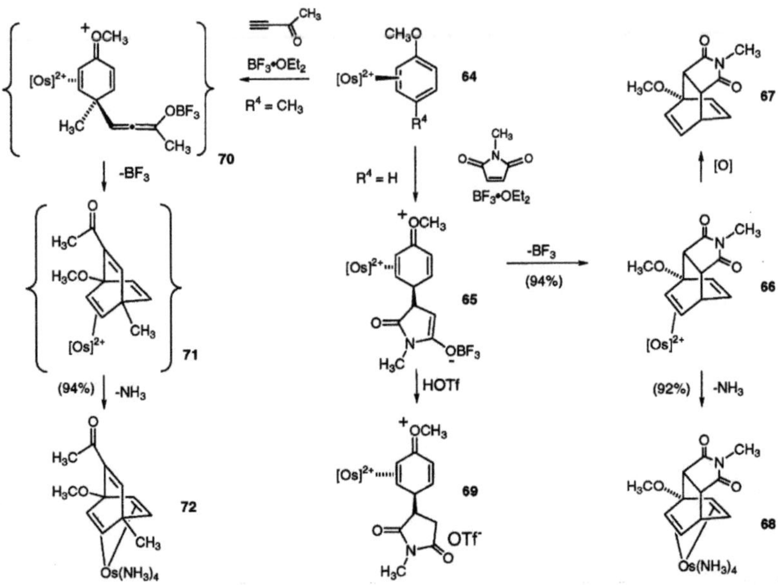

Fig. 15 [4+2] Cycloaddition reactions of η^2-anisole complexes via a 4*H*-anisolium intermediate and the formation of various bicyclo[2.2.2]octadiene and bicyclo[2.2.2.]octatriene complexes

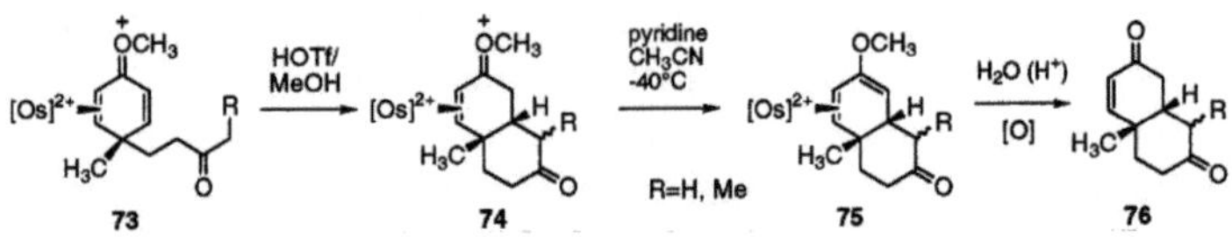

Fig. 16 The formation of functionalized *cis*-decalins from an η^2-anisole complex and an enone via a Michael-type cyclization

c. Michael-Aldol Ring Closures

When C3 of a 4*H*-anisolium complex is alkylated, the corresponding benzyl protons are highly acidic, rendering this position potentially nucleophilic (Fig. 17). For example, if the anisolium complex resulting from MVK addition to 6-methoxy-1,2,3,4-tetrahydro-naphthalene (**77**) is treated with pyridine, deprotonation occurs to generate the methoxytriene (**79**) [49]. Subsequent treatment of this material with TBSOTf induces an intramolecular aldol reaction to generate a tricyclic anisolium intermediate (**80**) that spontaneously eliminates the silanol to give alkene (**81**). Upon hydrolysis and oxidation, the free dienone (**82**) may be recovered.

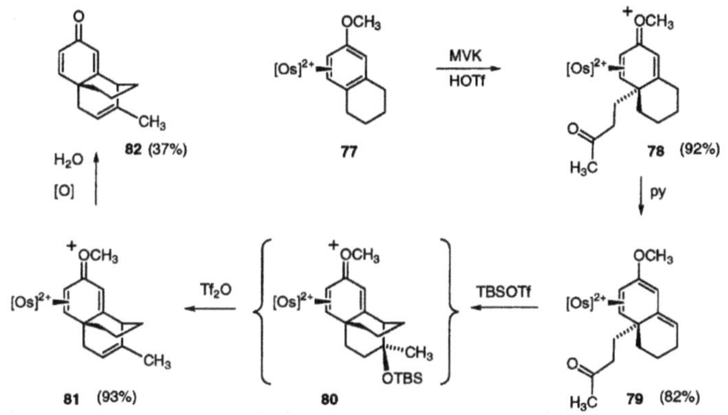

Fig. 17 The stereospecific formation of a tricyclic *cis*-decalin species via an intramolecular Aldol condensation

d. Diels-Alder Cycloaddition Reactions of 4-Methoxystyrenes

Styrenes usually do not participate in Diels-Alder reactions and when they do, they usually act as a dienophile. With a π base disrupting the aromatic character of the ring it becomes possible to use styrene as a diene precursor to decalins. Treatment of the anisole complex (**17**) with an acetal under acidic conditions generates a benzyl ether complex (Fig. 18; **83**). Elimination of alcohol generates a methylated *para*-quinone methide that may be stabilized as the pyridine adduct (**84**). Subsequent deprotonation gives the 4-methoxystyrene complex (**85**) in nearly quantitative yield. With the metal partially localizing the arene π system, the styrene complex resembles a vinylogous methoxydiene and readily undergoes Diels-Alder reactions, even in the absence of a Lewis acid. This cycloaddition reaction is stereo-and regioselective (e.g., **87**), with the addition occurring from the opposite face of the ring from metal coordination and with the

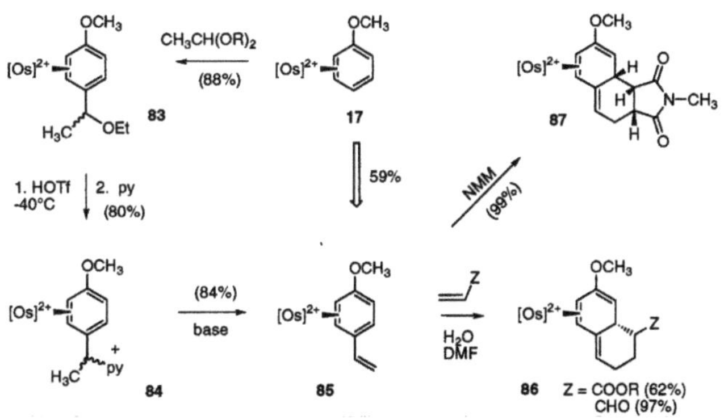

Fig. 18 The formation of a *para*-methoxystyrene complex from anisole and its use in Diels-Alder cycloaddition reactions

most electron-deficient carbon of the dienophile adding to the β carbon of the vinyl group. Given that the cycloaddition works for a variety of dienophiles and that the procedure for the formation of the styrene is generally applicable to acetals bearing an α proton, the reaction sequence outlined can provide a diverse set of functionalized decalin systems.

e. [2+2+2] Michael-Michael-Michael Ring Closures

When aniline or anisole complexes are combined with a large excess of a Michael acceptor in the presence of a Lewis acid, the enolate resulting from conjugate addition at C4 can, in some cases, be intercepted by another equivalent of the Michael acceptor, and the enolate resulting from this reaction can close at C3. In contrast to the high stereocontrol generally observed for the arene ring carbons, the newly formed ring is generally formed with poor stereocontrol, and as a result, the product mixture is usually complex. One exception comes from the reaction of the *N,N*-dimethylaniline complex (**88**) with α-methyl-γ-butyrolactone (Fig. 19). Even when this reaction is carried out with only 1 equivalent of electrophile, the only recovered products other than starting material are those resulting from the Michael-Michael-Michael sequence. For this system, one stereoisomer dominates (**90**) and its structure has been determined from X-ray diffraction data. The congested dienamine product (**90**) resists addition of carbon electrophiles at C2, but it does undergo protonation at C2 followed by reduction at C1 to yield the organic allylamine (**91**) after oxidative decomplexation. The stereoselectivity for the reduction of the iminium carbon is a rare exception to the general rule that all addition reactions occur at the opposite face of the arene ring from metal coordination. This anomaly is a consequence of the highly congested *exo* face for this compound.

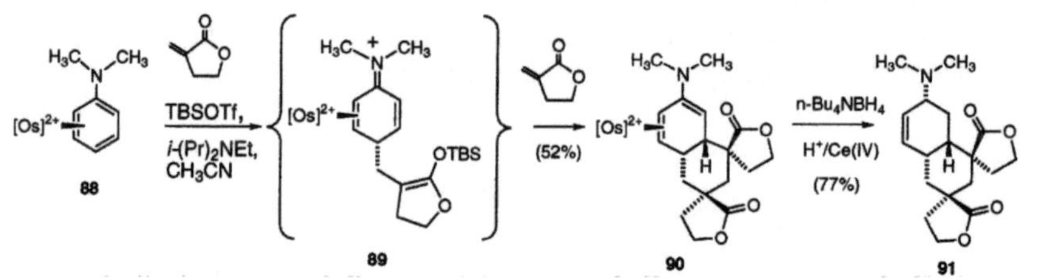

Fig. 19 A Michael-Michael-Michael ring-closure reaction sequence with an η^2-aniline complex

3.1.3
Addition Reactions to Benzene Hydrocarbons

The reactions outlined in the previous sections all utilized an arene with a heteroatom substituent that could stabilize the arenium intermediates resulting from electrophilic addition. Benzene, alkylated benzenes, and naphthalene are more difficult to activate because they lack this mode of stabilization. Osmium complexes of these hydrocarbons are readily hydrogenated, to cycloalkenes, but polar addition reactions require strong electrophiles used at reduced temperatures in order to avoid oxidation of the metal. The resulting benzenium complexes, which resemble η^3-allyl complexes, are unstable but may be trapped with the appropriate nucleophile.

3.1.3.1
Hydrogenation Reactions

With the π base partially localizing the π system of an arene, selective hydrogenation of benzene to cyclohexene is readily accomplished (**92**). When a methanol solution of the benzene complex is put under 1 atm of hydrogen in the presence of Pd°, the cyclohexene analog is produced in high yield (Fig. 20). Repeating this experiment with D_2 generates a single isomer (**93**) in which all the deuterium is located *anti* to the osmium [50]. Similarly, hydrogenation of the naphthalene complex **3** in the presence of Pd/C generates the 1,2-dihydronaphthalene complex **94** exclusively.

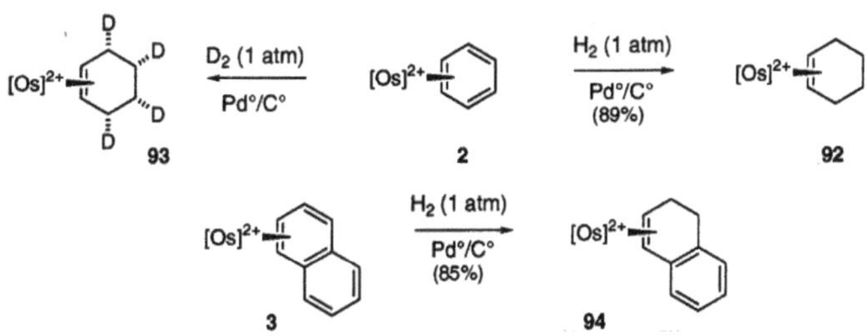

Fig. 20 Hydrogenation of benzene and naphthalene bound to pentaammineosmium(II)

3.1.3.2
Tandem Addition Reactions

In contrast to the reactivity pattern seen for heteroatom-substituted arene complexes, tandem addition reactions with benzene and naphthalene generally provide 1,4-addition products. This outcome is thought to be the result of C4 of the corresponding naphthalenium (resonance contributors **100–102**) or benzenium (resonance contributors **95–97**) having a greater partial positive charge because of its overlap/delocalization with the adjacent π system (Fig. 21) [21].

The complex $[Os(NH_3)_5(\eta^2\text{-benzene})]^{2+}$ reacts with both acetals and Michael acceptors to form benzenium intermediates [28]. These intermediates are difficult to isolate but are readily characterized at -40 °C. In contrast to the arenium complexes derived from heteroatom substituted benzenes, ^1H and ^{13}C NMR data indicate that the benzenium ring is trihapto-coordinated. The benzenium complexes react with a silyl ketene acetal, a silyl enol ether, or aryl lithium to yield substituted 1,4-cyclohexadiene complexes [21]. The 1,4-cyclohexadiene

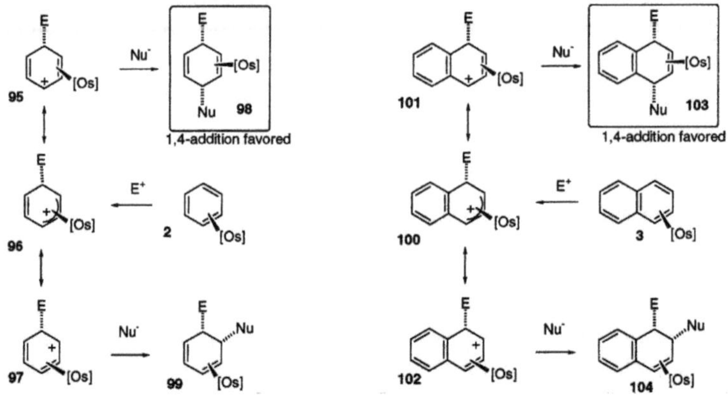

Fig. 21 Regiochemistry of a tandem addition sequence for benzene and naphthalene

can be removed from the metal by treatment with the appropriate oxidant (Table 4).

Alkylated benzene complexes of osmium(II) are more reactive than their parent toward electrophiles because the alkyl group can stabilize the arenium intermediate through hyperconjugation [28]. The toluene complex exists in solution as a 3:2 mixture of interconverting linkage isomers favoring the C5-C6 bound arene. The addition of either trifluoromethanesulfonic acid (HOTf) or 3-penten-2-one at C2 of the bound toluene generates the η^3-toluenium complex that can be trapped by 1-methoxy-2-methyl-1-trimethylsiloxypropene (MMTP), resulting in a single regioisomer. Upon oxidation with AgOTf, the diene is released (Table 4) [21]. The tandem addition of dimethoxymethane and MMTP to the toluene complex leads to a mixture of two 1,4-dienes after demetallation. If the tandem addition is performed at -40 °C, the ratio of diastereomers is 5:1, but if performed at -80 °C, this ratio becomes is 10:1. The regiochemistry of the electrophilic addition arises from the hyperconjugation of the toluene methyl group with the arenium resulting from addition at C2.

Similar to what is found for the toluene complex, electrophilic additions to the *ortho* and *meta* xylene complexes of pentaammineosmium(II) are highly regioselective for attack at C6 (Table 4) [21]. Electrophilic addition of HOTf or dimethoxymethane generates the complexed xylenium cation that can be trapped by MMTP to form the complexed diene. Demetallation using AgOTf releases the free diene. As with toluene and benzene, the electrophile and nucleophile add in a 1,4-fashion, with the electrophile adding to a carbon adjacent to a single methyl group.

A one-pot tandem addition/oxidative decomplexation methodology has also been developed for naphthalene which yields 1,4-disubstituted dihydronaphthalenes from osmium(II) naphthalene complexes. The process involves the regio- and stereospecific addition of an electrophile and a nucleophile to generate *cis*-1,4-dihydronaphthalene complexes. Oxidative decomplexation produces 1,4-dihydronaphthalenes in good overall yield (Table 5) [27].

Table 4 Formation of alkylated 1,4-cyclohexadienes from arenes

Arene	Electrophile	Nucleophile	Product	Overall yield
		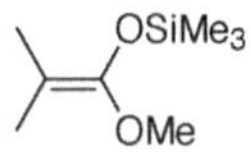	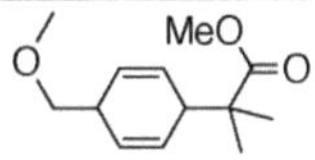	82%
		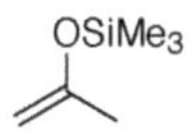	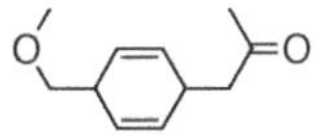	23%
		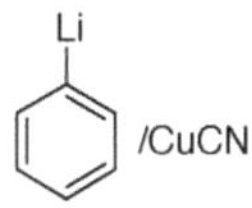	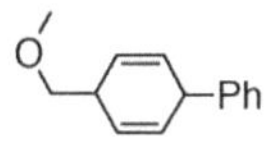	16%
		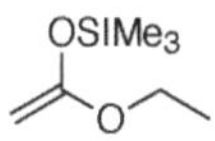	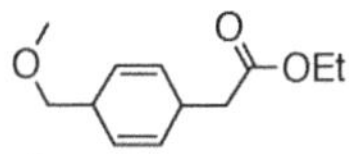	31%
	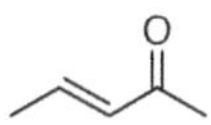	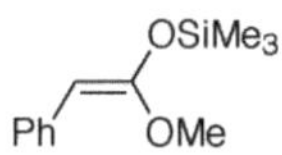	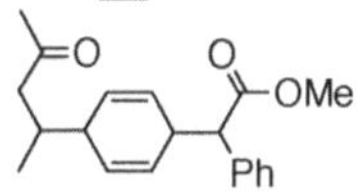	27%
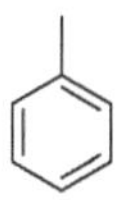		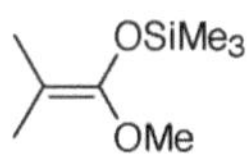	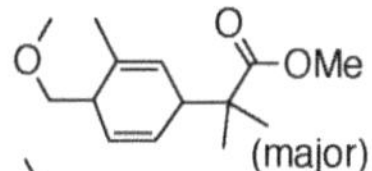	34%
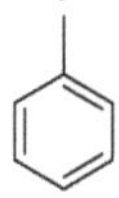	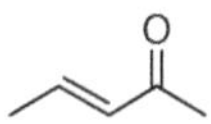	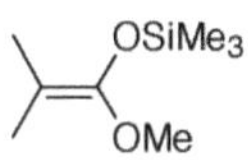	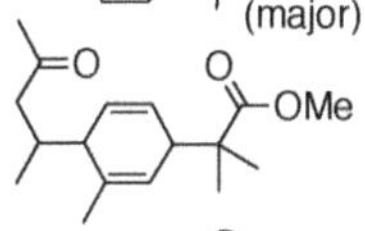	43%
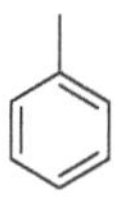	HOTf	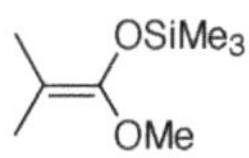	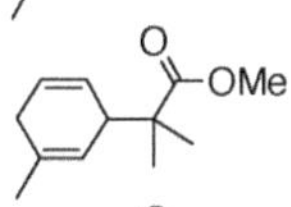	56%
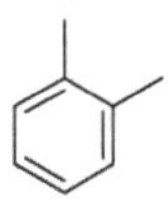	HOTf	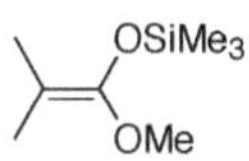	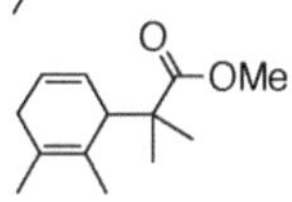	67%
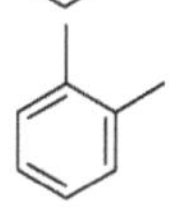		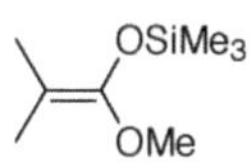	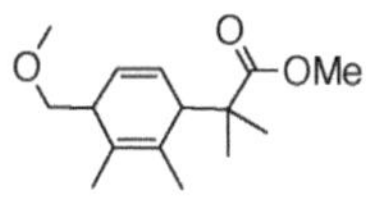	28%
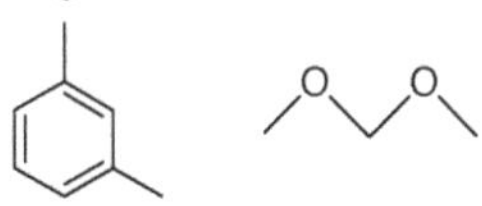		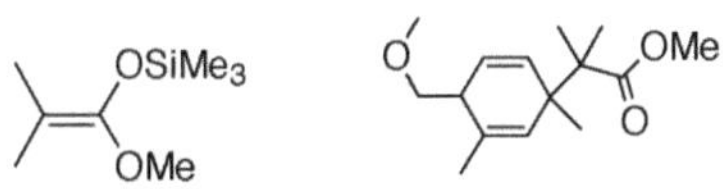		32%

Table 5 Overall yields for the tandem addition to naphthalene

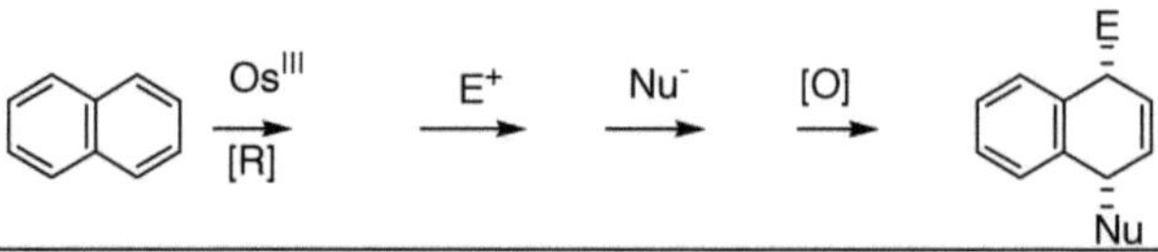

Electrophilic Substituent	Nucleophilic Substituent	Overall Yield
H	H	56%
H	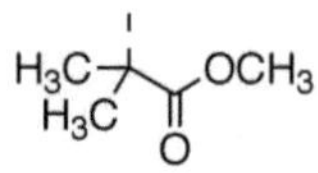	77%
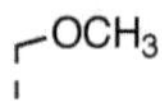	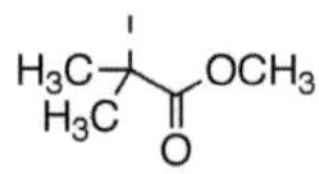	65%
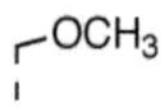	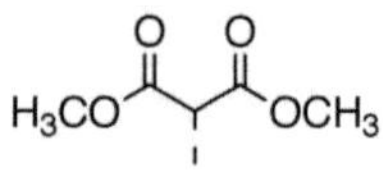	54%
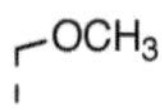	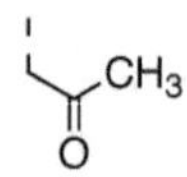	69%
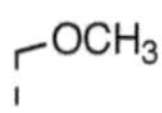	H	45%
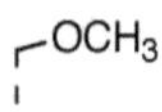	CH_3	40%
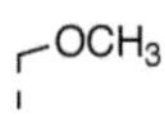	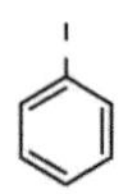	40%
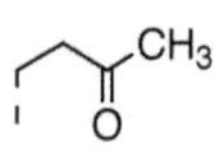	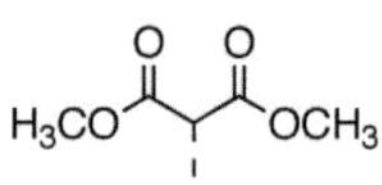	41%
	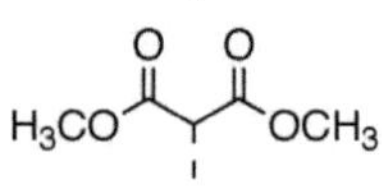	25%

3.2
Reactions Promoted by Rhenium(I) and Molybdenum(0)

The majority of organic reactions that have been developed to date for group VI and VII transition metal π bases are with the system {TpRe(CO(L)}. The limited number of examples of reactions with these systems compared to that of the pentaammineosmium system reflects their more recent discovery.

3.2.1
TpRe(CO)(L)

3.2.1.1
Preparation of Complexes

Metal fragments of the form {TpRe(CO)(L)} in which L=MeIm or NH_3 have the capability to bind substituted benzenes in an η^2-fashion. The fragment having L=MeIm has seen more application because of difficulties associated with the synthesis and isolation of aromatic complexes having L=NH_3 [16]. Anisoles, phenols, and naphthalenes all form thermally stable η^2-complexes when stirred in excess (10–25 eq.) with TpRe(CO)(MeIm)(η^2-benzene) in THF. The complexes are typically isolated by precipitation into hexanes. The benzene complex can be made from the direct reduction of the Re(III) precursor, TpRe(MeIm)$(Br)_2$, with Na^0 in benzene under one atmosphere of CO [51].

Naphthalene forms stable η^2 complexes with {TpRe(CO)(L)} when L=*t*-BuNC, PMe_3, py, DMAP, or MeIm. In toluene and an excess of naphthalene, the direct reduction of TpRe(L)Br_2 (L=py or DMAP) with Na^0 under a CO atmosphere will generate the py or DMAP naphthalene complex in ~40% yield [16]. Generation of the Re(I) PMe_3 and *t*-BuNC naphthalene systems is not as straightforward. They are synthesized by the oxidation of the corresponding Re(I) olefin complexes to Re(II) complexes, liberation of the olefin by heating, and reduction of the Re(II) species in the presence of naphthalene [16].

3.2.1.2
Diels-Alder Cycloaddition Reactions

The benzene complex **106** (TpRe(CO)(MeIm)(η^2-benzene)) undergoes an endo-selective Diels-Alder reaction with *N*-methylmaleimide to afford the bound bicyclo[2.2.2]octadiene complex **107** in 65% yield (Fig. 22) [52]. Oxidation yields the bicyclo[2.2.2]octadiene **108** and/or the bicyclo[2.2.2]octenone **109** depending upon the choice of oxidation conditions.

The more activated TpRe(CO)(MeIm)(η^2-anisole) complex has demonstrated a slightly broader range of cycloaddition reactivity. The anisole complex **110** cyclizes with DMAD to give the diastereomeric bound bicyclo[2.2.2]octatriene **111** isolated as a 1:1 ratio of coordination diastereomers (Fig. 23). Oxidation of these complexes liberates the corresponding triene **112** as well as the disubstituted anisole **113**, which presumably is generated with acetylene from the cycloreversion of **112**.

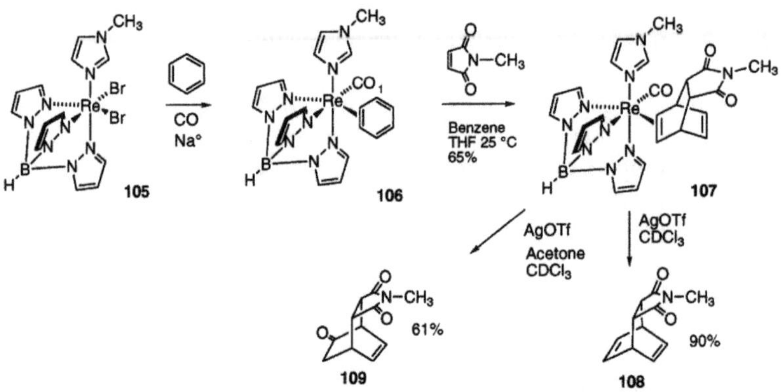

Fig. 22 Diels-Alder cycloaddition reaction of benzene and *N*-methylmaleimide promoted by rhenium(I)

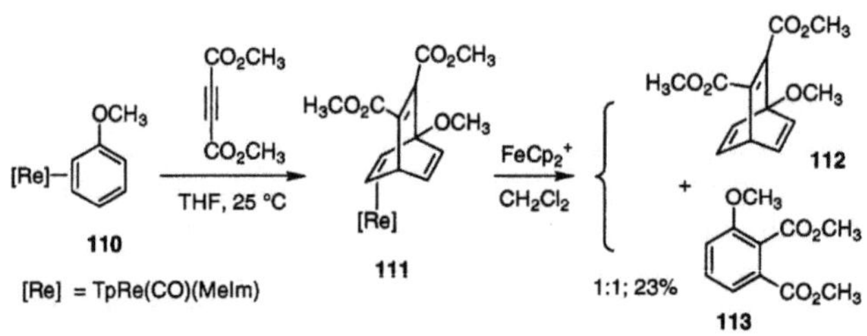

Fig. 23 Diels-Alder cycloaddition of anisole and DMAD promoted by rhenium(I)

3.2.1.3
Tandem Addition Reactions with Naphthalene

Naphthalene complexes of the form TpRe(CO)(L)(η^2-naphthalene) exist, in solution, as a mixture of diastereomers **A** and **B** (Table 6) [53]. Only when L=PMe_3 does the unbound ring of the naphthalene have a thermodynamic preference for quadrant **a** (I). For all other Ls, the preference is for the unbound ring to lie in quadrant **d** (II) (see above). When exposed to an acetonitrile solution of HOTf followed by MMTP, the naphthalene complex undergoes tandem electrophilic/nucleophilic addition reactions with complete regiocontrol. Oxidative decomplexation with AgOTf at room temperature liberates the dihydronaphthalenes **III** and **IV**. When L=PMe_3 the reaction favors the 1,4 addition product (**III**), a result similar to that observed with $[Os]^{2+}$-bound naphthalene. Complexes that do not have L=PMe_3 generate the 1,2 addition product (**IV**) with different degrees of selectivity [53]. The regioselectivity of the addition is highest for L=MeIm (25:1), but the overall yield is highest for L=py (89%). Of note, these reactions can be performed outside of the glovebox with ~500 mg of metal complex (Method B). A high degree of stereocontrol has also been demonstrated in these reactions, with both the electrophile and the nucleophile adding *anti* to the face of metal coordination.

Table 6 Regioisomeric ratios and yields of dihydronaphthalenes

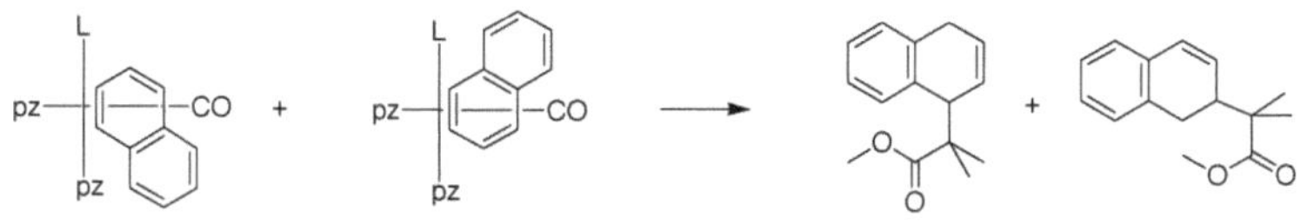

I II III IV

L	T (°C)	I : II	Method A* III : IV	Method A* Yield	Method B** III : IV	Method B** Yield
PMe_3	0	-	-	-	13:1	29%
	20	10:1	12:1	52%	12:1	24%
Pyridine	-40	-	1:15	63%	-	-
	0	-	-	-	1:7	80%
	20	1:3	1:8	89%	1:7	66%
DMAP	-40	-	1:25	19%	-	-
	20	1:1.5	1:10	50%	-	-
MeIm	0	-	-	-	1:25	55%
	20	1:5	1:23	83%	1:16	63%
NH_3	-40	-	1:7	~25%	-	-
	20	1:4	-	-	-	-

* = Reaction sequence performed in inert atmosphere. ** = Reaction sequence performed on benchtop, with ~500mg of metal complex, and blanketed under dinitrogen.

3.2.2
TpMo(NO)(L)

3.2.2.1
Preparation of Complexes

The complex TpMo(NO)(MeIm)(naphthalene) (**114**) has been prepared from the hexacarbonyl according to the sequence outlined in Fig. 24. The Tp ligand is installed by heating the hexacarbonyl in the presence of KTp, and the resulting tricarbonyl is then converted to the nitrosyl by action of Diazald. The compound $TpMo(NO)(CO)_2$ is then oxidized to form $TpMo(NO)Br_2$. From this point the auxiliary ligand is added in the presence of a reducing agent and arene to give **114**. The overall yield for this sequence of steps ranges from 30–40% [17].

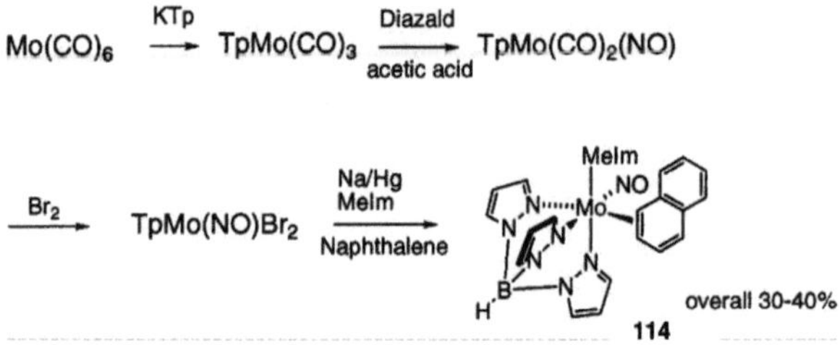

Fig. 24 Preparation of molybdenum(0) η^2-naphthalene complex

3.2.2.2
1,2-Tandem Addition Reactions with Naphthalene

In order to probe the tolerance of the molybdenum(0) aromatic complexes to electrophilic/acidic environments, a tandem addition sequence was attempted for the complex TpMo(NO)(MeIm)(η^2-naphthalene) (**114**) (Fig. 25) [17]. In a strategy similar to that used with the TpRe(CO)(MeIm)(η^2-naphthalene) analog,(see below) an acetonitrile solution (-35 °C) of **114** was exposed sequentially to triflic acid, 1-methoxy-2-methyl-1-trimethylsiloxypropene, and an amine base. The 1,2-dihydronaphthalene complex **116** was isolated in virtually quantitative yield. No evidence of free naphthalene or 1,4-addition product was observed. Stirring the reaction mixture with exposure to air resulted in an 80% overall yield of 2-(1,2-dihydro-naphthalen-2-yl)-2-methyl-propinoic acid methyl ester (**117**) following TLC purification [17].

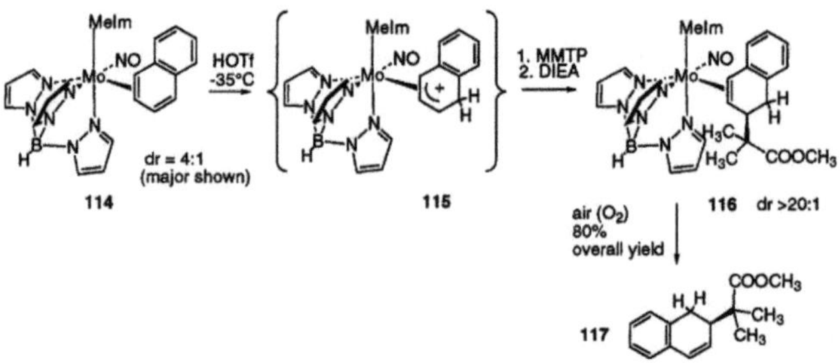

Fig. 25 1,2-Tandem addition to naphthalene promoted by molybdenum(0)

4
Reactions of η^2-Arene Complexes that Generate Enantio-Enriched Products

4.1
Lactate-Derived Phenyl Ether Complexes of Osmium(II)

In order to take advantage of the high degree of stereoselectivity observed in tandem additions to pentaammineosmium arene complexes, a chiral anisole derivative was prepared that demonstrated high coordination diastereoselectivity (>9:1) [54]. Complex **118** can be prepared in 83% yield and >90% de from a two-step sequence involving the coupling of phenol and (*S*)-(-)-methyl lactate followed by complexation (Fig. 26). In addition to the η^2-bond common to other pentaammineosmium(II) complexes, kinetics studies of **118** provide evidence of a hydrogen-bonding interaction between the acidic ammine ligands and the ester carbonyl (see Fig. 26) [54]. The result of these interactions is a thermodynamic differentiation between the arene faces. If the metal binds to the *re* face of the arene, the methyl group adjacent to the methine group is placed into a sterically disfavored position near an *ortho* ring proton. If the metal binds to the *si* face of the arene, the proton of the methine group is forced into the same position. As the latter scenario results in a smaller steric repulsion, this form is thermodynamically preferred. Thus, the metal binds preferentially to the *si* face.

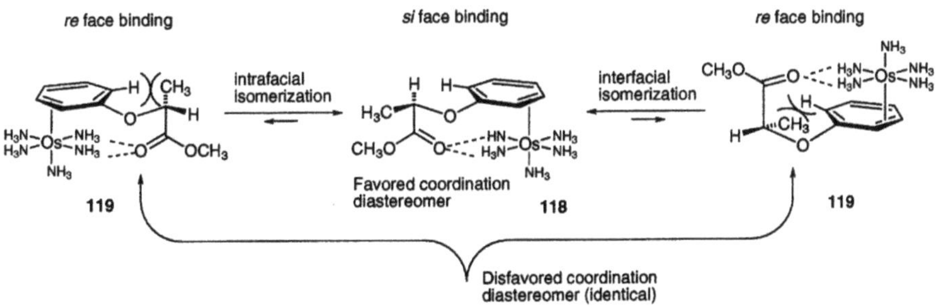

Fig. 26 Preferential binding of one enantioface of an arene induced by a lactate chiral auxiliary

The chiral anisole derivative **119** has been used in the synthesis of several asymmetric functionalized cyclohexenes (Fig. 27) [55]. In a reaction sequence similar to that employed with racemic anisole complexes (see above), **119** adds an electrophile and a nucleophile across C4 and C3, respectively, to form the cyclohexadiene complex **120**. The vinyl ether group of **120** can then be reduced by the tandem addition of a proton and hydride to C2 and C1 respectively, affording the olefin complex **121**. Direct oxidation of **121** liberates the cyclohexenes of the type **122** having the initial asymmetric auxiliary still intact. Alternatively, the auxiliary may be cleaved under acidic conditions to afford η^3-allyl complexes, which can undergo reactions with nucleophiles regioselectively at C1. Oxidative decomplexation liberates the cyclohexenes **123–127**.

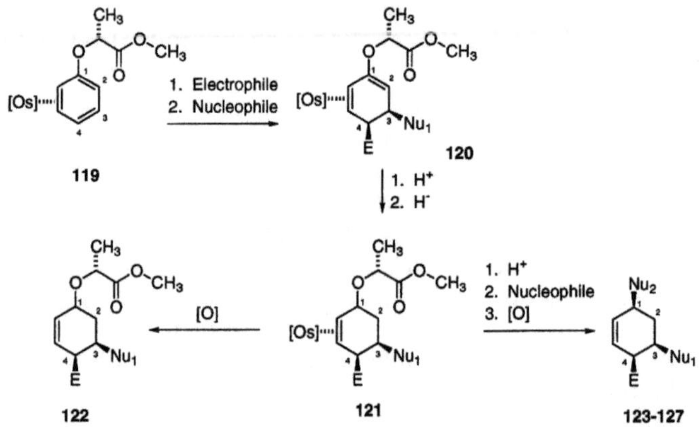

Compound	E	Nu_1	Nu_2	%Yield	ee(%)
123	H	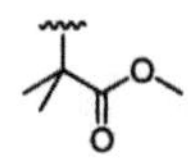	-	46	97
124	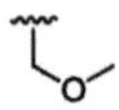	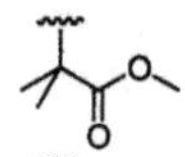	-	41	97
125	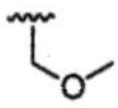	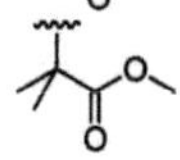	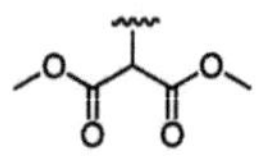	33	81
126	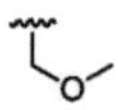	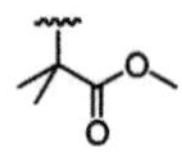	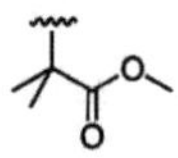	28	90
127	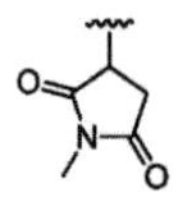	H	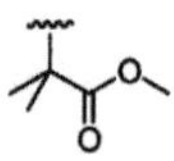	19	93

Fig. 27 Formation of enantio-enriched cyclohexenes from a chiral anisole precursor

4.2 Resolved TpRe(CO)(L)(L′)

A methodology for resolving TpRe(CO)(1-methylimidazole)(η^2-benzene) has been developed utilizing (1*R*)-α-pinene. Each enantiomer of the {TpRe(CO)(MeIm)} system can be obtained with ee=94% by taking advantage of differing rates of pinene substitution for the two diastereomers of TpRe(CO)(MeIm)(η^2-(*R*)-α-pinene).

This methodology was tested on the tandem addition of a proton and silyl ketene acetal to naphthalene (Fig. 28). A sample of racemic benzene complex **106** was stirred with an excess of (*R*)-α-pinene for 30 h to generate the matched and mismatched pinene complexes **128** and **129**, respectively. The resulting mixture was then combined with naphthalene and again stirred for 30 h. A 1:1 mixture of the naphthalene complex (*R*)-**130** and the matched pinene complex (*S*,*R*)-

128 was isolated in 48% yield. This mixture was then subjected a sequence of steps in which the naphthalene ligand first protonates (HOTf/MeOH), then reacts with 1-methoxy-2-methyl-1-trimethylsiloxypropene (MMTP), then is oxidatively decomplexed. The resulting 2-alkylated 1,2-dihydronaphthalene (*S*)-**132** was isolated in 63% yield based on available **130.** The enantiomeric ratio was found to be 93:7 (86% ee). Of note, when the reaction was repeated with the stronger acid HOTf·CH_3CN the enantiomer ratio for (*S*)-**132** was only 80:20, even though the diastereomer ratio for (*R*)-**131** was still >20:1. This observation suggests that acid may play a role in epimerizing the rhenium stereogenic center, and care must be taken to keep the acid strength and concentration as low as possible when using this rhenium system for asymmetric organic reactions.

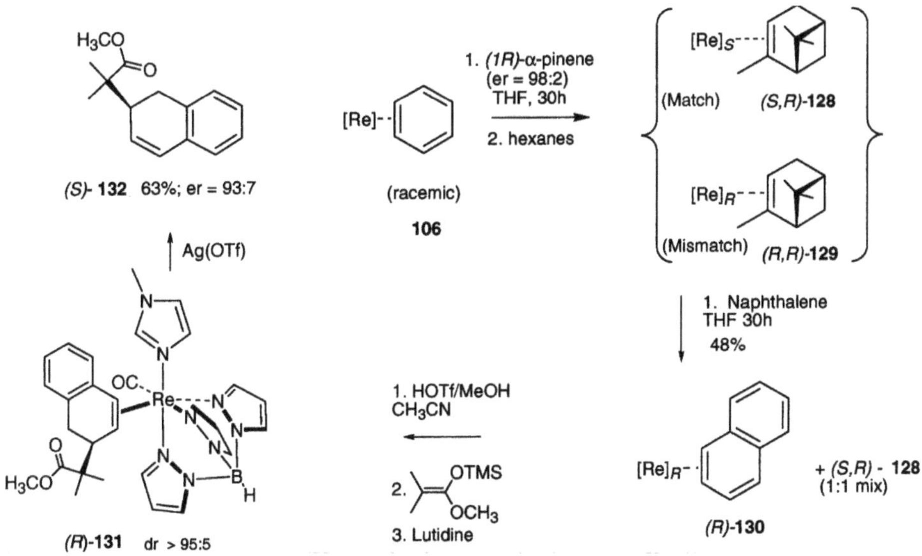

Fig. 28 Resolution of a rhenium(I) π base and its use in the formation of an enantio-enriched dihydronaphthalene

5
Concluding Remarks

The synthetic methodology outlined in this chapter is fundamentally different than that described elsewhere in this book. It is based on a transition metal acting as an electron-donor (i.e., a π base) for the aromatic ligand. Thus, reactions with electrophiles are promoted, and the resulting arenium systems are stabilized such that a nucleophile may then be added. In this regard, the reactions outlined in this chapter complement the more established η^6-arene methodology. Whereas the chromium η^6-arene complexes have been known for over 40 years, viable η^2-arene complexes have been known for the past 15, and only recently has an alternative to osmium emerged. Ultimately the impact that π bases have on conventional organic synthesis will depend heavily on practical issues such as handling and cost. The advent of rhenium and molybdenum dearomatization agents represents a significant step in the alleviation of these concerns.

Acknowledgement is made to the donors of The Petroleum Research Fund, administered by the ACS, for partial support of this research (ACS-PRF#36638-AC1). This work was also supported by the NSF (CHE0111558 and 9974875) and the NIH (NIGMS; R01-GM49236).

References

1. Mander LN (1991) Synlett 134
2. Mander LN (1991) In: Trost BM, Flemming I (eds) Comprehensive organic synthesis, vol 8. Pergamon, p 341
3. Hegedus LS (1999) Transition metals in the synthesis of complex organic molecules. University Science Books, Sausalito
4. Paradisi C (1991) In: Trost BM, Fleming I, Semmelhack MF (eds) Comprehensive organic synthesis, vol 4. Pergamon, Oxford, p 423
5. Semmelhack MF (1995) In: Abel EW, Stone FGA, Wilkinson G (eds) Comprehensive organometallic chemistry II, vol 12. Pergamon, Oxford, pp 979–1015
6. Harman WD (1997) Chem Rev 97:1953
7. Park S-HK, Geib SJ, Cooper NJ (1997) J Am Chem Soc 119:8365
8. Chin RM, Dong L, Duckett SB, Partridge MG, Jones WD, Perutz RN (1993) J Am Chem Soc 115:7685
9. Sweet JR, Graham WAG (1983) J Am Chem Soc 105:305–306
10. Tagge CD, Bergman RG (1996) J Am Chem Soc 118:6908
11. Bach I, Porschke K-R, Goddard R, Kopiske C, Kruger C, Rufinska A, Seevogel K (1996) Organometallics 15:4959–4966
12. Reinartz S, White PS, Brookhart M, Templeton JL (2001) J Am Chem Soc 123:12724
13. Scott F, Krüger C, Betz P (1990) J Organomet Chem 387:113
14. Striejewske WS, Conry RR (1998) Chem Commun 555
15. Brooks BC, Gunnoe TB, Harman WD (2000) Coord Chem Rev 3
16. Meiere SH, Brooks BC, Gunnoe TB, Carrig EH, Sabat M, Harman WD (2001) Organometallics 20:3661
17. Meiere SH, Keane JM, Gunnoe TB, Sabat M, Harman WD (2002) (in press)
18. Trindle C, Sacks G, Harman WD (2002) Int J Quantum Chem (2003) (in press)
19. Tellers DM, Skoog SJ, Bergman RG, Gunnoe TB, Harman WD (2000) Organometallics 19:2428–2432
20. Harman WD, Taube H (1987) J Am Chem Soc 109:1883
21. Ding F, E. KM, Sabat M, Harman WD (2002) J Am Chem Soc (in press)
22. Harman WD, Sekine M, Taube H (1988) J Am Chem Soc 110:5725
23. Kolis SP, Kopach ME, Liu R, Harman WD (1997) J Org Chem 62:130
24. Gonzalez J, Sabat M, Harman WD (1993) J Am Chem Soc 115:8857
25. Kopach ME, Harman WD, Hipple WG (1992) J Am Chem Soc 114:1737
26. Kopach ME, Kelsh LP, Stork K, Harman WD (1993) J Am Chem Soc 115:5322
27. Winemiller MD, Harman WD (1998) J Am Chem Soc 120:7835
28. Winemiller WD, Kopach ME, Harman WD (1997) J Am Chem Soc 119:2096
29. Winemiller MD, Kelsch BA, Sabat M, Harman WD (1997) Organometallics 16:3672
30. Brooks BC, Meiere SH, Friedman LA, Carrig EH, Gunnoe TB, Harman WD (2001) J Am Chem Soc 123:3541
31. Harman WD, Taube H (1988) J Am Chem Soc 110:5403
32. Sekine M, Harman WD, Taube H (1988) (????) 27:3604
33. Harman WD, Sekine M, Taube H (1988) J Am Chem Soc 110:2439
34. Meiere SH, Harman WD (2001) Organometallics 20:3876

35. Meiere SH, Ding F, Friedman L, Sabat M, Harman WD (2002) J Am Chem Soc (in press)
36. Harman WD, Dobson JC, Taube H (1989) J Am Chem Soc 111:3061
37. Brauer DJ, Kruger C (1977) Inorg Chem 16:884
38. Harman WD (1987) PhD Dissertation, Stanford University
39. van der Heijden H, Orpen AG, Pasman P (1985) J Chem Soc Chem Commun 1576
40. Gunnoe TB, Sabat M, Harman WD (1998) J Am Chem Soc 120:8747
41. Gladysz JA, Boone BJ (1997) Angew Chem Int Ed 36:551
42. Lay PA, Magnuson RH, Taube H (1986) Inorg Syn 24:269
43. Harman WD, Schaefer WP, Taube H (1990) J Am Chem Soc 112:2682
44. Kopach ME, Harman WD (1994) J Am Chem Soc 116:6581
45. Kolis SP, Gonzalez J, Bright LM, Harman WD (1996) Organometallics 15:245
46. Kopach ME, Kolis SP, Liu R, Robertson JW, Chordia MD, Harman WD (1998) J Am Chem Soc 120:6199
47. Kopach ME, Harman WD (1994) J Org Chem 59:6506
48. Stokes S (2002) MSc Dissertation, University of Virginia
49. Kolis SP, Epp KM, Liu R, Harman WD (1998) J Am Chem Soc 120:6205
50. Harman WD, Taube H (1988) J Am Chem Soc 110:7906
51. Meiere SH, Brooks BC, Gunnoe TB, Sabat M, Harman WD (2001) Organometallics 20:1038
52. Chordia MD, Smith PL, Meiere SH, Sabat M, Harman WD (2001) J Am Chem Soc 123:10756
53. Valahovic MT, Gunnoe TB, Sabat M, Harman WD (2002) J Am Chem Soc 124:3309
54. Chordia MD, Harman WD (1998) J Am Chem Soc 120:5637
55. Chordia MD, Harman WD (2000) J Am Chem Soc 122:2725

Topics Organomet Chem (2004) 7: 129–156
DOI 10.1007/b12824

(Arene)Cr(CO)$_3$ Complexes: Cyclization-, Cycloaddition- and Cross-Coupling-Reactions

Motokazu Uemura

Department of Chemistry, Faculty of Integrated Arts and Sciences,
Osaka Prefecture University, Sakai, 599-8531 Osaka, Japan
E-mail: uemura@ms.cias.osakafu-u.ac.jp

Abstract (Arene)tricarbonylchromium complexes have been utilized in organic synthesis based on the significant properties of the arene chromium complexes. This chapter describes the cyclization, cycloadditions, and cross-coupling reactions of the arene chromium complexes. The application of these reactions to a key step of natural products synthesis is also mentioned briefly.

Keywords Cyclization · Dipole cycloaddition · Cross-coupling · Axially biaryls · Radical-mediated coupling · Planar chirality

1
Introduction

(Arene)chromium complexes have some characteristic properties due to an electron-withdrawing ability and steric bulkiness of tricarbonylchromium fragment. Reagents approach usually from the opposite face to the tricarbonylchromium fragment, since one face of the arene ring is blocked with the chromium group. Furthermore, the (arene)chromium complexes can exist as two enantiomeric forms based on a planar chirality, when the arene ring is substituted with different substituents at *ortho*- or *meta*-positions. Therefore, these planar chiral chromium complexes could be used as chiral source in asymmetric reactions. In this chapter, cyclization, cycloaddition, and cross-coupling reactions using the arene chromium complexes are focused.

2
Cyclization

2.1
Benzylic Anion- and Cation-Mediated Cyclization

It is well known that the tricarbonylchromium-complexed benzylic anions and cations are stabilized due to overlapping between d-orbital of the chromium and p-orbital of the benzylic carbon [1]. Tricarbonylchromium complexes of α-tetralone and α-indanone having a carbonyl group at the side chain underwent a deprotonation of the *exo*-benzylic protons by treatment with base to give the stereo-controlled tricyclic compounds (Eqs. 1 and 2) [2]. In these cases, Robinson annulation products were formed in less than 10% yield. Also, base treatment of benzyl ether chromium complex having a chlorine at the side chain **3** gave cyclization product as a diastereomeric mixture (Eq. 3) [3].

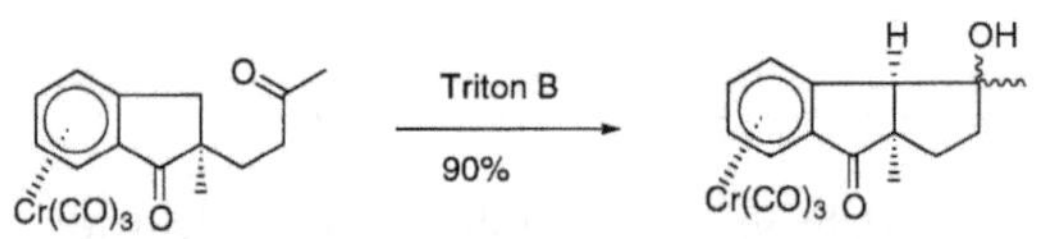

1

(1)

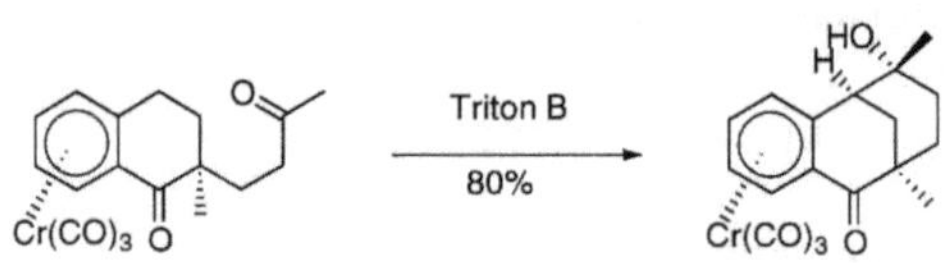

2

(2)

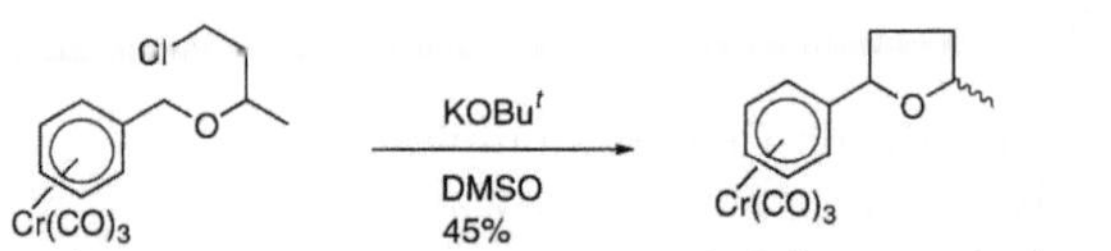

3

3 : 2 diastereomeric mixture

(3)

The chromium complexes of styrene and the related arenes can be susceptible to nucleophilic addition at the β-position of the double bond. The conjugate addition of nucleophiles at the β-position generates the tricarbonylchromium-stabilized benzylic carbanion and the generated benzylic carbanion could be further trapped with electrophiles. An intramolecular cyclization to the chromium complexed styrene double bond gave tricyclic compound (Eq. 4) [4].

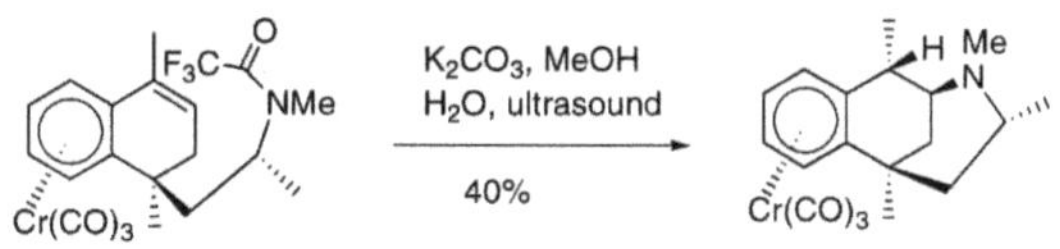

(4)

Similarly, chromium-complexed benzylic cations are also stabilized and organic reactions based on the benzylic cation species have been developed. For example, planar chiral *o*-substituted benzaldehyde dimethylacetal chromium complexes **4** were treated with 3-buten-1-ol in the presence of $TiCl_4$ to give tetrahydropyran derivatives with high diastereoselectivity (Eq. 5) [5]. The chromium-complexed benzylic oxonium ion **6** would be also generated and subsequent intramolecular cyclization afforded the cyclization product **7**. Furthermore, the chromium-complexed benzyl alcohol derivative having electron-rich arene ring at the side chain produced tetrahydroisoquinoline skeleton by treatment with Lewis acid with stereochemical retention at the benzylic position (Eq. 6) [6].

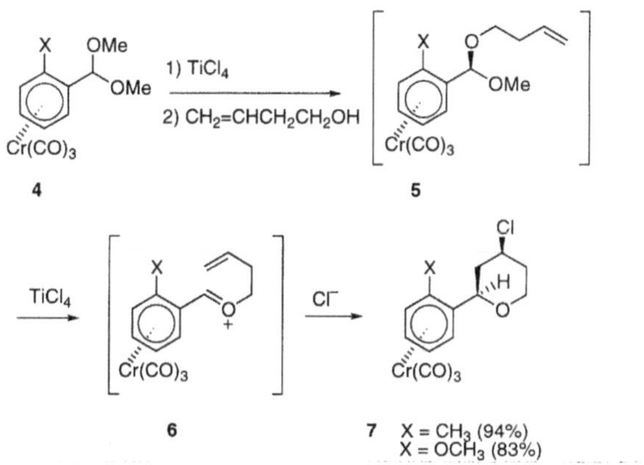

4 **5** **6**

7 X = CH_3 (94%)
X = OCH_3 (83%)

(5)

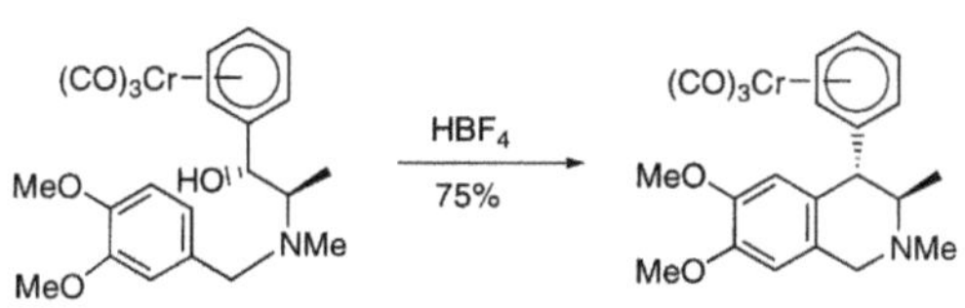

8

>98% de

(6)

2.2
Radical-Mediated Cyclization

The chromium-complexed benzylic radicals are also known and were employed in a variety of useful organic reactions. One electron reductive coupling of the tricarbonylchromium complexes of *o*-substituted benzaldehydes and benzaldimines with samarium iodide gave 1,2-diols and diamines, respectively, in good yields with extremely high *threo* diastereoselectivity (Eq. 7) [7]. Therefore, enantiomerically pure 1,2-diols and diamines were easily obtained by using the planar chiral chromium complexes of benzaldehyde and benzaldimine.

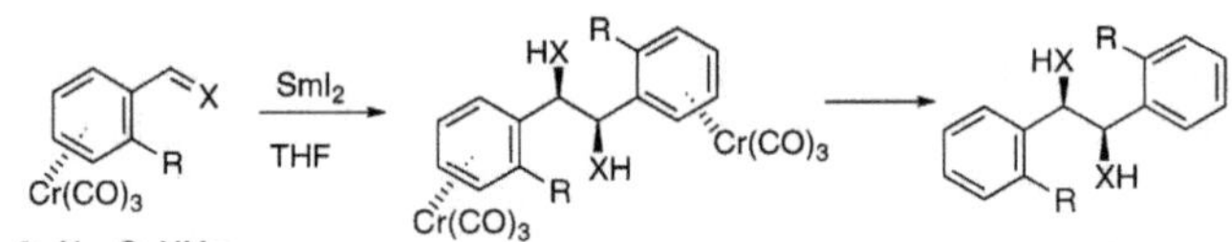

9 X = O, NMe
enatiomerically pure

enantiomerically pure
1.2-diol. diamine

(7)

An intramolecular reductive coupling of mono-chromium complexes of the biphenyls having carbonyl or imine groups at the both *ortho*-positions gave the *trans*-diols, diamines or amino alcohols without formation of any stereoisomers (Eq. 8) [8].

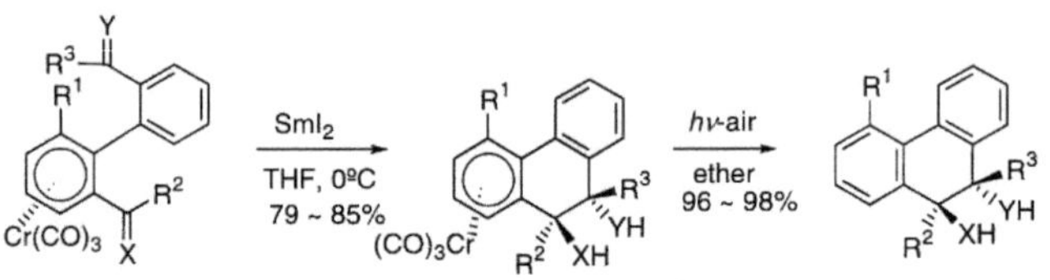

10

X = O, NPh; Y = O, NPh
R^1 = H, OMe, R^2 = H, Me, R^3 = H, Me

(8)

The generated chromium-complexed benzyl radical species could be trapped with α,β-unsaturated enones and esters giving cyclization products. Thus, treatment of *o*-substituted benzaldehyde chromium complexes with SmI_2 in the presence of α,β-unsaturated ester gave γ-lactones with high diastereoselectivity (Eq. 9) [9].

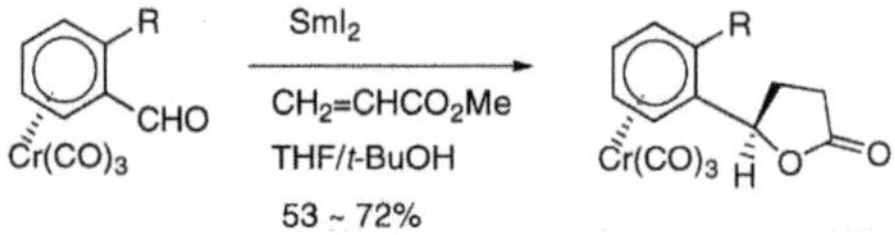

(9)

The high stereoselectivity for the preparation of γ-lactone derivatives is contributed to a conformation of the *o*-substituted benzaldehyde chromium complexes, in which a carbonyl oxygen of the benzaldehyde chromium complexes is oriented *anti* to the *ortho*-substituent due to a stereoelectronic effect, and samarium iodide attacks to the *anti*-oriented carbonyl from an opposite side to $Cr(CO)_3$ fragment (Scheme 1). The generated benzyl cation **11** incorporates a

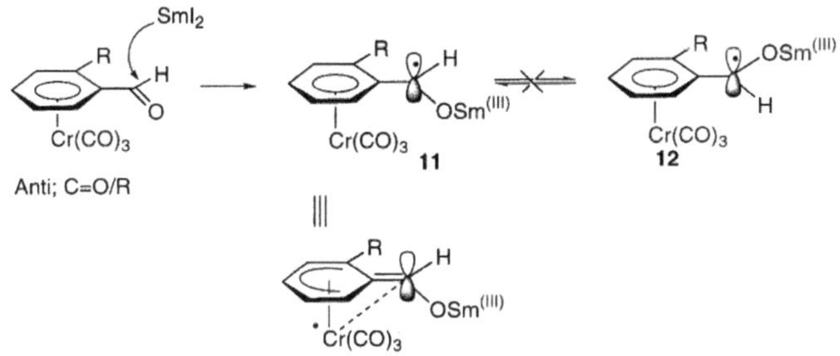

Scheme 1

substantial amount of an exocyclic double bond character **13**. This, in turn, implies that a rotation about Cα-C*ipso* bond giving **12** will be restricted. Then, the generated stereochemically stable ketyl radical intermediate **11** was trapped with α,β-unsaturated esters giving the γ-lactones.

As further extension of radical-mediated cyclization, (-)-steganone was stereoselectively synthesized by an intramolecular reductive coupling between aldehyde and α,β-unsaturated γ-lactone moiety of biphenyl mono $Cr(CO)_3$ complex **14** (Eq. 10) [10].

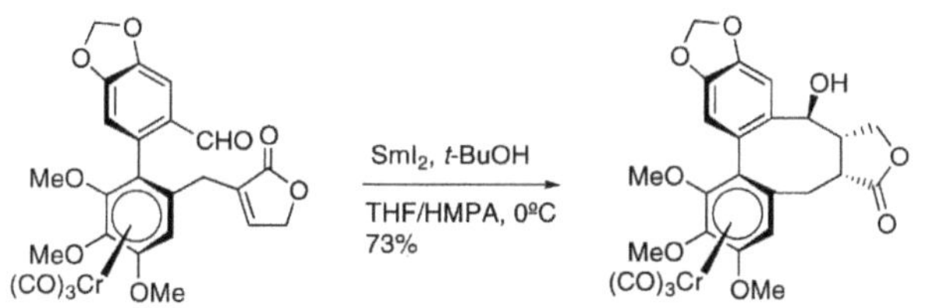

14

(10)

Furthermore, aryl radicals generated from chromium-uncomplexed aryl bromides of 2-aryl-2,3-dihydro-4-pyridinones attack to α,β-unsaturated enones to give the indolizidine and quinolizidine derivatives depending on the carbon chain with high diastereoselectivity (Eq. 11) [11]. The high selectivity is due to the minimization of A-strain between the aryl methyl group and the dihydropyridinone ring. The preferred conformation of transition state is the dihydropyridinone *N* close to the $Cr(CO)_3$ fragment, in which one face of the enone is blocked by the tricarbonylchromium fragment.

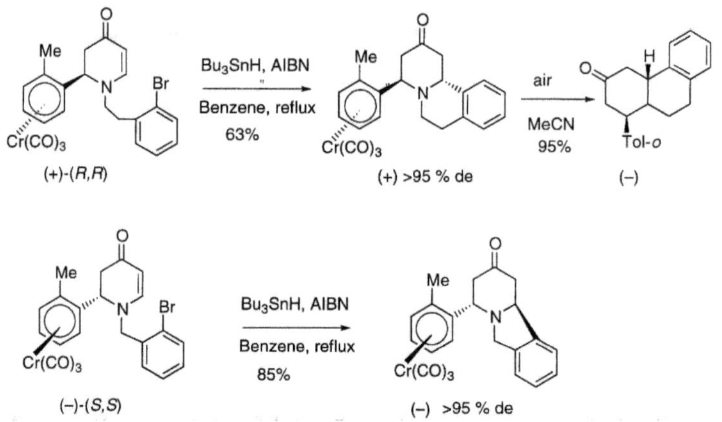

(11)

Interestingly, the ketyl radicals generated from the carbonyl group at the side chain of 6-7-dimethoxy tetraline chromium complex **15** attacked on the chromium-complexed arene ring with loss of the methoxyl group to afford the cyclization chromium complex **16** as a single diastereomer (Eq. 12) [12].

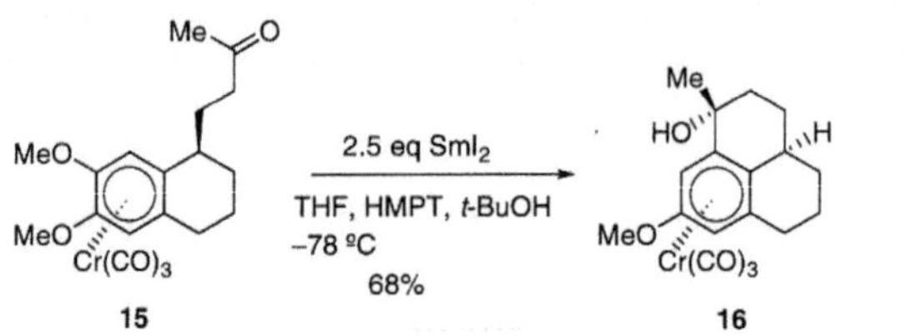

(12)

A plausible reaction mechanism is shown in Scheme 2. The generated ketyl radical **17** adds to the chromium-complexed arene ring from the face opposite to the $Cr(CO)_3$ fragment. Further electron reduction and protonation give an anionic η^4-intermediate **19**, which lead to the cyclization product **16** by elimination of MeOH.

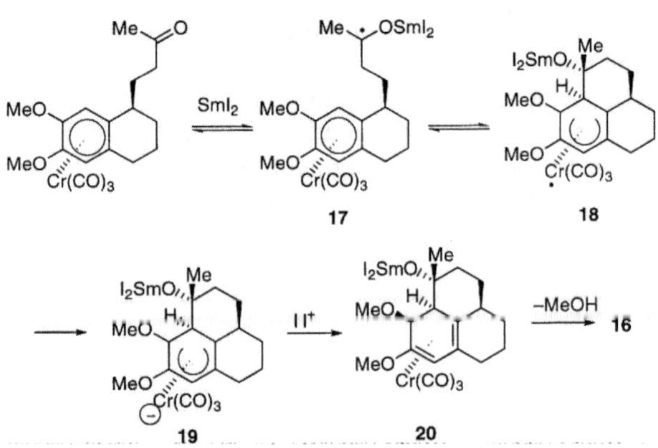

Scheme 2

The corresponding imine derivative **21** produced similar cyclization products as a diastereomeric mixture under the same conditions (Eq. 13) [13].

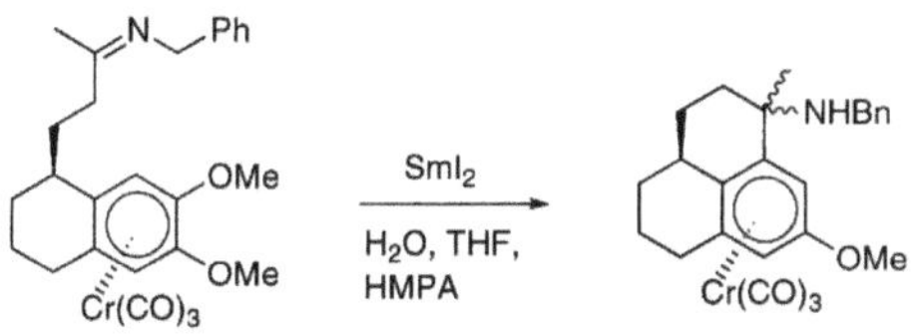

(13)

21 2 : 1 mixture

However, the chromium complex of dimethoxy dihydronaphthalene derivative **22** gave *5-endo-trig* cyclization product **23**, which is generally based on unfavorable process (Eq. 14) [12a]. The corresponding dihydrotetraline chromium complex without the methoxy group afforded hydrophenylene derivative **24** via attack of the generated ketyl radical intermediate at the chromium-coordinated arene ring (formally alkylative Birch reduction product) along with the formation of hydrobenzindene derivative.

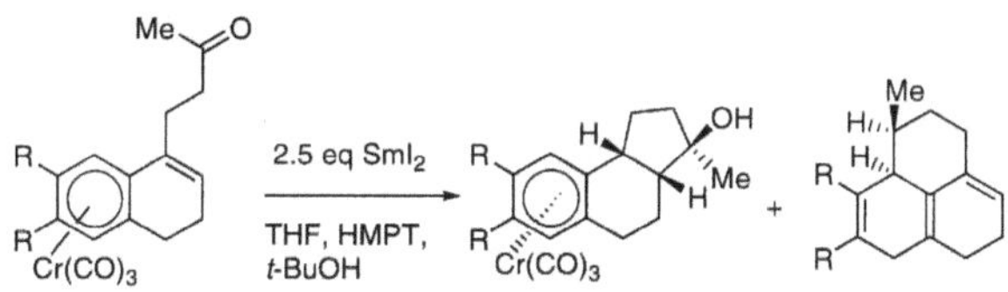

(14)

22 **23** **24**

	23	24
R = H	32%	32%
R = OMe	66%	–

3 Cycloaddition

3.1 Cycloaddition to Tricarbonylchromium-Complexed Benzaldehydes

Planar chiral *ortho* substituted benzaldehyde chromium complexes are useful compounds for a variety of asymmetric reactions. For example, planar chiral tricarbonylchromium complexes of *o*-substituted benzaldehydes were reacted with Danishefsky's diene in the presence of Lewis acid at room temperature to afford the chromium-complexed 2,3-dihydro-4-pyranones **25** with high diastereoselectivity (Eq. 15) [14]. The high diastereoselectivity of the formation of cycloaddition products **25** is also contributed to an *exo*-side approach of the diene to *anti*-oriented carbonyl oxygen of the planar chiral *ortho* substituted benzaldehyde chromium complexes. When the reaction takes place at lower temperature, aldol-type condensation product **26** was obtained along with the formation of pyranones. The open intermediate **26** was easily transformed to the corresponding cycloaddition product after stirring at room temperature [14].

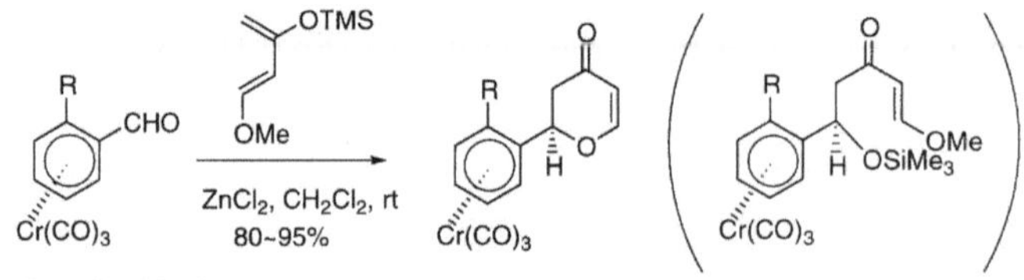

(15)

R = OMe, Me, Cl **25** 98 %de **26**

The planar chiral *ortho* substituted benzaldehyde chromium complexes were also reacted with tosyl methylisocyanide (TosMic) in the presence of K_2CO_3 in MeOH at 0 °C to give *trans*-oxazolines with high diastereoselectivity (Eq. 16) [15]. The corresponding benzaldehydes without chromium complexation resulted in lower selectivity. Decomplexation and $LiAlH_4$ reduction afforded amino alcohols. Similarly, ethyl isocyanoacetate afforded the corresponding *trans*-oxazolines with high diastereoselectivity by the reaction with planar chiral benzaldehyde chromium complexes in the presence of LDA in THF at -78 °C (Eq. 17) [16]. The reaction conditions using KCN instead of LDA as base (EtOH at room temperature) gave diastereomeric mixture of *trans*- and *cis*-oxazolines. The *trans* oxazolines obtained from the planar chiral benzaldehyde chromium complexes afforded optically pure α-amino-β-hydroxy acids by decomplexation followed by treatment with acid.

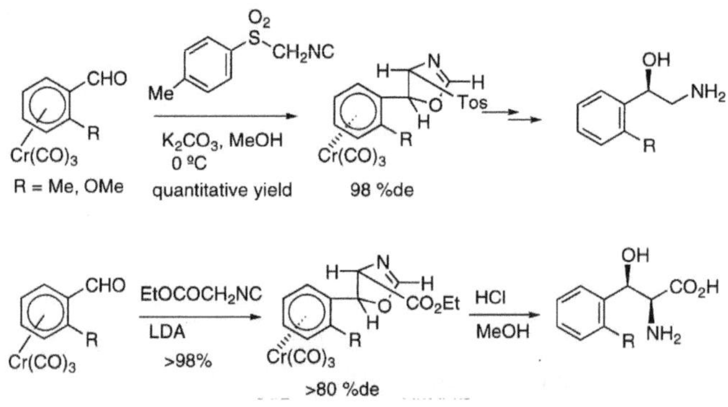

(16)

R = Me, OMe quantitative yield 98 %de

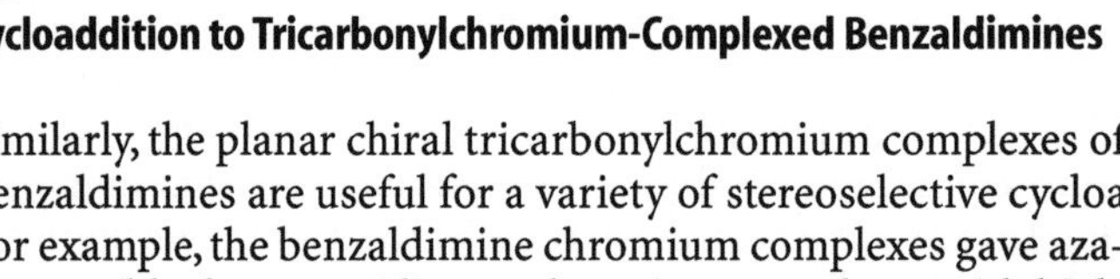

(17)

>80 %de

3.2 Cycloaddition to Tricarbonylchromium-Complexed Benzaldimines

Similarly, the planar chiral tricarbonylchromium complexes of *ortho* substituted benzaldimines are useful for a variety of stereoselective cycloaddition reactions. For example, the benzaldimine chromium complexes gave aza-Diels-Alder product, 2,3-dihydro-4-pyridinone chromium complexes with high diastereoselectivity by reaction with Danishefsky's diene (Eq. 18) [11, 17, 18]. The high diastereoselectivity of the cycloadducts is also based on the preferred *anti*-conformation of the starting benzaldimine chromium complexes as well as the planar chiral benzaldehyde chromium complexes. The cycloaddition of imines having arene chromium complex at the remote position with Danishefsky's diene underwent smoothly in good yields, but the diastereoselectivity was low (Eq. 19) [17].

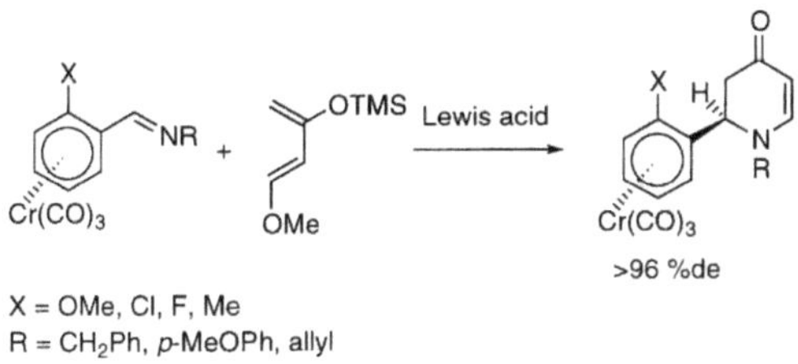

(18)

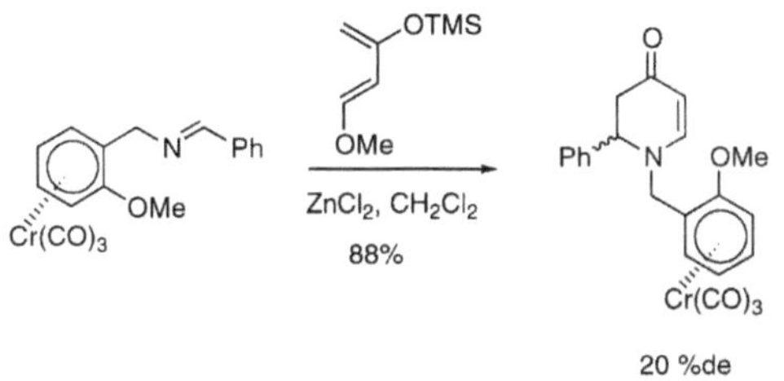

(19)

Similarly, ketene generated from acid chloride by treatment with triethylamine reacted with tricarbonylchromium-complexed benzaldimines to afford β-lactam derivatives via [2+2] cycloaddition with high diastereoselectivity. Thus, the cycloaddition of benzaldimine chromium complexes with ketenes generated from acid chloride at 0 °C in the presence of triethylamine afforded *cis*-β-lactam as a single diastereomer **27** (Eq. 20) [19]. Remote positioned imine having the planar chiral arene chromium complex was also reacted with ketene to afford β-lactam complex as diastereomeric mixture (Eq. 21) [19].

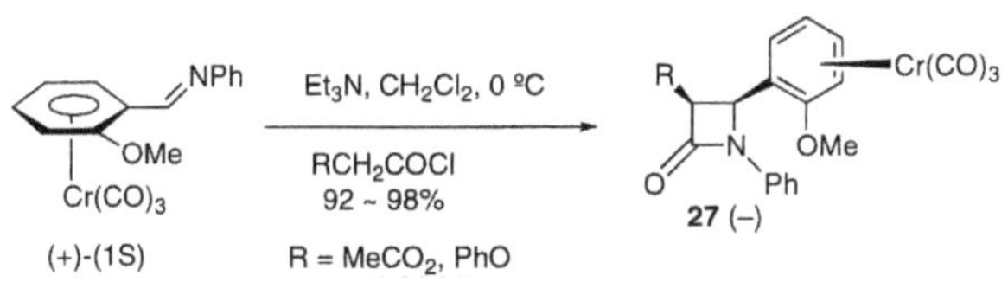

(20)

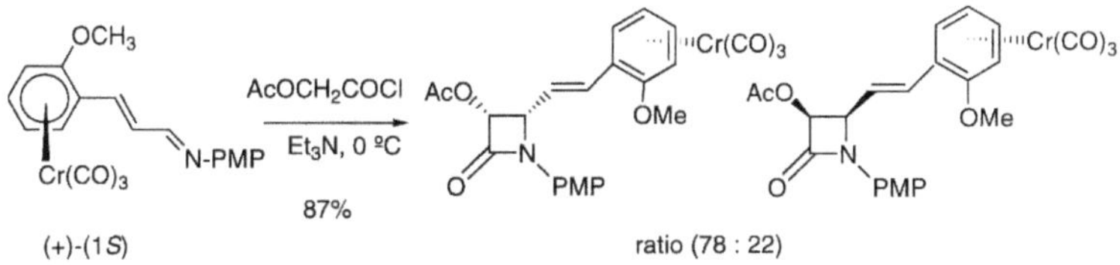

(21)

Diastereoselective [2+2] cycloaddition of *o*-fluorobenzaldehyde chromium complex with ketene generated from acetoxyacetyl chloride and subsequent intramolecular displacement of the chromium-complexed fluorine atom gave tricyclic *cis*-β-lactam **29** (Eq. 22). Surprisingly, the tricarbonylchromium fragment of the tricyclic compound **29** was found to be the opposite side to the hydrogens of the β-lactam ring [20]. A stereochemical inversion takes place during the nucleophilic addition step, and the tricyclic compound **29** might be formed via β-lactam ring opening/reclosing.

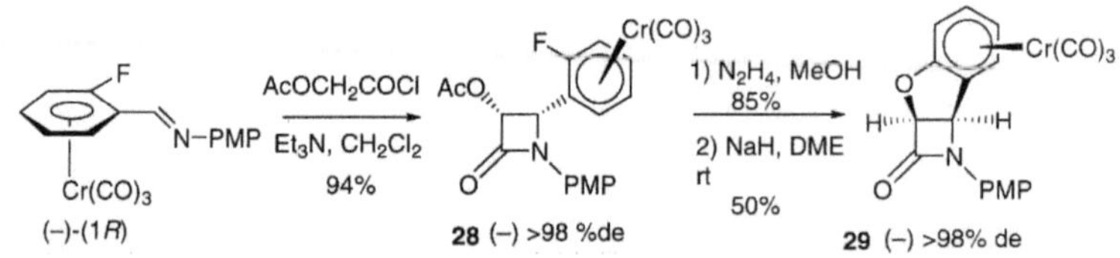

(22)

Azomethine ylides linked to the planar chiral arene chromium complexes were reacted with methyl acrylates to give highly diastereoselective [3+2] cycloaddition pyrrolidines [21]. The regioselectivity in the cycloadditions is largely controlled by the nature of Lewis acid. Thus, *N*-lithiated azomethine ylides, generated from α-aminoesters **30** by treatment with triethylamine in the presence of LiBr, reacted with methyl acrylates to give arylpyrolidines **31** as a single diastereomer in good yields (Eq. 23). The use of $TiCl(OPr\textit{i})_3$ instead of LiBr as Lewis acid resulted in formation of a reversed regioselective cycloaddition product **32** (Eq. 24). The change in regioselectivity presumably arises from different coordination structures of the Ti and LiBr to the carbonyl as shown in proposed transition states **33** and **34** (Scheme 3).

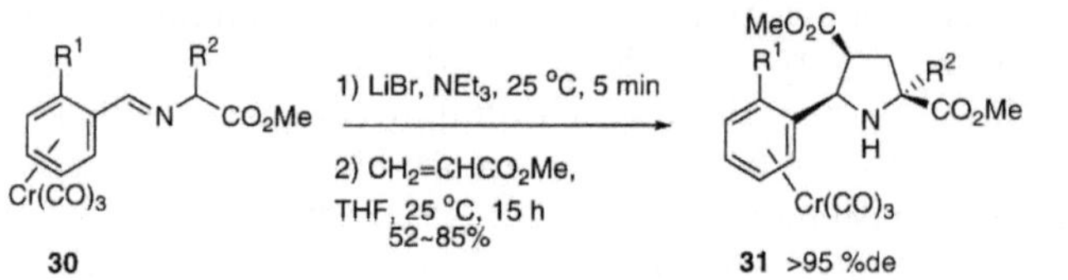

(23)

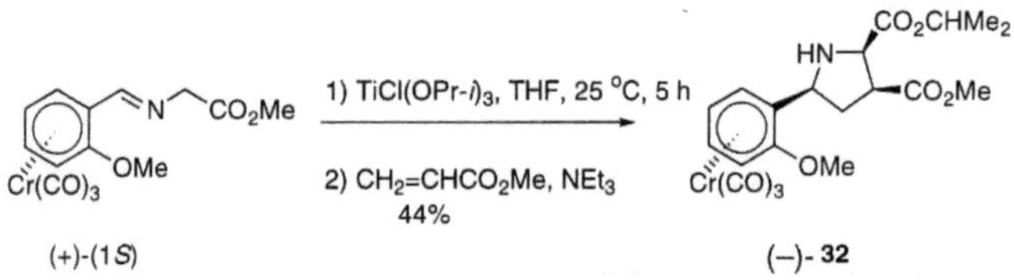

(24)

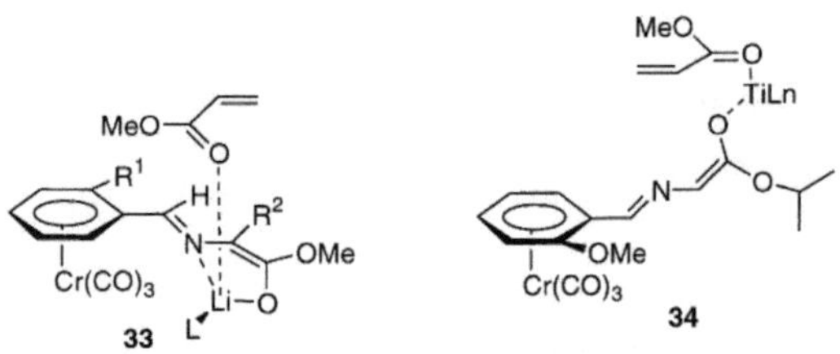

Scheme 3

1,3-Dipolar addition of tricarbonylchromium-complexed nitrone **35** with electron-rich olefin was also achieved for the preparation of *cis*-3,5-disubstituted isooxazolidine **36** much more stereoselectively than in the corresponding chromium uncomplexed arenas (Eq. 25) [22]. Also, an intramolecular version was easily observed in the 1,3-dipolar addition of the chromium-complexed nitrone with high *cis*-selectivity (Eq. 26) [23].

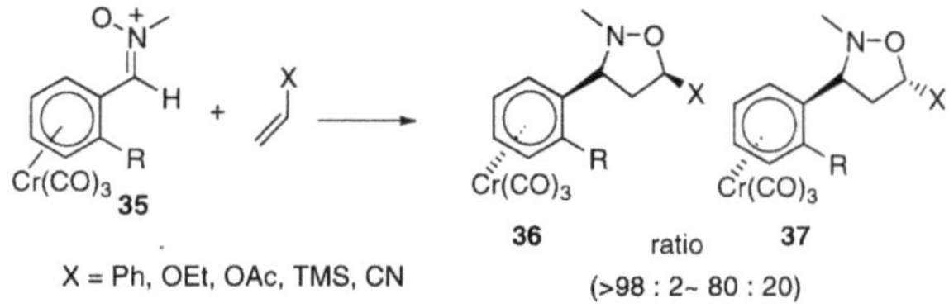

(25)

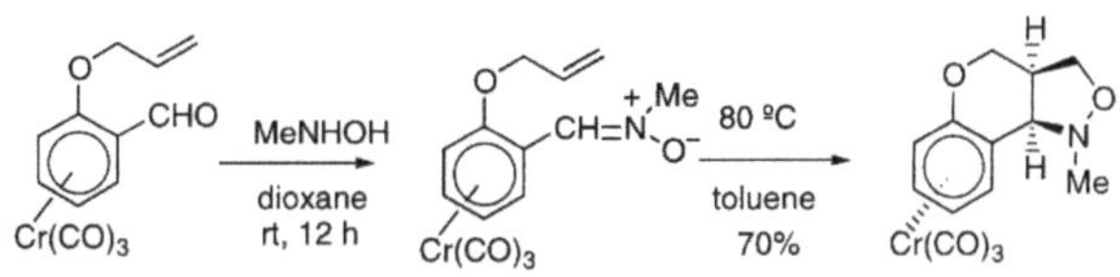

(26)

Also, the planar chiral benzaldimine chromium complexes undergo other type cycloadditions. Treatment of the planar chiral *N*-aryl benzylideneamine chromium complexes with ester enolates gave β-lactams in good yield (Eq. 27) [24]. When the imine was reacted with lithium enolate generated from ethyl ester of *N*-benzoyl alanine, the *trans-cis* stereoselectivity was low (1:1). However, optically active β-lactams could be prepared via separation of diastereomers; each diastereomer afforded the corresponding β-lactams after decomplexation. Similarly, Reformatsky condensation of α-bromo esters with benzaldimine chromium complexes and zinc under ultrasound irradiation produced optically pure β-amino esters **39** and β-lactams **40** as a single diastereomer, respectively, in a various ratio (Eq. 28) [25]. Further treatment of the β-amino esters **39** with LDA gave the corresponding β-lactams **40**.

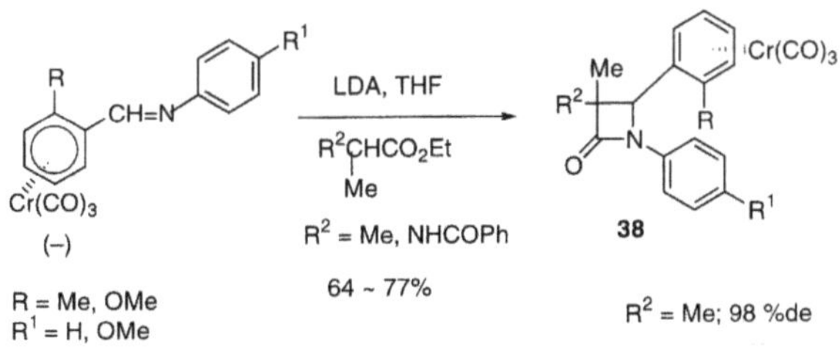

(27)

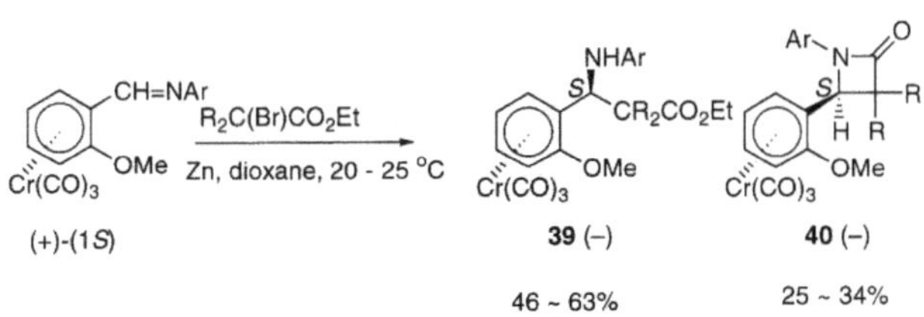

(28)

An intramolecular cycloaddition of aza-diene **41** with arene chromium complex in the presence of Lewis acid afforded the chromium-complexed tetrahydroquinoline derivatives in good yields with high diastereoselectivity (Eq. 29) [26].

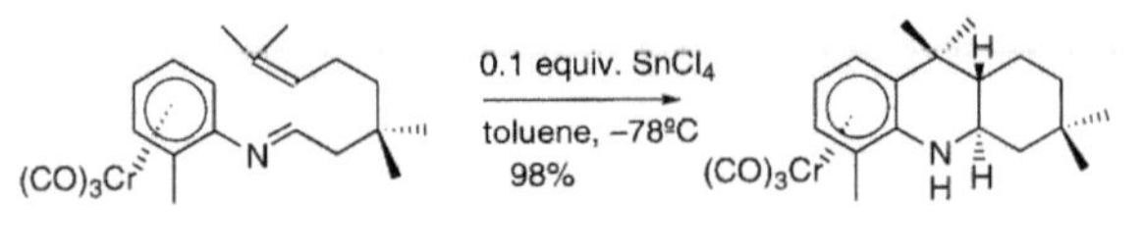

41 42

(29)

3.3 Cycloaddition to Tricarbonylchromium-Complexed Styrenes and Remote Positioned Olefin

Similarly, the styrene chromium complexes undergo cycloaddition reactions with high diastereoselectivity. Thus, treatment of the planar chiral styrene chromium complexes with 3,5-dichloro-2,4,6-trimethylbenzonitriloxide produced optically active 3,5-disubstituted 4,5-dihydroisoxazoles (Eq. 30) [27].

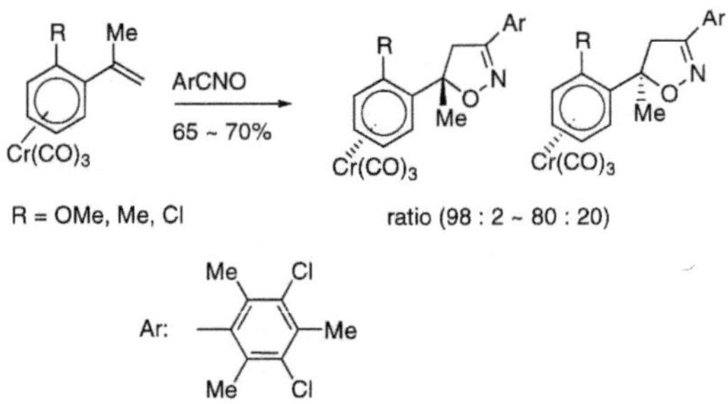

(30)

Diels-Alder reaction with cyclopentadiene gave the corresponding adduct (Eq. 31) [28]. (Dihydronaphthalene)chromium complex gave addition product by reaction with diazomethane with high diastereoselectivity [29]. The acrylate having chiral arene chromium complex underwent Diels-Alder reaction with high selectivity in the presence of Lewis acid (Eq. 32) [30].

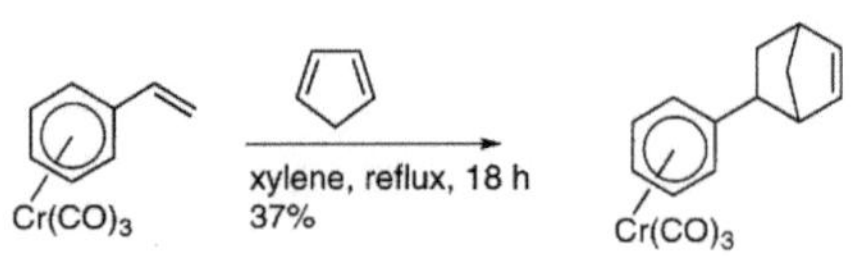

(31)

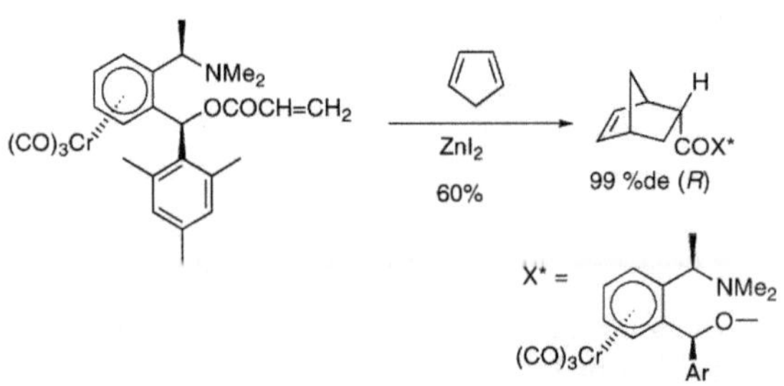

(32)

3.4
Cycloaddition to Tricarbonylchromium-Complexed *o*-Quinodimethanes

1,2-Dihydrocyclobutabenzenes are widely used as precursors of the highly reactive *ortho*-quinodimethanes under thermal conditions, which are synthetically important intermediates for cycloadditions. In addition, the cyclobutabenzene can be complexed to the tricarbonylchromium fragment. Therefore, the effect of chromium coordination on the dihydrocyclobutabenzene/cycloaddition is significant for the preparation of optically active products. When tricarbonylchromium complexes of *syn*- and *anti*-1-ethoxycyclobutabenzene were heated in benzene, an interconversion between *syn*- and *anti*-isomers was observed as an equilibrium mixture via reversible ring opening/ring closure, analogous to the reaction of the chromium-free ligand (Eq. 33) [31]. This result indicates the formation of (*o*-quinodimethane)Cr(CO)$_3$ intermediate. Therefore, heating of **43** in *trans*-1,2-bis(trimethylsilyl)ethane at 160 °C gave the cycloaddition products **46** and **47** in a ratio of 17:1. The stereochemistry of the major diastereomer **46** was that expected from the addition of dienophile to the *anti*-face of (*o*-quinodimethane)Cr(CO)$_3$ intermediate **44**. However, the high temperature is required for the thermal ring opening of cyclobutabenzenes giving *o*-quinodimethane intermediate. Milder conditions for the generation of the *o*-quinodimethane chromium complex intermediate are significant for development of this chemistry.

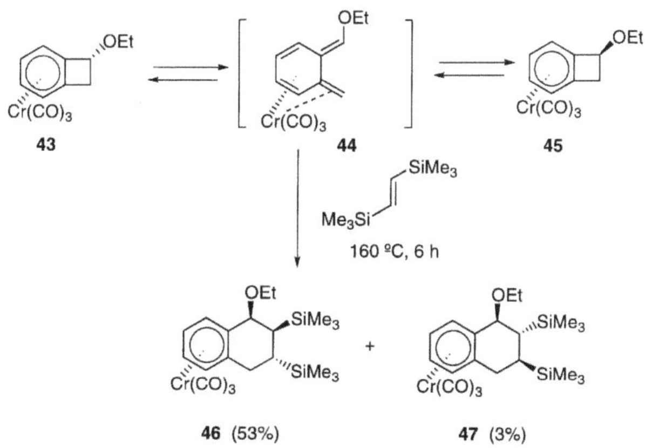

(33)

An electron-donating group at sp^3 carbons at the cyclobutane ring lowers the energy barrier of ring opening towards *o*-quinodimethane formation [32]. Taking advantage of this fact, the ring opening reaction was much more accelerated when 1-hydroxy or acetoxycyclobutabenzene underwent ring opening by treatment with *n*-butyl lithium at temperature below 0 °C to *o*-quinodimethane intermediate [33]. The corresponding Cr(CO)$_3$ complexes, like free ligand, undergo facile anion accelerated ring opening to reactive *o*-quinodimethane intermediates. Thus, in the presence of a reactive dienophile, treatment of (*syn*-1-acetoxycyclobutabenzene)Cr(CO)$_3$ in THF with *n*-BuLi at low temperature afforded the expected tetralol chromium complexes as diastereomeric mixture in a various ratio (Eq. 34) [34, 35]. Dimethyl fumarate as dienophile resulted in 1,2-*cis* tetraline chromium complex as a single diastereomer. And methyl acrylate gave a diastereomeric mixture of two cycloaddition products with predominant formation of 1,2-*cis* isomer. The observed stereochemistry might be explained by ring opening reaction resulting in (*E*)-configuration of the lithium enolate via an *endo* transition state. On the other hand, vinyl sulfones afforded exclusively 1,2-*trans* isomers (Eq. 35). Thus, the hydroxy substituent is located on the face opposite to the chromium, and the sulfonyl group is located on the same face as chromium. *N*-Methyl maleimide was also reacted to give the expected cycloaddition product, albeit low yield (Eq. 36).

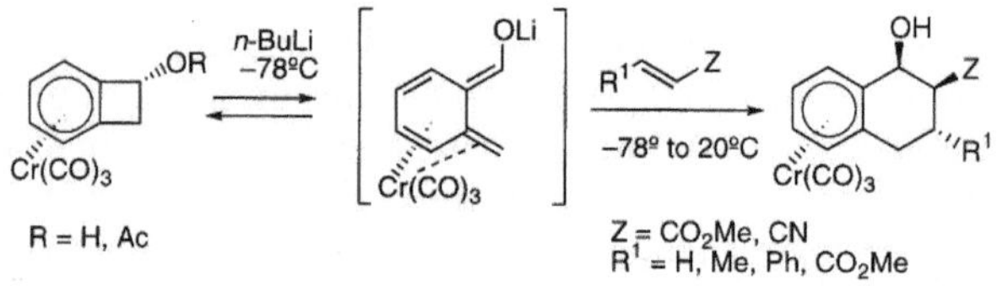

(34)

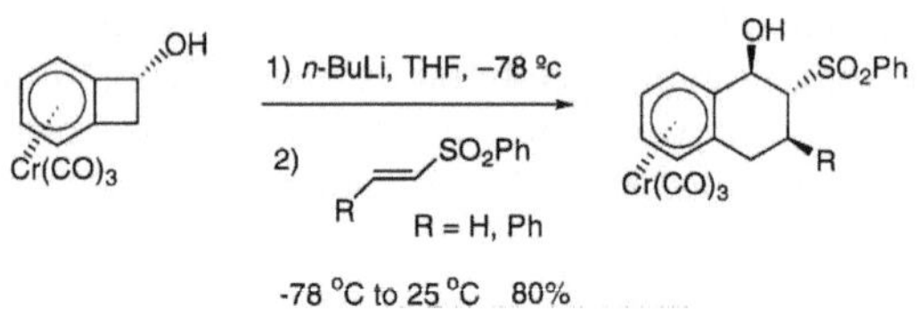

(35)

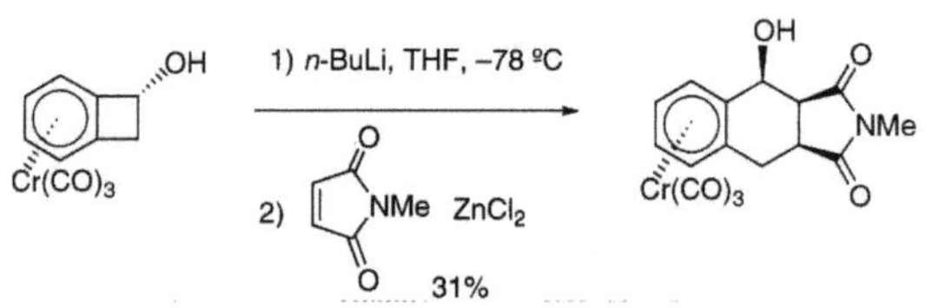

(36)

Tricarbonylchromium complex of 1-*endo*-hydroxybenzocyclobutene having 1-*exo*-hexenyl group would be expected to afford intramolecular cyclization product **50** via the generated (*o*-quinodimethane)$Cr(CO)_3$ intermediate as a further extension of this methodology, but a proximal ring opened product, (benzyl hexenyl ketone)$Cr(CO)_3$ **49**, was obtained without formation of expected cycloaddition product **50** (Eq. 37) [36]. The hexenyl dienophile incorporated in the side chain was not sufficiently reactive in the desired cycloaddition. In the presence of more electron deficient dimethyl fumarate, the hexenyl incorporated chromium complex **48** was treated with *n*-BuLi to give an intermolecular cycloaddition product **51** via formation of *o*-quinodimethane intermediate.

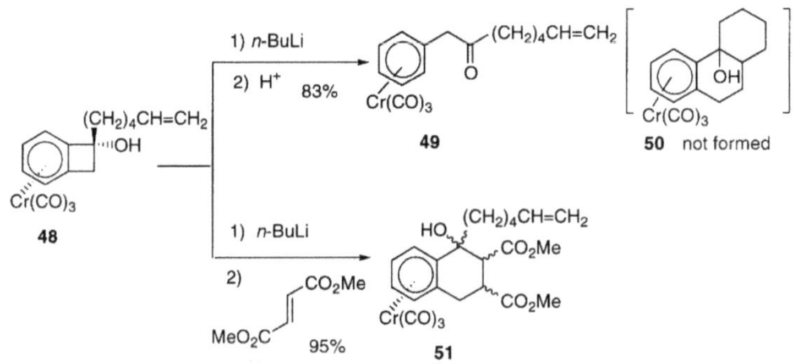

(37)

3.5 Anionic Oxy-Cope Rearrangement

1,2-Dioxobenzocyclobutene has also attracted interest for organic synthesis. In particular, much attention has been paid to the addition of carbon nucleophiles to the carbonyl group for transformation to naphthoquinone or indanone derivatives. Double additions of excess vinyllithium to (1,2-dioxobenzocyclobutene)chromium complex **52** at -78 °C gave the tricarbonylchromium complex of 5,10-dioxobenzocyclooctene **53** in 87% yield without formation of simple di-adduct (Eq. 38) [37]. Obviously, the product **53** was formed by a double addition of vinyllithium and subsequent anionic oxy-Cope rearrangement.

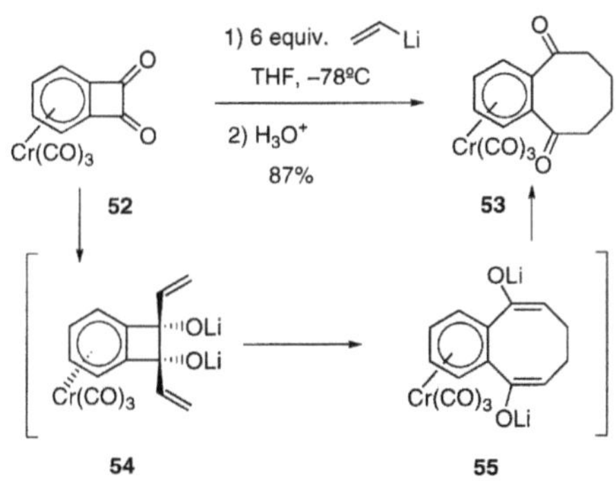

(38)

Similarly, treatment of (1,2-dioxobenzocyclobutene)chromium complex **52** with excess of 2-propenyllithium, 1-lithio-1-phenylethene and cyclopentenyllithium at -78 °C followed by hydrolysis with diluted aqueous hydrochloric acid afforded similar type benzocyclooctenediones as diastereomeric mixture along with formation of cyclopentaanellated indanones (Eq. 39) [37].

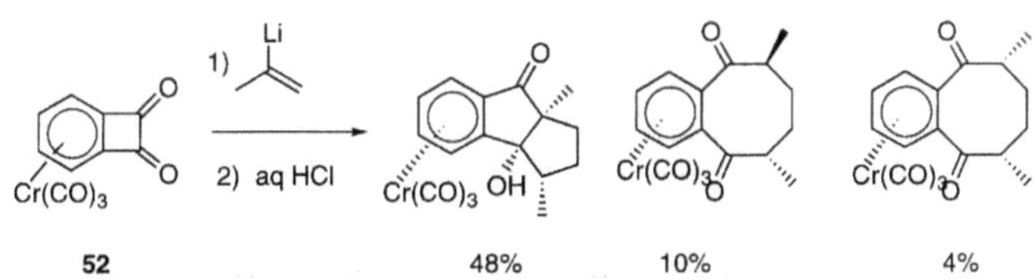

(39)

4
Coupling Reaction

4.1
Cross-Coupling of (Aryl Halide)$Cr(CO)_3$ Complexes

Palladium(0)-catalyzed cross-coupling of aryl halides or aryl triflates with organometals is well known [38]. It is also found that a coordination of an electron-withdrawing tricarbonylchromium fragment to the aryl halides accelerates an oxidative addition of the arene-halogen bond to the palladium(0) due to the decrease of π-electron density, and even chlorobenzene chromium complex undergoes oxidative addition to the Pd(0) at ambient temperature [39]. Thus, the palladium(0)-catalyzed carbonylation, cross-coupling and Heck reaction using aryl halide chromium complexes have been easily achieved. For example, the palladium(0)-catalyzed cross-coupling of tricarbonyl(chlorobenzene)chromium with vinyl- or tetraalkyl stannane gave vinylbenzene or alkylbenzene after oxidative demetallation in good yields (Eq. 40) [40]. Also, palladium-catalyzed carbonylation of chlorobenzene tricarbonylchromium gave the corresponding esters, aldehydes, amides, or α-oxo-amides according to the reaction conditions [41]. A combination of sodium alkoxide and alkyl formate instead of carbon monoxide could be used for the palladium-catalyzed ester preparation from chlorobenzene chromium complex under mild conditions [42]. Similarly, Heck reaction with methyl vinyl ketone afforded the α,β-unsaturated ketone (Eq. 42) [40].

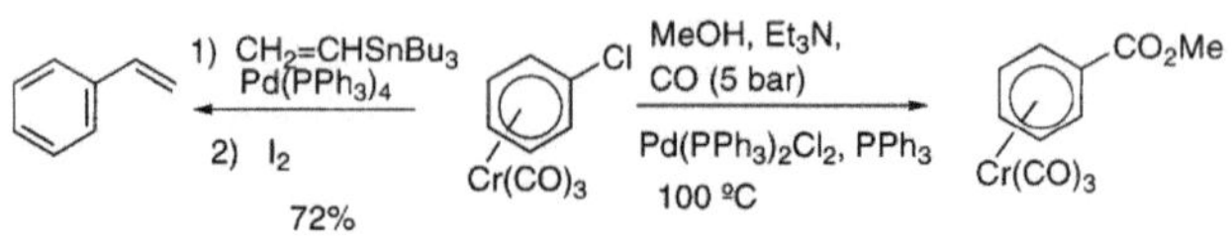

(40)

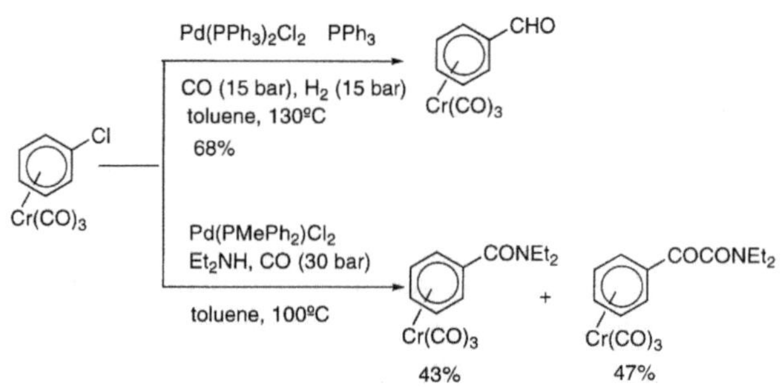

(41)

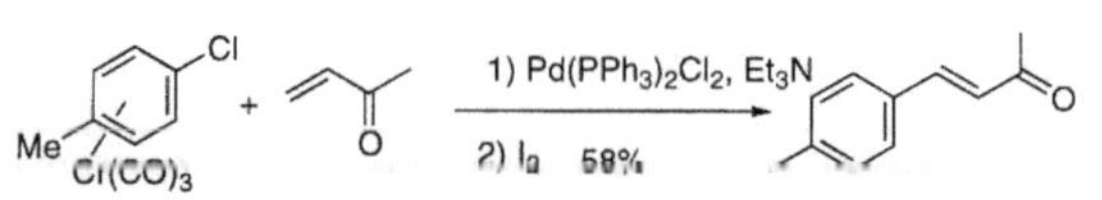

(42)

Also, palladium-catalyzed Sonogashira coupling of chlorobenzene chromium complexes with alkynes gave coupling products in good yields (Eq. 43) [43]. Propargyl alcohols afforded the corresponding Sonogashira coupling products **57** by the palladium-catalyzed coupling in the presence of CuI and Et_3N, while terminal aryl secondary propargyl alcohols gave the α,β-unsaturated enone chromium complexes **58** instead of expected propargyl alcohols under the same reaction conditions (Eq. 44) [43]. The chalcone chromium complexes **58** would be formed by base-catalyzed isomerization of the initially formed chromium-complexed propargyl alcohols. An electron-withdrawing ability of the tricarbonylchromium fragment plays a key role for the base-catalyzed isomerization. An exchange of one of CO tripod on the chromium ligand to an electron donating phosphine group renders the metal more electron rich, but (chlorobenzene)$Cr(CO)_2PPh_3$ underwent the palladium-catalyzed Sonogashira coupling even with terminal aryl secondary propargyl alcohols without formation of the corresponding α,β-unsaturated enones under the usual conditions [44]. This mild enone synthesis enables the development of novel one pot synthesis of pyrazoline skeleton. Thus, addition of *N*-methylhydrazine is compatible with the reaction conditions after the coupling. After the formation of chalcone chromium complexes, the hydrazine enters the Michael-addition and following cyclocondensation sequence giving rise to the formation of 2-pyrazolines (Eq. 45) [45].

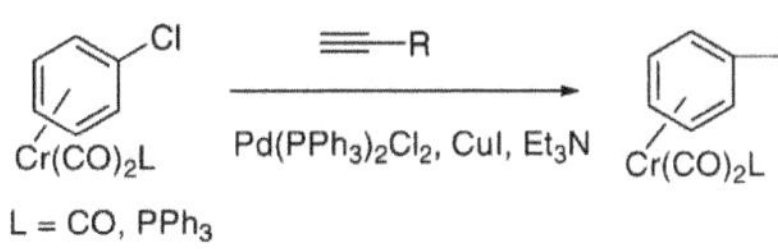

(43)

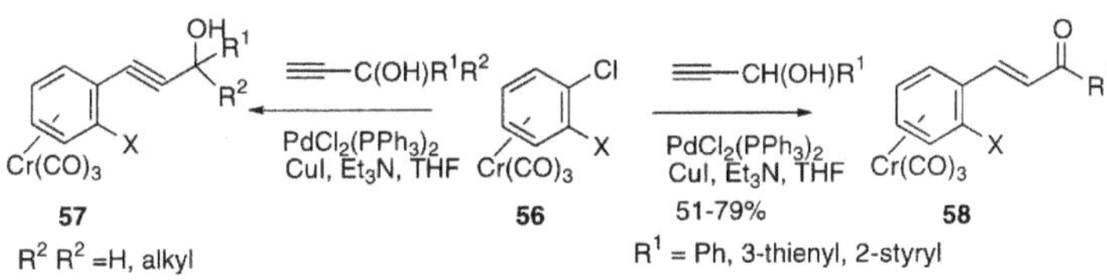

(44)

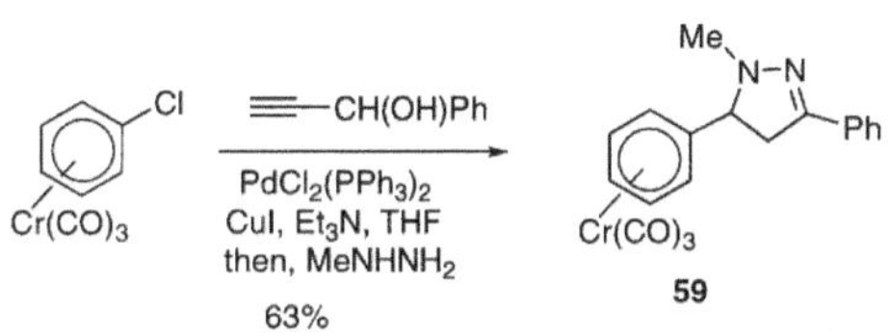

(45)

The chromium-complexed propargyl alcohols are useful compounds for further organic transformation. For example, the treatment of **60** with thionyl chloride or chlorodiethylphosphite to give the chloro- or phosphorylallenyl substituted arene chromium complexes in good yields (Eq. 46) [46]. The phosphoryllallenyl substituted η^6-arene chromium complexes have been developed in further cyclization reactions (Eq. 47).

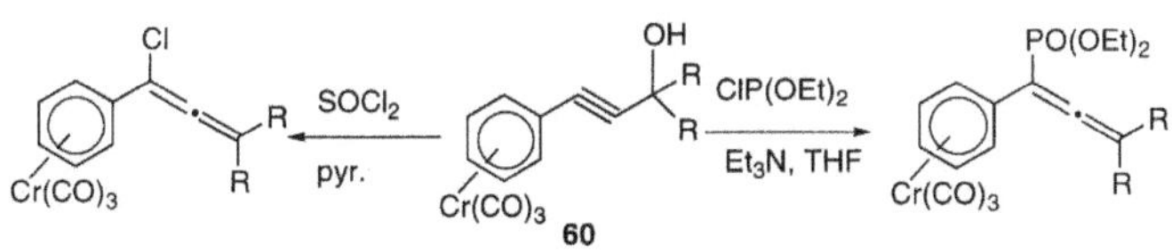

(46)

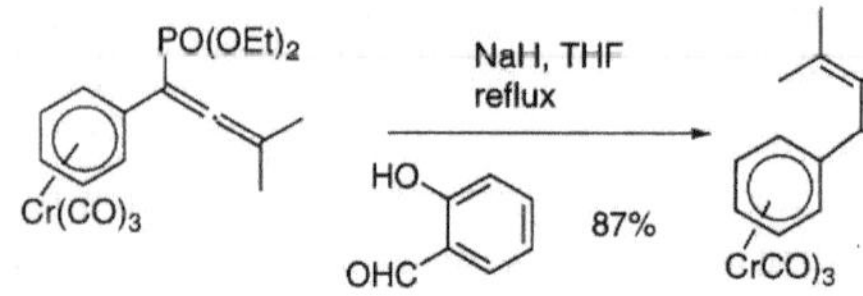

(47)

The tricarbonylchromium-complexed phenylacetylene underwent also Sonogashira coupling with various arylbromides or aryliodides (Eq. 48) [48].

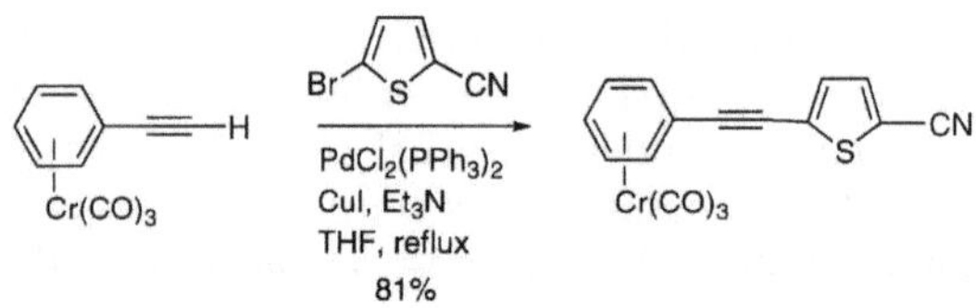

(48)

Intramolecular Heck type reaction undergoes smoothly with high stereoselectivity in good yields (Eq. 49) [49]. The diastereofacial selectivity of the alkene carbopalladation is only controlled by the planar chirality of the arene chromium complex. The chromium complex **61** yielded the product **62** with an *anti*-relationship between the $Cr(CO)_3$ fragment and CH_2CO_2Me group by palladium-catalyzed reaction in the presence of triethylamine and methanol under CO atmosphere (Eq. 50) [50]. Although the transition state geometry of the Heck-reaction favors an eclipsed arrangement of the alkene to the Pd-C(Ar) bond, a proposed geometry with an equatorial benzylic substituent and a coordination to the alkenes such as to form. A coordination to the opposite alkene face and Pd coordination coplanar with the arene plane would provide a rationale for the observed diastereoselectivity. The proposed transition state A would lead to severe steric congestion between the $Cr(CO)_3$ fragment and the one of the other Pd-ligands (Scheme 4).

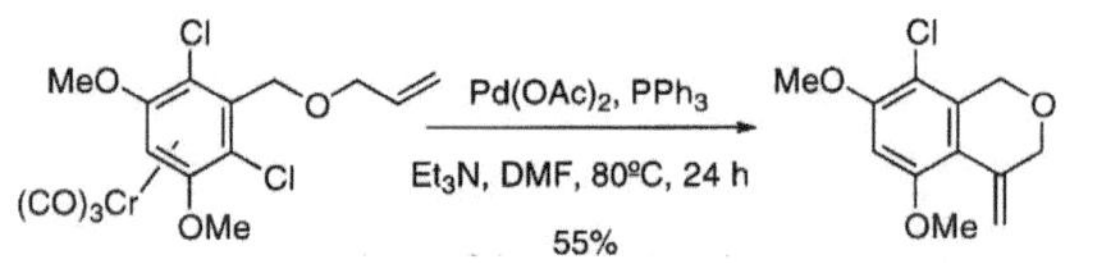

(49)

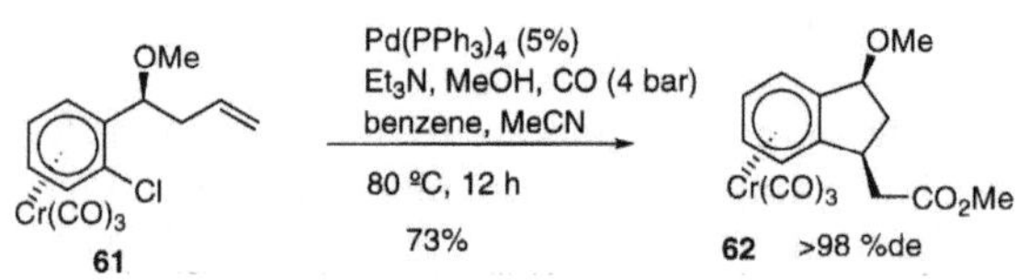

(50)

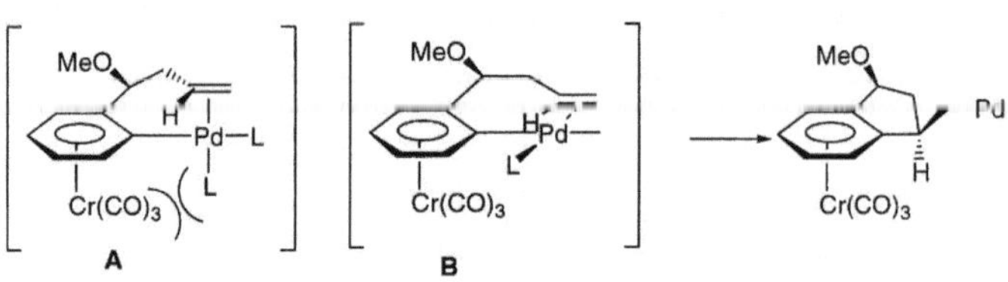

Scheme 4

Planar chiral arene chromium complexes are useful compounds as chiral ligands or auxiliaries in asymmetric reactions [51]. The usual method for the preparation of the optically active (arene)chromium complexes is a resolution via recrystallization or column chromatography of the corresponding diastereomers derived from racemic (arene)chromium complexes and suitable chiral reagents [52]. Biocatalysts have been also used for the optical resolution of racemic arene chromium complexes [53]. Furthermore, the diastereoselective and enantioselective *ortho* lithiation of arene chromium complexes have been employed for the preparation of optically active chromium complexes [54]. Diastereoselective chromium complexation with distinction of the arene face is also useful method for the preparation of the optically active arene chromium complexes [55]. However, in these methods, stoichiometric amount of the chiral reagents is necessary for the preparation of optically active arene chromium complexes. The use of catalytic amount of the chiral reagents is an attractive method, and the catalytic asymmetric desymmetrization of prochiral 1,2-dichlorobenzene chromium complex was achieved in the presence of a chiral palladium catalyst. Thus, the cross-coupling of (1,2-dichlorobenzene)Cr(CO)$_3$ with alkenyl- and arylmetals in the presence of a chiral phosphine-palladium catalyst gave mono-coupling product up to 69% ee (Eq. 51) [56]. Furthermore, enantioselectivity of mono-methoxycarbonylation product was enhanced by a sequent kinetic resolution connected to the formation of the bis-methoxycarbonylation product by the reaction of CO in MeOH and triethylamine in the presence of chiral palladium P,N-ligand (Eq. 52) [57].

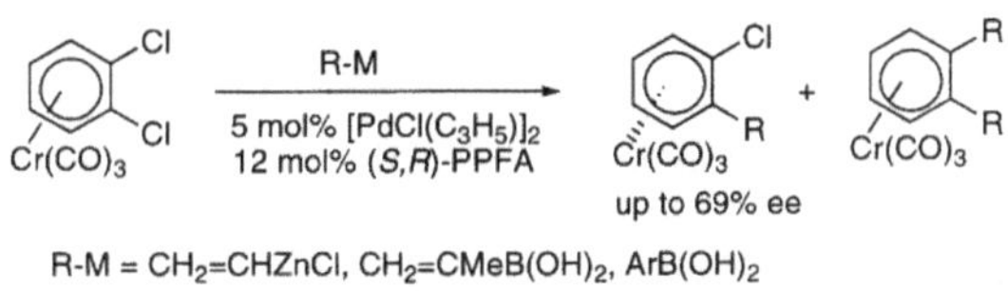

(51)

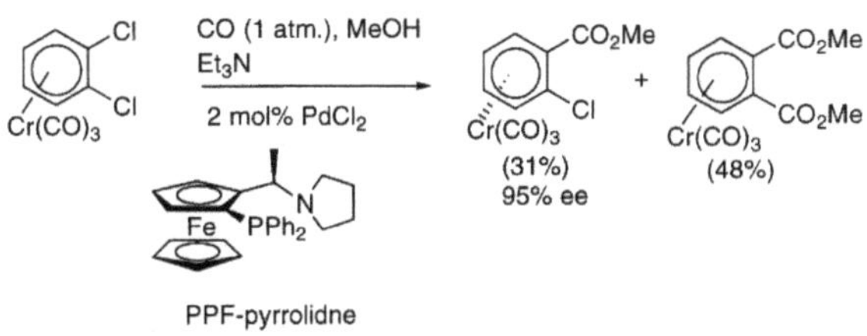

(52)

4.2
Axially Chiral Biaryls

Aryl metals such as arylboronic acids, aryl Grignard and zinc reagents were also coupled with aryl halide chromium complexes to give mono Cr(CO)$_3$ complexes of biphenyls [58]. Among these aryl metals, arylboronic acids resulted in higher yields of the coupling products, while aryl stannane did not afford the desired coupling products. Axially chiral biaryls could be stereoselectively prepared by Suzuki-Miyaura cross-coupling of the planar chiral arylhalide chromium complexes with *o*-substituted arylboronic acids. The axially chirality of the biaryls was controlled by the steric size of *ortho* substituents using arylboronic acids and the reaction conditions. Thus, (2-methoxy-6-substituted bromobenzene)$Cr(CO)_3$ complexes were heated with *o*-methylphenylboronic acid in the presence of $Pd(PPh_3)_4$ and aqueous Na_2CO_3 in MeOH at 70~75 °C for 30 min to give *syn*-configurated biphenyl chromium complexes **63** without formation of the corresponding *anti*-diastereomers (Eq. 53) [59]. Similarly, other *o*-substituted phenylboronic acids and α-naphthylboronic acid produced predominantly thermodynamically unstable *syn* cross-coupling products as major compounds in spite of a severe non-bonding steric interaction between the tricarbonylchromium fragment and the *ortho* substituents derived from phenylboronic acids. However, *o*-formylphenylboronic acid afforded exclusively *anti* cross-coupling products **64** under the same reaction conditions. In some case, carbonyl inserted benzophenone chromium complexes were obtained.

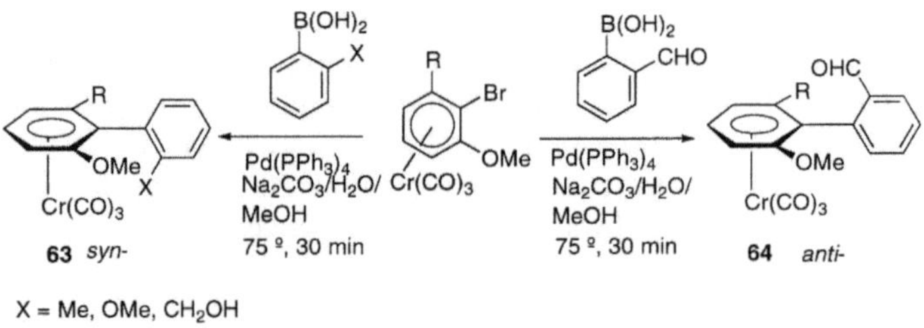

(53)

The *syn*-configured cross-coupling products were easily isomerized to the thermodynamically more stable *anti* biaryl chromium complexes by following two methods; the first method is modification of steric size of *ortho* substituent to less one. For example, an oxidation of *syn*-configurated *o*-hydroxymethyl to formyl group with DMSO/Ac_2O gave an *anti*-formyl biphenyl chromium complex along with axial isomerization (Eq. 54) [59b,c].

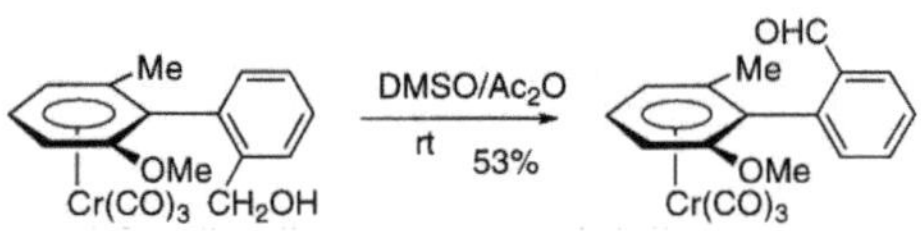

(54)

The second procedure for an axial isomerization is thermal conditions. Refluxing of *syn* biphenyl chromium complexes **65** in high boiling aromatic solvent, e.g., toluene, xylene, mesitylene, for 2 h gave the *anti*-biphenyl chromium complexes **66** (Eq. 55) [59b,c].

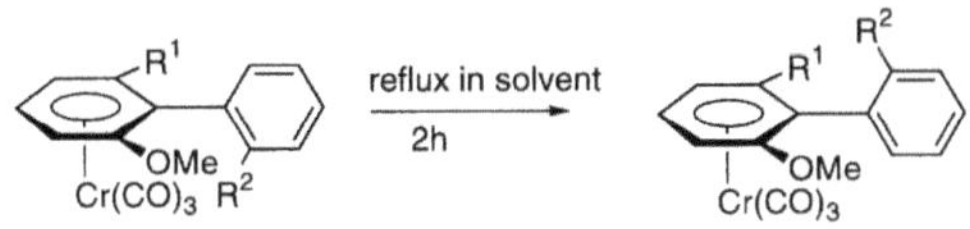

65 *syn-* **66** *anti-*

(55)

R^1	R^2	solvent	**65** (*syn*) : **66** (*anti*)	yield (%)
CHO	Me	toluene	2 : 98	(98)
Me	OMe	toluene	42 : 58	(98)
Me	OMe	xylene	9 : 91	(98)
Me	Me	xylene	80 : 20	(98)
Me	Me	mesitylene	2 : 98	(79)

A plausible reaction mechanism for the palladium(0)-catalyzed diastereoselective cross-coupling reaction is proposed as follows (Scheme 5) [59]. Two *cis* diorganopalladium(II) intermediates **67** and **68** with square planar configurations would be the transient species. A transmetallation step between Ar(Cr)PdBr and arylboronic acid to give both intermediates **67** and **68** is an equilibration step, and the subsequent reductive elimination would be the rate-determining step in the cross-coupling reaction. The palladium intermediate **67** has a severe steric interaction between the substituent R at the ring B and the sterically bulky L substituent. Both arene rings are coupled via an overlap of p-orbitals, avoiding the severe non bonding interactions between R and triphenylphosphine groups in the less hindered intermediate **68**, and then, the R substituent rotates toward the Cr(CO)$_3$ moiety giving the *syn*-(S^*,S^*)-configured products **63**. The formation of thermodynamically stable *anti*-(S^*,R^*)-complexes **64** by the coupling with *ortho*-formylphenylboronic acid might be attributed to an axial isomerization under reaction conditions.

67 **68**

Scheme 5

Since some *syn*-configurated chromium-complexed biaryls can easily undergo the axial isomerization under thermal conditions as mentioned above, the thermodynamically stable *anti*-isomers could be expected to be obtained directly by the cross-coupling at the elevated reaction temperature. Thus, when the cross-coupling of planar chiral (arene)chromium complex **69** with *ortho* methyl phenylboronic acid was carried out in *refluxing aqueous xylene* for 2 h, the expected *anti*-configurated biphenyl **70** was exclusively obtained without formation of the corresponding *syn*-biphenyl (Eq. 56) [59b,c]. However, the *syn*-configured biaryl chromium complex was exclusively obtained under heating in aqueous MeOH at 70–75 °C as mentioned above (Eq. 53). Similarly, the arene chromium complex **69** was coupled with α-naphthylboronic acid in *refluxing xylene* to give the *anti*-configured biaryl **71**.

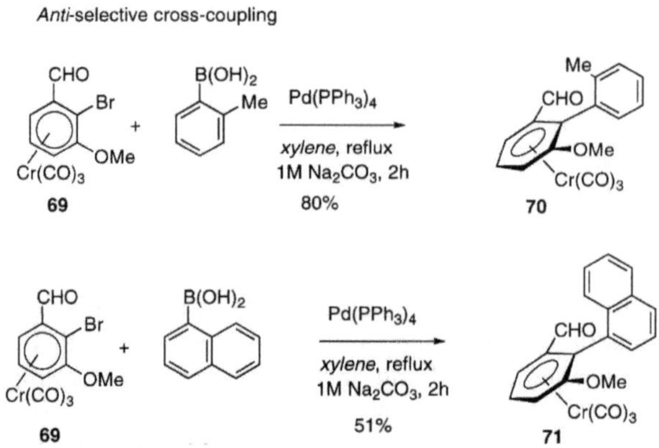

(56)

The palladium(0)-catalyzed diastereoselective cross-coupling of the planar chiral (arylhalide)chromium complexes with arylboronic acids provides a promising approach to the synthesis of both enantiomerically pure biaryls starting from a single optically pure arene chromium complex as a chiral source by following two methods. The first procedure is the cross-coupling of a *single planar chiral arene chromium complex* with phenylboronic acids having different *ortho*-substituents under *kinetically controlled reaction conditions* (Eq. 57). Thus, enantiopure (+)-(2-bromo-3-methoxybenzaldehyde ethyleneacetal)chromium complex **72** gave stereoselectively the *syn*-configurated (+)-biphenyl chromium complex **73** by palladium(0)-catalyzed cross-coupling with *o*-methylphenylboronic acid in the presence of sodium carbonate under reflux in aqueous methanol. The *syn*-coupling product **73** was converted to (-)-(*R*)-2-methoxy-2′-methyl-6-(1,3-dioxolanyl)biphenyl **74** by photo-oxidation. In contrast, the cross-coupling of **72** with *o*-formylphenylboronic acid produced the *anti*-configurated complex **75** under the same conditions. Antipode (+)-(*S*)-biphenyl **76** was obtained by conversion of the formyl group of the *anti*-biphenyl complex **75** to a methyl group. In this way, both enantiomers of axially chiral biphenyls were stereoselectively prepared using two different *ortho*-substituted phenylboronic acids [59c].

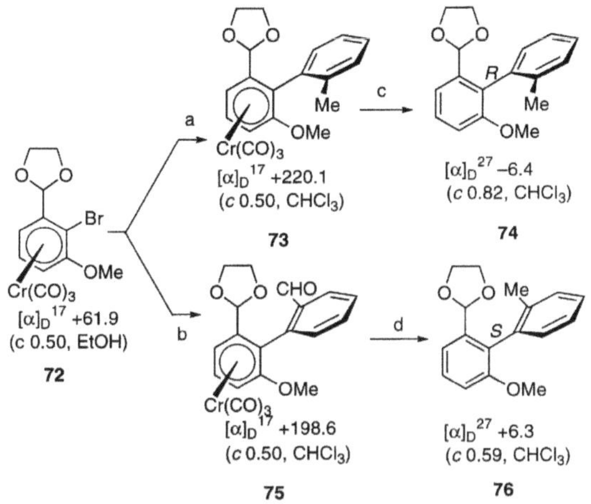

(57)

(a) 2-methylphenylboronic acid, $Pd(PPh_3)_4$, Na_2CO_3, MeOH, H_2O, 75 °C, 30 min (81%); (b) 2-formylphenylboronic acid, $Pd(PPh_3)_4$, Na_2CO_3, MeOH, H_2O, 75 °C, 30 min (52%); (c) *hn*-O_2, ether (90%); (d) 1) $LiAlH_4$, ether (62%); 2) MsCl, pyr (38%); 3) $LiAlH_4$, ether (51%); 4) *hν*-O_2, ether (90%)

The second method for the preparation of both enantiomers from a single planar chiral arene chromium complex was based on the combination of the diastereoselective cross-coupling and subsequent axial isomerization (Eq. 58) [57c]. Enantiopure (-)-tricarbonyl(2-bromo-3-methoxybenzaldehyde)chromium **77** was reacted with 2-methylphenylboronic acid under refluxing *aqueous methanol* to give the *syn* (+)-(*R*,*R*)-tricarbonyl[(1,2,3,4,5,6-η)-2-methoxy-2′-methyl-6-formylbiphenyl]chromium **78**. When the palladium-catalyzed cross-coupling was carried out in refluxing aqueous *xylene* instead of methanol, the *anti*-product **80** was obtained in 75% yield. Furthermore, refluxing of the *syn*-coupling product **78** under *xylene* gave the thermodynamically stable *anti* (-)-(*R*,*S*)-biphenyl chromium complex **80** with axial isomerization. Both *syn*-and *anti*-biphenyl chromium complexes **78** and **80** were converted to the axially chiral (*R*)- and (*S*)-biphenyls **79** and **81**, respectively, by reduction with $NaBH_4$, acetylation and subsequent photo-oxidation.

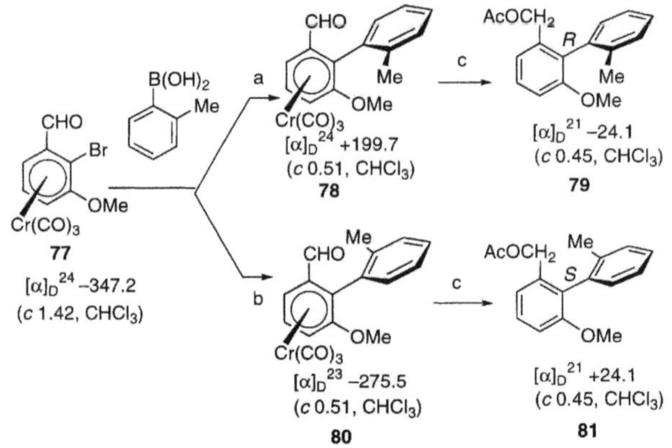

(58)

(a) $Pd(PPh_3)_4$, Na_2CO_3, MeOH, H_2O, 75 °C, 30 min (80%); (b) $Pd(PPh_3)_4$, Na_2CO_3, xylene, H_2O,140 °C, 30 min (75%); (c) 1) $NaBH_4$, MeOH (98%); 2) Ac_2O, py (90%); 3) $h\nu$-O_2, ether (52%)

These diastereoselective cross-coupling of planar chiral aryl halide chromium complexes with arylboronic acids have been applied to the synthesis of biologically active natural products having the axial chirality as a key step. An enantiomerically pure (-)-tricarbonyl(3,4,5-trimethoxy-2-bromobenzyl alcohol)chromium **82** with 4,5-methylenedioxy-2-formylphenylboronic acid in the presence of $Pd(PPh_3)_3$ in aqueous MeOH gave *anti*-configurated biaryl chromium complex **83** without formation of the corresponding *syn*-isomer (Eq. 59). The *anti*-coupling product **83** was stereoselectively converted to (-)-steganone [61].

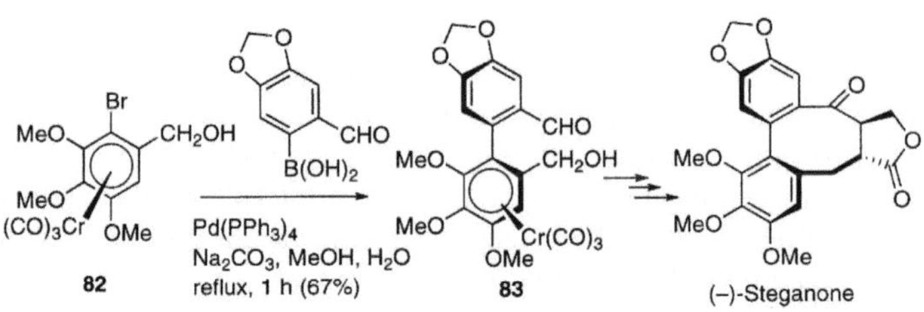

(59)

Vancomycin has attracted multidisciplinary interest for the clinical use and has been enlisted as drug of the last resort for the treatment of infections due to methicillin-resistant *Staphyococcus aureus*. One essential problem in the total synthesis of vancomycin is the stereoselective construction of the axially chiral A-B biaryl ring system. Diastereoselective cross-coupling of arylhalide chromium complex with *o*-substituted arylboronic acid is useful method for the A-B ring. Both enantiomers of planar chiral arylbromide chromium complex could be stereoselectively converted to the key intermediate of axially chiral A-B ring system depending on the nature of *ortho*-substituent of arylbromide chromium complex [62]. The formyl group of enantiomerically pure (+)-(2-bromo-3,5-dimethoxybenzaldehyde)chromium complex **84** was initially converted to monosilyl protected diol chromium complex **85** with sterically bulky substituent. The chromium complex **85** was coupled with *ortho* methoxyphenylboronic acid **86** under the conditions of $Pd_2(dba)_3$, $(o\text{-tolyl})_3P$ in a mixture of toluene/MeOH/1 Mol/l aqueous Na_2CO_3, 80 °C, 10 min to give the *syn*-coupling product **87** without formation of the corresponding *anti*-isomer (Eq. 60). The coupling product **87** was further converted to **88**, a key intermediate of axially chiral A-B ring system of vancomycin [63]. In contrast, *ent*-(-) complex **84** was coupled with identical phenylboronic acid **86** under the same conditions to give the *anti*-coupling product **89** without formation of the corresponding *syn*-isomer. The formyl group of the *anti*-product **89** was converted to the identical key intermediate **88** by stereoselective construction of the benzylic chiral center with introduction of a nitrogen atom.

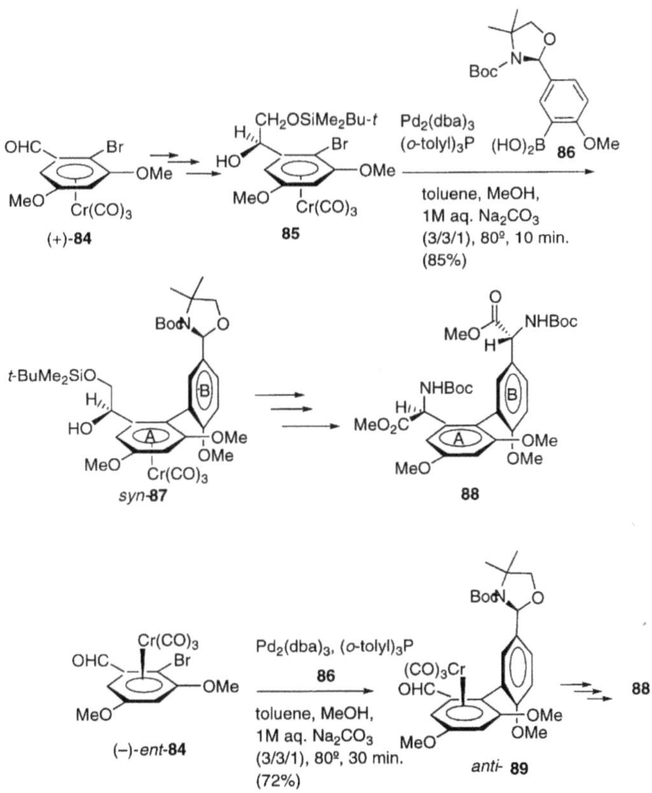

(60)

The corresponding enantiomer of A-B ring system **92** was synthesized by a similar method [64]. Strecker reaction of chromium-complexed benzaldimine takes place for construction of amino acid function at the benzylic position. Thus, planar and axially chiral disulfinilimine complex **91** obtained by reaction of the *anti*-coupling product **90** and (+)-(*S*)-sulfinylamine was treated with Et_2AlCN to afford nitrile complex which was converted to antipode vancomycin A-B ring system **92** (Eq. 61).

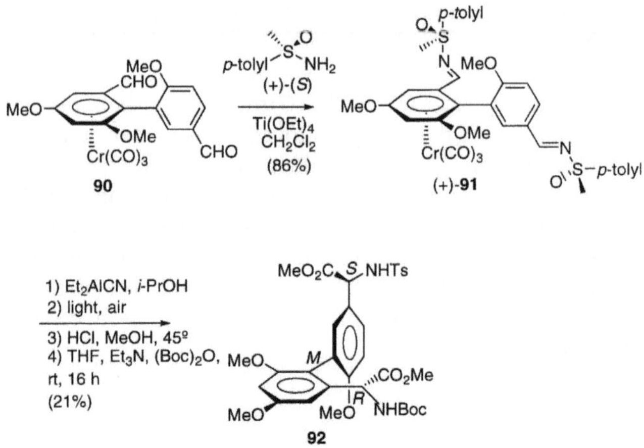

(61)

Axially chiral biaryls can be also prepared by nucleophilic substitution of tricarbonylchromium complexes of 2-methoxy-benzoates with aryl Grignard reagents [65].

References

1. (a) Davies SG, Coote SJ, Goodfellow CL (1991) In: Liebeskind LS (ed) Advances in metal-organic chemistry, vol 2. JAI Press, Greenwich, Conneticut, p 1; (b) Collmann JP, Hegedus L, Norton JR, Finke RG (1987) Principles and applications of organotransition metal chemistry. University Science, Mill Valey, CA; (c) Davies SG, McCarthy TD (1995) In: Wikinsons G, Stone PGA, Abel EW (eds) Comprehensive organometallic chemistry II, vol 12. Pergamon, Oxford, p 1039; (d) Davies SG, Donohoe TJ (1993) Synlett 323
2. (a) Meyer A, Hofer O (1980) J Am Chem Soc 102:4410; (b) Jaouen G, Meyer A (1976) Tetrahedron Lett 3547; (c) Jaouen G (1977) Ann NY Acad Sci 295:59
3. Blagg J, Davies SG, Holman NJ, Laughton CA, Mobbs BE (1986) J Chem Soc Perkin Trans 1581
4. Sainsbury M, Williams CS, Naylor A, Scopes DIC (1990) Tetrahedron Lett 31:2763
5. Davies SG, Newton RF, Williams JMJ (1989) Tetrahedron Lett 30:2967
6. Coote SJ, Davies SG, Midderriss D, Naylor A (1989) J Chem Soc Perkin Trans 1:2223
7. (a) Taniguchi N, Kaneta M, Uemura M (1996) J Org Chem 61:6088; (b) Taniguchi N, Uemura M (1997) Synlett 51; (c) Taniguchi N, Uemura M (1998) Tetrahedron 54:12775
8. Taniguchi N, Hata T, Uemura M (1999) Angew Chem Int Ed 38:1232
9. (a) Taniguchi N, Uemura M (1997) Tetrahedron Lett 38:7199; (b) Merlic CA, Walsh JC (1998) Tetrahedron Lett 39:2083
10. Monovich LG, Huérou YL, Rönn M, Molander GA (2000) J Am Chem Soc 122:52
11. Kündig EP, Xu LH, Romanens P, Bernarfinelli G (1996) Synlett 270
12. (a) Schmalz HG, Siegel S, Bats JW (1995) Angew Chem Int Ed Engl 36:2383; (b) Schmalz HG, Siegel S, Schwalz A (1996) Tetrahedron Lett 37:2947
13. Hoffmann O, Schmalz HG (1998) Synlett 1426
14. Baldoli C, Del Buttero P, Di Ciolo M, Maiorana S, Papagni A (1996) Synlett 258
15. Solladié-Cavallo A, Quazzotti S, Colonna S, Manfredi A (1989) Tetrahedron Lett 30:2933
16. Colonna S, Manfredi A, Solladié-Cavallo A, Quazzotti S (1990) Tetrahedron Lett 31:6185

17. Baldoli C, Del Buttero P, Di Ciolo M, Maiorana S, Papagni A (1996) Synlett 258
18. Ratni H, Crousse B, Kündig EP (1999) Synlett 626
19. Baldoli C, Del Buttero P, Licandro E, Maiorana S, Papagni A (1994) Tetrahedron Asymmetry 5:809
20. Del Buttero P, Baldoli C, Molteni G, Pilati T (2000) Tetrahedron Asymmetry 11:1927
21. Schnell B, Bernardinelli G, Kündig EP (1999) Synlett 348
22. (a) Mukai C, Cho WJ, Kim IJ, Hanaoka M (1990) Tetrahedron Lett 31:6893; (b) Mukai C, Kim IJ, Cho WJ, Kido M, Hanaoka (1993) J Chem Soc Perkin Trans 1 2495
23. Baldoli C, Del Buttero P, Licandro E, Maiorana S, Papagni A (1995) Tetrahedron Asymmetry 6:1711
24. Baldoli C, Del Buttero P (1991) J Chem Soc Chem Commun 982
25. Baldoli C, Del Buttero P, Licandro E, Papagni A, Pilati T (1996) Tetrahedron 52:4849
26. Laschat S, Noe R, Riedel M, Krüger C (1993) Organometallics 12:3738
27. Baldoli C, Del Buttero P, Maiorana S, Zecchi G, Moret M (1993) Tetrahedron Lett 34:2529
28. Knox CR, Leppard DG, Pauson PL, Watts WE (1972) J Organomet Chem 34:347
29. Vebrel J, Tonnard F, Carrie R (1987) Bull Soc Chim Fr 1057
30. Uemura M, Hayashi Y, Hayashi Y (1993) Tetrahedron Asymmetry 11:2291
31. (a) Kündig EP, Leresche J (1993) Tetrahedron 49:5595; (b) Kündig EP, Bernardinelli G, Leresche J, Romanens P (1990) Angew Chem Int Ed Engl 29:407
32. Oppozer W (1978) Synthesis 793
33. (a) Choy W, Yang H (1988) J Org Chem 53:5796; (b) Choy W (1990) Tetrahedron 46:2281
34. (a) Wey HG, Butenschön H (1991) Angew Chem Int Ed Engl 30:880; (b) Brands M, Wey HG, Krömer R, Krüger C, Butenschön H (1995) Liebigs Ann 253
35. (a) Kündig EP, Bernardinelli G, Leresche J (1991) J Chem Soc Chem Commun 1713; (b) Kündig EP, Leresche J (1993) Tetrahedron 49:5599; (c) Kündig EP, Leresche J, Saudan L, Bernardinelli G (1996) Tetrahedron 52:7363
36. Brands M, Wey HG, Butenschön H (1991) J Chem Soc Chem Commun 1541
37. (a) Brands M, Goddard R, Wey HG, Butenschön H (1993) Angew Chem Int Edn Eng 32:267; (b) Brands M, Bruckmann J, Krüger C, Butenschön H (1994) J Chem Soc Chem Commun 999; (c) Brands M, Wey HG, Bruckmann J, Krüger C, Butenschön H (1996) Chem Eur J 2:182; (d) Butenschön H (1999) Synlett 680
38. For representative references: (a) Knight DW (1991) In: Trost BM, Fleming I (eds) Comprehensive organic synthesis. Pergamon, New York, p 449; (b) Suzuki A (1998) In: Diederich F, Stang PJ (eds) Metal-catalyzed cross-coupling reactions, chap 2. Wiley-VCH, Weinheim; (c) Stille JK (1986) Angew Chem Int Ed Engl 25:508; (d) Negishi E (1982) Acc Chem Res 15:340; (e) Diederich F, Stang PJ (eds) (1997) In: Diederich F, Stang PJ (eds) Metal-catalyzed cross-coupling reactions. Wiley-VCH, Weinheim
39. Some representative references: (a) Clough JM, Mann IS, Widdowson DA (1987) Tetrahedron Lett 28:2645; (b) Mutin R, Lucas C, Thivolle-Cazat J, Dufaud V, Dany F, Basset JM (1988) J Chem Soc Chem. Commun 896; (c) Dany F, Mutin R, Lucas C, Dufand V, Thivolle-Cazat J, Basset JM (1989) J Mol Catal 51:L15
40. Scott WJ (1987) J Chem Soc Chem Commun 1755
41. (a) Mutin R, Lucas C, Thivolle-Cazat J, Dufaud V, Dany F, Basset JM (1988) J Chem Soc Chem Commun 896; (b) Dufaud V, Thivolle-Cazat J, Basset JM, Mathieu R, Jaud J, Waissermann J (1991) Organometallics 10:4005
42. (a) Carpentier JF, Castanet Y, Brocard J, Mortreux A, Petit F (1991) Tetrahedron Lett 32:4705; (b) Carpentier JF, Castanet Y, Brocard J, Mortreux A, Petit F (1992) Tetrahedron Lett 33:2001
43. (a) Müller TJJ, Lindner HJ (1996) Chem Ber 129:607; (b) Müller TJJ, Ansorge M, Lindner HJ (1996) Chem Ber 129:1433; (c) Müller TJJ (1997) Tetrahedron Lett 38:1025
44. Ansorge M, Müller TJJ (1999) J Organomet Chem 585:174
45. Müller TJJ, Ansorge M, Aktah D (2000) Angew Chem Int Ed 39:1253

46. Müller TJJ, Ansorge M (1997) Chem Ber Recueil 130:1135
47. (a) Müller TJJ, Ansorge M (1998) Tetrahedron 54:1457; (b) Ansorge M, Polborn K, Müller TJJ (1999) Eur J Inorg Chem 225
48. Tranchier JP, Chavignon R, Prim D, Auffrant A, Plyta ZF, Rose-Munch F, Rose E (2000) Tetrahedron Lett 41:3607
49. (a) Bräse S (1999) Tetrahedron Lett 40:6757; (b) Ratni H, Crousse B, Kündig EP (1999) Synlett 626
50. (a) Crousse B, Xu LH, Bernardinelli G, Kündig EP (1998) Synlett 658; (b) Kündig EP, Ratni H, Crousse B, Bernardinelli G (2001) J Org Chem 66:1852
51. Bolm C, Muniz K (1999) Chem Soc Rev 28:51
52. Some representative references: (a) Mandelbaum A, Neuwirth Z, Caïs N (1963) Inorg Chem 2:902; (b) Dabard R, Meyer A, Jaouen G (1969) CR Acad Sci Paris Ser C 268:201; (c) Falk H, Schögl K, Steyrer W (1966) Monatsch Chem 97:1029; (d) Rosca S, Nenitzescu CD (1970) Rev Roum Chim 15:259; (e) Solladié-Cavallo A, Solladié G, Tsamo E (1979) J Org Chem 44:4189; (f) Solladié-Cavallo A, Solladié G, Tsamo E (1985) Inorg Syn 23:85; (g) Davies SG, Goodfellow CL (1989) J Chem Soc Perkin Trans 1:192; (h) Davies SG, Goodfellow CL (1990) J Chem Soc Perkin Trans 1:393; (i) Bromley LA, Davies SG, Goodfellow CL (1991) Tetrahedron Asymmetry 2:139
53. Some references: (a) Nakamura K, Ishihara K, Ohono A, Uemura M, Nishimura H, Hayashi Y (1990) Tetrahedron Lett 31:3603; (b) Yamazaki Y, Hosono K (1990) Tetrahedron Lett 31:3895; (c) Top S, Jaouen G, Gillois J, Buldoli C, Maiorana S (1988) J Chem Soc Chem Commun 1284
54. Some representative references: (a) Price DA, Simpkins NS, MacLeod AM, Watt AP (1994) J Org Chem 59:1961; (b) Heppert JA, Aube J, Thomas-Miller ME, Milligan ML, Takusagawa F (1990) Organometallics 9:727; (c) Blagg J, Davies SG, Goodfellow CL, Sutton KH (1987) J Chem Soc Perkin Trans 1:1805; (d) Kondo Y, Green JR, Ho J (1991) J Org Chem 56:7199; (e) Han JW, Son SK, Chung YK (1997) J Org Chem 62:8264
55. Some representative references: (a) Schmalz HG, Arnold M, Hollander J, Bats JW (1994) Angew Chem Int Ed Engl 33:109; (b) Brocard J, Lebibi J, Prelinski L, Mahmoudi M (1986) Tetrahedron Lett 27:6325; (c) Uemura M, Kobayashi T, Isobe K, Minami T, Hayashi Y (1986) J Org Chem 51:2859; (d) Alexakis A, Mangeney P, Marek I, Rose-Munch F, Rose E, Semra A, Robert F (1992) J Am Chem Soc 114:8288
56. (a) Uemura M, Nishimura H, Hayashi T (1993) Tetrahedron Lett 34:107; (b) Uemura M, Nishimura H, Hayashi T (1994) J Organomet Chem 473:129
57. Gotov B, Schmalz HG (2001) Org Lett 3:1753
58. Uemura M, Nishimura H, Kamikawa K, Nakayama K, Hayashi (1994) Tetrahedron Lett 35:1909
59. (a) Uemura M, Kamikawa K (1994) J Chem Soc Chem Commun 2697; (b) Kamikawa K, Watanabe T, Uemura M (1996) J Org Chem 61:1375; (c) Kamikawa K, Uemura M (2000) Synlett 938
60. Kamikawa K, Watanabe T, Uemura M (1995) Synlett 1040
61. (a) Uemura M, Daimon A, Hayashi Y (1995) J Chem Soc Chem Commun 1943; (b) Kamikawa K, Watanabe T, Daimon A, Uemura M (2000) Tetrahedron 56:2325
62. Kamikawa K, Tachbana A, Sugimoto S, Uemura M (2001) Org Lett 3:2033
63. Nicolau KC, Li H, Boddy CNC, Ramanjulu JM, Yue TY, Natarajan S, Chu XJ, Bräse S, Rünsam F (1999) Chem Eur J 5:2584
64. Wilhelm R, Widdowson DA (2001) Org Lett 3:3079
65. Kamikawa K, Uemura M (1996) Tetrahedron Lett 37:6359

Topics Organomet Chem (2004) 7: 157–179
DOI 10.1007/b94494

Natural Products Synthesis

Hans-Günther Schmalz · Battsengel Gotov · Andreas Böttcher
Institut für Organische Chemie, Universität zu Köln, Greinstrasse 4, 50939 Köln, Germany
E-mail: schmalz@uni-koeln.de, battsengel@mcscdecola.mn, andreas.boettcher@uni-koeln.de

Abstract Planar-chiral η^6-arene-$Cr(CO)_3$ complexes represent highly valuable building blocks for the diastereo- and enantioselective synthesis of complex natural products and related bioactive compounds. Highly original and competitive overall syntheses of various classes of natural products, such as sesquiterpenes, diterpenes, alkaloids and compounds with axial chirality, have been developed. In certain cases, the whole strategy is based on arene chromium chemistry and the various chemical and stereochemical effects of the metal unit are exploited in several subsequent transformations. Cationic Cp-ruthenium complexes allow arylether formation by S_NAr reactions and have found application in the synthesis of glycopeptide antibiotics.

Keywords Arene chromium complexes · Arene ruthenium complexes · Enantioselective total synthesis · Planar-chiral metal complexes · Natural products

1 Introduction

Among the known types of transition metal arene complexes, only η^6-benzene-$Cr(CO)_3$ derivatives have received broad recognition by synthetic chemists. This is because such complexes are rather air-stable and easy to handle while the

$Cr(CO)_3$ group activates the arene ligand in a characteristic fashion allowing a variety of transformations which cannot be achieved with the free arenes [1–3]. Most useful are the facilitated attack of nucleophiles (even allowing for efficient dearomatization reactions) [4, 5], the pronounced stabilization of both positive [6] and negative [7] charge in benzylic position [8], the increased acidity of aromatic and benzylic hydrogens [9], and the activation of C_{Ar}-X bonds towards Pd-catalyzed coupling reactions [10]. Fortunately, the various *chemical effects* (see Fig. 1) are associated with pronounced *stereochemical effects*, which mainly result from the effective shielding of the complexed π-face of the ligand by the bulky $Cr(CO)_3$ tripod and from the remarkable configurative stability of reactive intermediates in benzylic position [11].

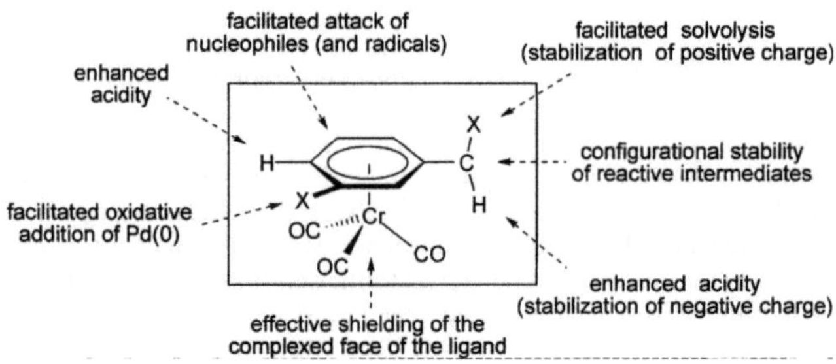

Fig. 1 Important effects of the $Cr(CO)_3$ unit on the reactivity of the complexed arene ligand

Another stereochemical feature is that (achiral) arene ligands bearing two non-identical substituents in *ortho* or *meta* position give rise to chiral complexes (Fig. 2) which can be viewed as structures possessing a plane of chirality [12, 13].

A *ent*-**A**

Fig. 2 The planar-chiral 1-tetralone-$Cr(CO)_3$ complex **A** and its enantiomer *ent*-**A**

As a consequence, arene-$Cr(CO)_3$ complexes offer new and unique opportunities for the stereoselective multi-step synthesis of complex molecules. By focusing on applications in the *enantioselective total synthesis of natural products* (and relevant analogs), this review intends to highlight the state of the art of synthetic arene chromium chemistry. It will be shown that highly original and competitive overall syntheses can be achieved especially in those cases, where the whole strategy is based on arene chromium chemistry and the chemical and stereochemical effects of the $Cr(CO)_3$ unit can be exploited in several subsequent transformations. Accordingly, this review covers enantioselective total syntheses of complex natural products, in which stoichiometric arene chromium chemistry plays a *significant* role. Only a few, particularly important racemic syntheses are mentioned. Not included are (1) the modification (derivatization) of aromatic natural products through temporary complexation (i.e., partial syntheses) and (2) the use of arene-$Cr(CO)_3$ complexes in the preparation of simple synthetic building blocks.

At the end of this chapter, a few cationic arene-RuCp complexes will be mentioned as well, because such complexes possess a unique potential for arylether formation and have thus found convincing application in total synthesis as well.

2
Synthesis of Sesquiterpenes

One of the first eye-catching synthetic applications of arene-chromium chemistry was the synthesis of the *spiro*-sesquiterpenes (±)-acorenone and (±)-acorenone B (*rac*-**7**) disclosed by Semmelhack and Yamashita in 1980 [14]. These authors twice exploited the *meta*-selective nucleophile addition to anisole-$Cr(CO)_3$ derivatives (Scheme 1). Starting from complex *rac*-**1**, such a reaction is first used for the regioselective introduction of an acyl sidechain to give **2** after oxidative workup. A few steps later, the nitrile *rac*-**4** (obtained from *rac*-**3** by complexation and separation of the diastereomeric products by preparative HPLC) is deprotonated to form the *spiro* addition product *rac*-**5**, from which the enone *rac*-**6** is obtained after protonation and hydrolysis of the initially formed dienol ether. The final conversion of *rac*-**6** into acorenone B (*rac*-**7**) efficiently proceeds over five steps and involves a diastereoselective hydrogenation of an *exo*-methylene group.

While representing an historic milestone in applied π-complex chemistry, the synthesis, however, does not meet modern criteria of synthetic efficiency, a major drawback being the low diastereoselectivity in the formation of complex

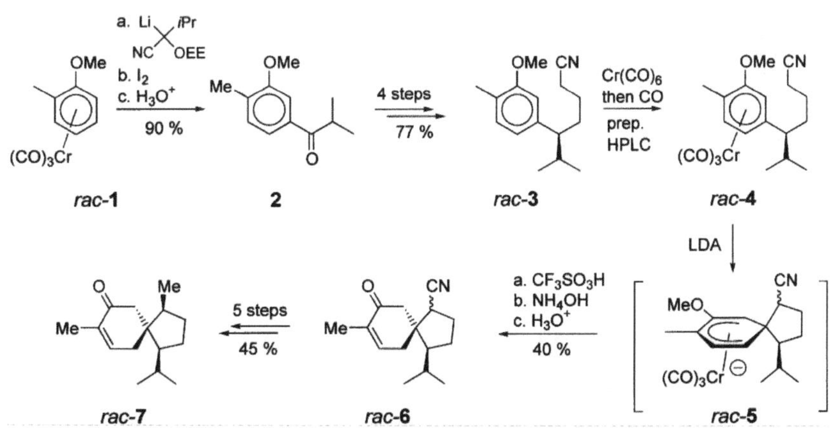

Scheme 1 Total synthesis of (±)-acorenone B (*rac*-7) according to Semmelhack

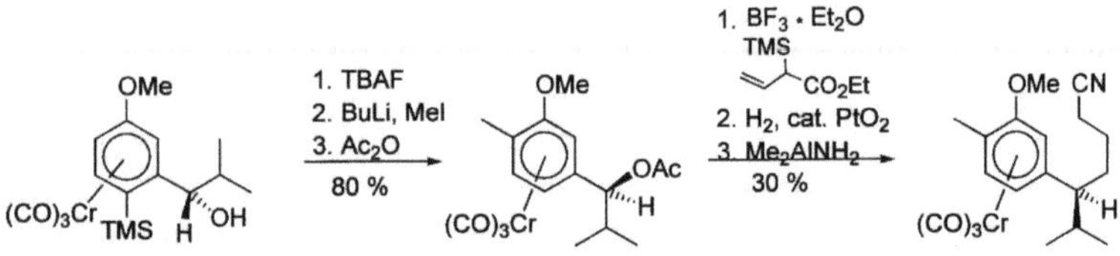

rac-**8** *rac*-**9** *rac*-**4**

Scheme 2 Preparation of the acorenone B intermediate *rac*-**4** according to Uemura

rac-**4** (d.r.=60:40). This problem was tackled by Uemura et al., who developed an alternative access to this key intermediate as shown in Scheme 2 [15]. Here, complex *rac*-**8** serves as a building block, which is diastereoselectively obtained by OH-directed complexation. After removal of the TMS-group, regioselective *ortho*-methylation and *O*-acetylation leads to *rac*-**9**, which on treatment with a Lewis acid in the presence of an allyl silane (to trap the intermediate carbocation) stereospecifically affords an alkylated product. Double-bond hydrogenation and interconversion of the ester into a nitrile functionality then yields the acorenone intermediate *rac*-**4** as a single diastereomer.

Hydroxylated aromatic calamenenes represent a group of structurally rather simple sesquiterpenes, which exhibit interesting biological activities, for instance, as anti-infective compounds. Here, arene chromium chemistry offers some highly useful opportunities for the stereo-controlled set-up of the two sidechains in benzylic position of the tetraline scaffold, as was demonstrated by Uemura and coworkers in the synthesis of (racemic) *cis*- and *trans*-7-hydroxycalamenenes *rac*-**13** and *rac*-**15**, respectively (Scheme 3) [3, 16, 17]. Starting from *rac*-**10**, the tetralone complex *rac*-**11** is obtained via (diastereoselective) Friedel-Crafts cyclization. While *trans*-configured products are accessible from *rac*-**11** by ionic hydrogenation of alcohols (e.g., *rac*-**12**) resulting from alkyl-lith-

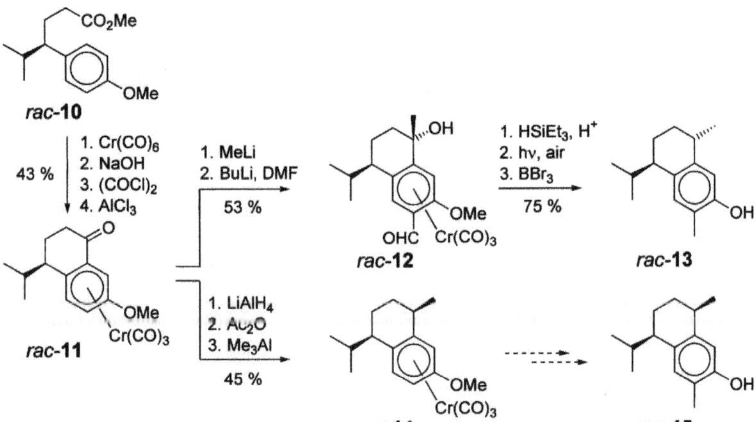

Scheme 3 Synthesis of (racemic) *cis* and *trans* 7-hydroxycalamenes according to Uemura

ium addition, *cis*-products arise from reaction of the acetylated reduction products with trimethyl aluminum. In both cases, an intermediate benzylic cation is trapped from the less hindered face of the ligand. These early syntheses, however, suffer from the fact that they open access only to racemic products.

Possibly the first application of non-racemic (planar-chiral) arene chromium complexes as synthetic building blocks is the *enantioselective* total synthesis of anti-infective 7,8-dihydroxycalamenenes reported by Schmalz and co-workers in the early 1990s (Scheme 4) [18, 19]. The key building block **18** is prepared in >99% *ee* from simple 5,6-dimethoxytetralone **16** in an efficient three step sequence. First, enantioselective reduction using the CBS protocol affords the corresponding tetralol (>99% ee after recrystallization) [20] which is then diastereoselectively complexed (d.r.≥9:1) to afford the *endo* tetralol complex **17** as a pure diastereomer after simple flash filtration in high yield. Of course, this compound is also accessible with opposite absolute configuration (*ent*-**17**) if the enantiomeric CBS-catalyst is used [21]. The OH-function, which served as a directing group in the complexation step, is then either oxidized (see below) or removed by ionic hydrogenation to give enantiomerically pure **18** in high yield.

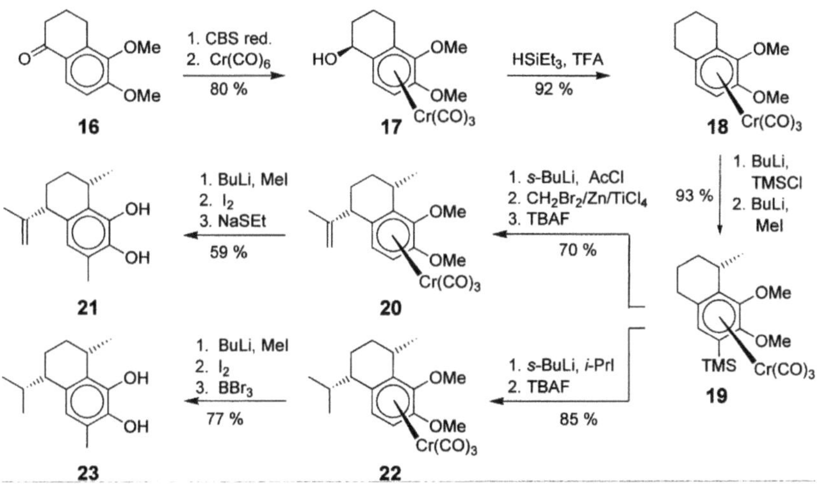

Scheme 4 Synthesis of 7,8-dihydroxycalamenenes **21** and **23** according to Schmalz

The further synthesis of (1*S*,4*S*)-7,8-dihydroxy-11,12-dehydrocalamenene (**21**) and its saturated analog **23** is now accomplished in a highly efficient manner: After protection of the most acidic aryl position through silylation, benzylic deprotonation/methylation proceeds with complete regio- and diastereoselectivity to afford complex **19**. The second benzylic alkylation is then achieved by deprotonation of **19** (with *sec*-butyllithium) and reaction with an appropriate alkyl- or acyl halide. After desilylation, aromatic methylation, oxidative decomplexation (iodine) and ether cleavage the desired target molecules **21** and **23**, respectively, arise as pure enantiomers in a high overall yield.

Remarkably, benzylic stereo-control is also possible in the "acyclic" series due to the significant configurative stability of the reactive intermediates. An example is the short enantioselective synthesis of (+)-ar-curcumene (**27**) (Scheme 5) [22]. The key step involves an electron transfer-mediated benzylic umpolung of the (central-chiral) complex **25** by means of the one electron reducing agent LiDBB (lithium di-*tert*-butyl diphenyl). Obviously, both the formation of the (planar-chiral) reactive intermediate **26** and the subsequent alkylation proceed in a stereo-controlled manner with overall retention of configuration.

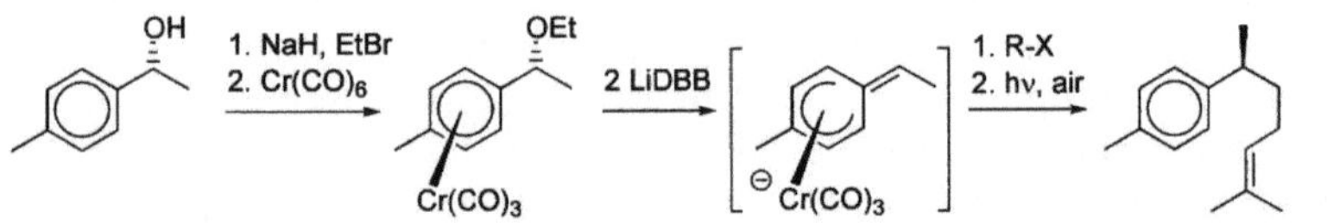

Scheme 5 Synthesis of (+)-ar-curcumene **27** according to Schmalz

3 Synthesis of Diterpenes

A group of challenging target molecules with important biological activities are the pseudopterosins [23] and the *seco*-pseudopterosins [24], marine diterpene glycosides with strong anti-inflammatory properties (Fig. 3).

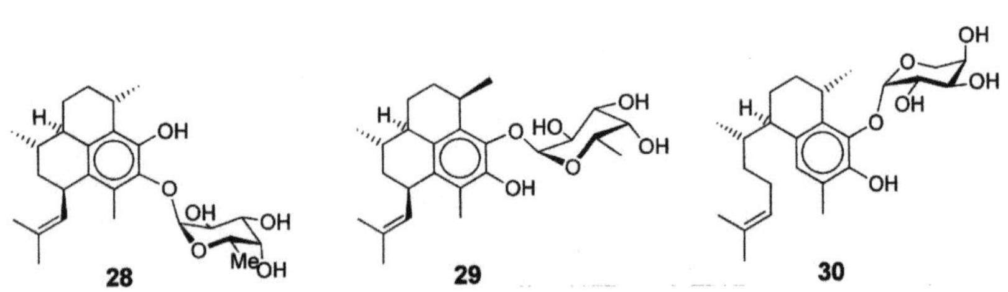

Fig. 3 Structure of pseudopterosin E (**28**), pseudopterosin G (**29**), and *seco*-pseudopterosin A (**30**)

As demonstrated by Schmalz and coworkers, arene chromium chemistry offers unique and highly efficient entries to the aglycones of such compounds. The most successful approach [25, 26] follows the retrosynthetic analysis shown in Scheme 6, where the pseudopterosin aglycone (**31**) derives from a *seco*-compound of type **32**. Such intermediates can be traced back (via **33**) to the planar-chiral complex **34** carrying the absolute stereochemical information.

The elaborated synthetic route is shown in Scheme 7 [25, 26]. Treatment of the enantiopure building block **34** (obtained from **17** by oxidation) with an excess of isopropenyl lithium results in diastereoselective nucleophilic addition to

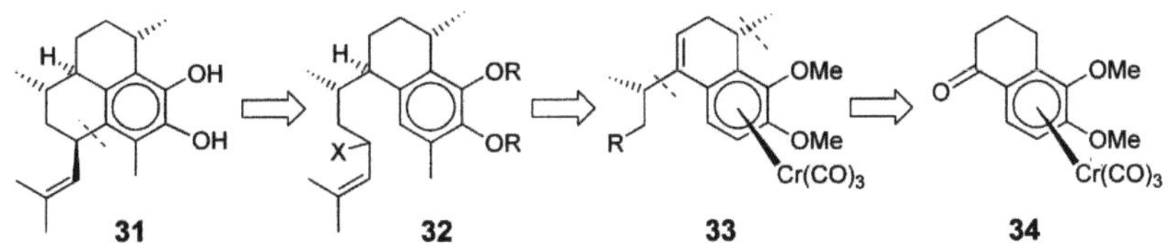

Scheme 6 Retrosynthetic analysis for the pseudopterosin aglycone **31** according to Schmalz

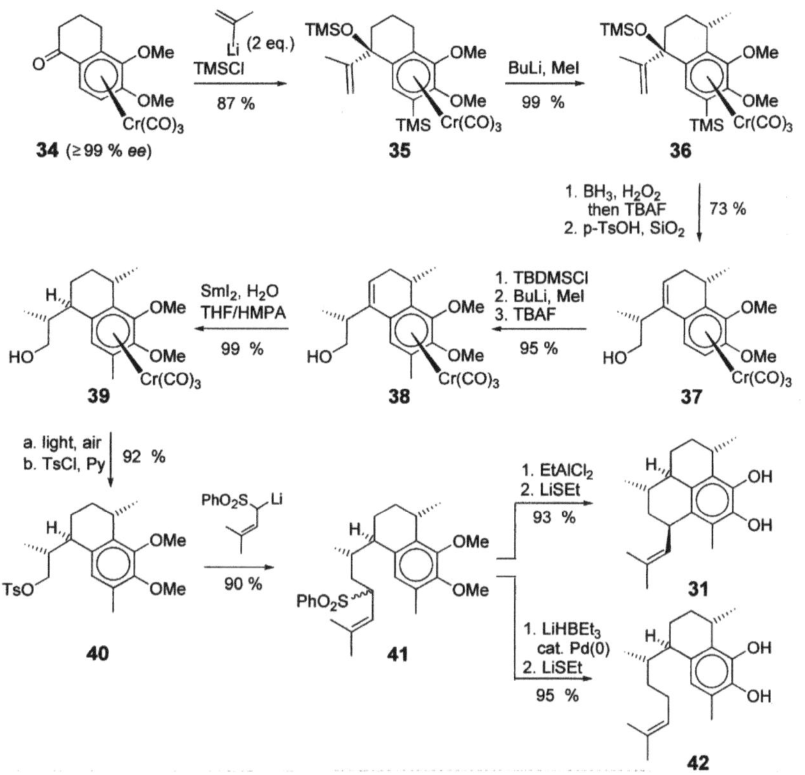

Scheme 7 Total synthesis of the pseudopterosin aglycone **31** and the *seco*-pseudopterosin aglycone **42** according to Schmalz

the ketone and in regioselective *ortho*-lithiation. On quenching the di-anionic intermediate with TMS-Cl complex **35** is obtained in high yield. The subsequent regio- and diastereoselective deprotonation/methylation proceeds particularly smoothly to give **36**, which is then converted by diastereoselective hydroboration, desilylation and selective elimination of the benzylic OH group to the olefin **37**. Aromatic methylation is then achieved under temporary protection of the primary OH-group. Unexpectedly, the aspired hydrogenation of intermediate **38** cannot be realized under common conditions. However, treatment of **38** with SmI_2 in THF/HMPT in the presence of water [27] gives the desired product **39** in almost quantitative yield. After oxidative decomplexation, activation of the OH-function by tosylation and alkylation of the obtained tosylate (**40**) with a lithiated sulfone affords the key-intermediate **41**. This is diastereoselectively transformed to the pseudopterosin aglycone **31** by $EtAlCl_2$-mediated cyclization (d.r.≥16:1) followed by nucleophilic cleavage of the methyl ethers. Alternatively, Pd-catalyzed allylic reduction with $LiBHEt_3$ gives access to the *seco*-pseudopterosin aglycone **42**.

The overall synthesis (Scheme 7) is in excellent competition in terms of yield and selectivity. The expenditure connected to the (stereoselective) introduction of the metal fragment in the beginning pays off tremendously, as both the chemical and stereochemical effects of the $Cr(CO)_3$ unit are exploited in several key transformations. Actually, all new (lasting) stereocenters are established with virtually complete diastereoselectivity under the influence of the planar-chiral complex substructure.

A different strategy was used by the same group in a synthesis of the 18-nor-*seco*-pseudopterosin aglycone (**47**), also starting from the chiral chromium complex **34** (Scheme 8) [28]. First, an exocyclic double bond is generated through Peterson olefination (→**43**). After protection of the acidic aromatic *ortho*-position by silylation and benzylic deprotonation/methylation, the key intermediate (**44**) undergoes conjugate addition of homoprenyllithium. The resulting benzylic anion is diastereoselectively protonated during aqueous work-up to afford the desired *trans*-configured product (d.r.=10:1). Following the established methodology, complex **45** is converted into the target structure in only three further steps. The whole synthetic sequence (**34**→**47**) consists of less than ten steps and affords the desired aglycone in a striking overall yield of 51%.

The helioporins, a group of marine diterpenes with antiviral and cytotoxic properties, are structurally closely related to the (*seco*-) pseudopterosins and all possess a characteristic benzodioxole unit [29]. Their stereostructure was assigned as shown in Fig. 4 with a *cis*-relationship of the two benzylic sidechains at the tetralin nucleus.

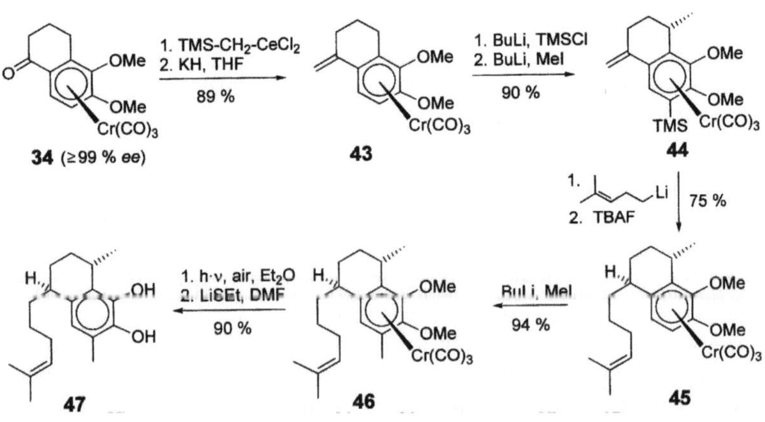

Scheme 8 Synthesis of the 18-nor-*seco*-pseudopterosin aglycone **47** according to Schmalz

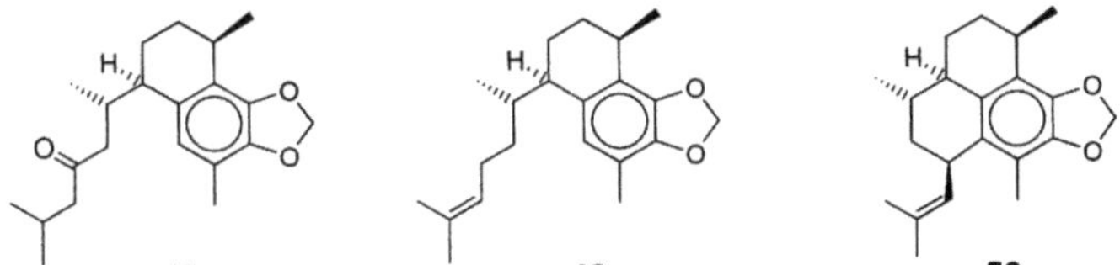

Fig. 4 Structure of helioporin B (**48**), helioporin D (**49**), and helioporin E (**50**) as assigned by Higa

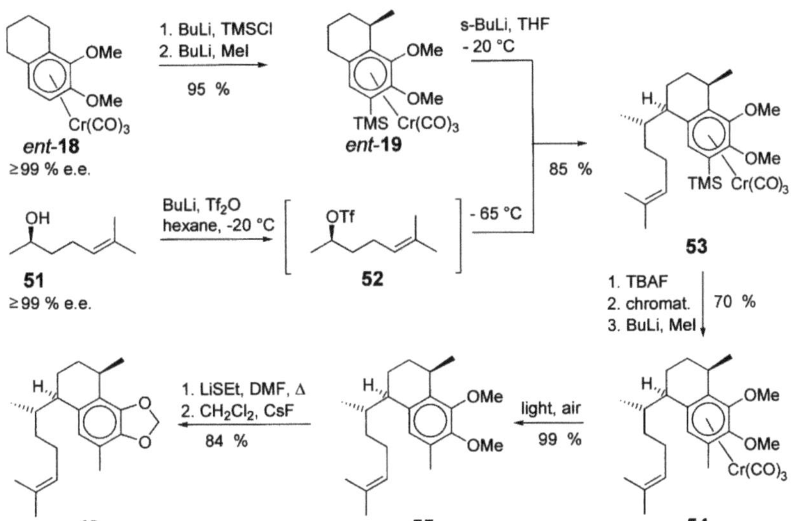

Scheme 9 Total synthesis of putative helioporin D (**49**) according to Schmalz

The first synthesis of a member of this class of marine diterpenes was disclosed by Schmalz and coworkers in 1998 (Scheme 9) [30]. Following a strategy related to the one used before in the synthesis of *cis*-calamenenes (see also Scheme 4), 5,6-dimethoxytetralin-$Cr(CO)_3$ (*ent*-**18**, ≥99% e.e.) is employed as a chiral scaffold and all sidechains are introduced by regio- and diastereoselective alkylation. In a key-step, the benzylic anion derived from *ent*-**19** is alkylated with the sensitive triflate **52** prepared in-situ from (*R*)-sulcatol (**51**). This stereoconvergent process affords the coupling product in high yield (d.r.=92:8). After desilylation the undesired diastereomer is removed by chromatography. The sequence is concluded by aromatic methylation (→**54**), decomplexation (→**55**), ether cleavage and dioxol formation. Thus, only eight linear steps are necessary for the conversion of the chiral building block *ent*-**18** into the target molecule **49** (putative helioporin D) in 47% overall yield.

Interestingly, the NMR data of compound **49** did not match those of the natural product helioporin D. This observation led to the revision of the stereostructure of helioporin D (and consequently of all other helioporins), which actually has the same configuration as the *seco*-pseudopterosin aglycone **42** [31, 32]. This was proven by conversion of **42** (Scheme 7) into natural helioporin D by benzodioxol formation [31].

During a recent attempt to synthesize helioporin B (*epi*-**48**) following a strategy related to the one shown in Scheme 8, Dehmel and Schmalz observed an unexpected *endo*-selectivity of the conjugate nucleophilic addition and obtained only 11-*epi*-helioporin B [33]. However, as the relative configuration of this compound corresponds exactly to the one of natural serrulatane diterpenes, advantage was taken of the unexpected finding in a concise synthesis of (+)-20-methoxy-serrulat-14-en-7,8-diol **60** (Scheme 10) [34]. Starting from the chiral building block *ent*-**34** (*E*-selective) ethylidenation of the ketone was achieved by addition of a vinyl-cerium reagent and subsequent ionic hydrogenation under allylic rearrangement. After silylation (→**56**), benzylic methoxymethylation and aromatic methylation, the key intermediate **57** was obtained as a pure isomer. Treatment of **57** with lithio-methylphenylsulfone in a 1,4-dioxane/HMPA mixture proceeded with high *endo* selectivity. After oxidative decomplexation, the conjugate addition product **58** is obtained, which is finally converted to the target compound **60** in three additional steps.

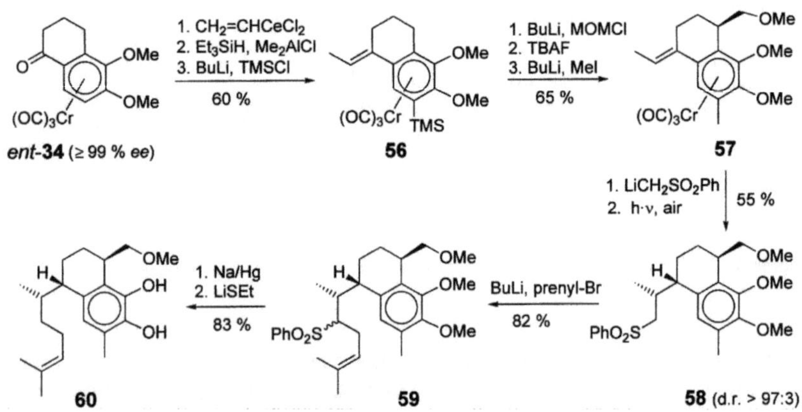

Scheme 10 Total synthesis of the serrulatane diterpene **60** according to Schmalz

Even carried out only in the racemic series, Uemura's synthesis of (±)-dihydroxyserrulatic acid (*rac*-**66**) [35, 36] should be mentioned here, because it nicely demonstrates another option to control the relative configuration of the stereocenters in serrulatenes (Scheme 11). Starting from the mono-ketal of dihydro-1,4-naphthoquinone, complex *rac*-**61** was prepared and converted into the acetate *rac*-**62** by diastereoselective reduction and acetylation. On treatment of this compound with crotyltrimethylsilane in the presence of a Lewis acid, benzylic alkylation proceeded with a reasonable diastereoselectivity (d.r.=3:1) to afford *rac*-**63**. The ketone function was now used to introduce an *endo*-methyl substituent by addition of methyl lithium followed by ionic hydrogenation. Having thus

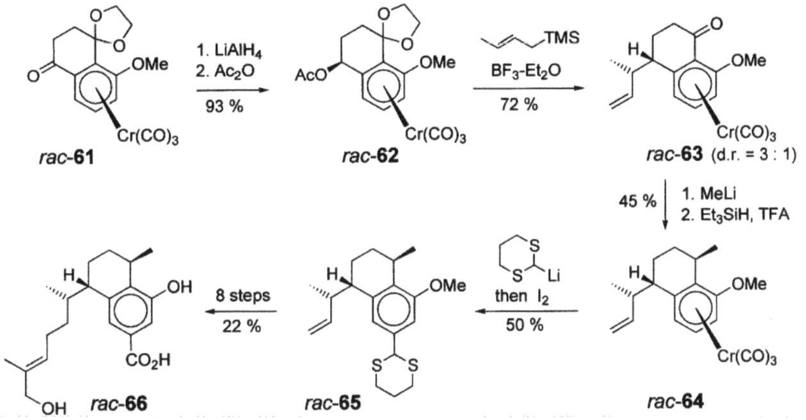

Scheme 11 Synthesis of (±)-dihydroxyserrulatic acid (*rac*-**66**) according to Uemura

established all three stereocenters (*rac*-**64**), the missing C_1 substituent is introduced by (*meta*-selective) nucleophilic addition. After oxidative work-up the decomplexed product *rac*-**65** is obtained, which is finally converted to the target compound in another eight steps.

4 Synthesis of Alkaloids

Semmelhack and his coworkers investigated the use of arene chromium chemistry for the regioselective functionalization of indoles. In a formal synthesis of teleocidin A [37] the sidechain at C-7 is regioselectively introduced by nucleophilic addition employing the $Cr(CO)_3$-complexed 4-aminoindole derivative *rac*-**67** and a cyano-stabilized lithioalkane (Scheme 12). After oxidative quench the desired C-7 alkylated compound (*rac*-**68**) is obtained in good yield. Reduction of the cyano-group to the corresponding aldehyde, Wittig-methylenation and removal of the SEM protecting group gives a 4-(methylamino)indole, which is subjected to a S_N2-type N-alkylation employing an enantiopure triflate to set up the l-valine part of the target structure. The indole derivative **69** (mixture of diastereomers) thus obtained had previously been converted to teleocidin A [38].

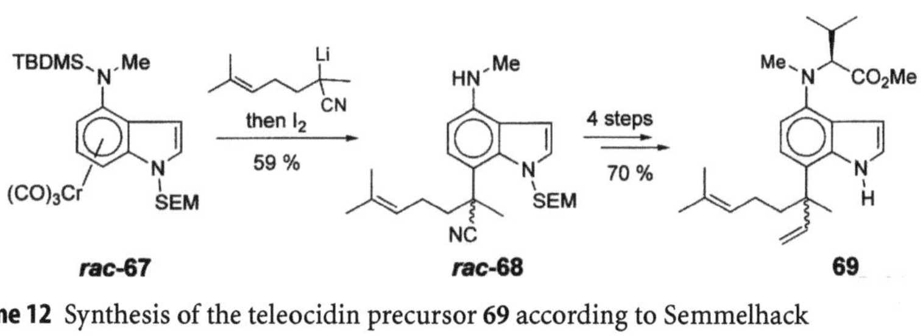

Scheme 12 Synthesis of the teleocidin precursor **69** according to Semmelhack

Another example for applied indole-$Cr(CO)_3$ chemistry from the Semmelhack laboratory uses an intramolecular variant of the well-established nucleophilic addition-oxidation methodology in the synthesis of clavicipitic alcohol (**73**) (Scheme 13) [39]. The synthesis starts from natural l-tryptophane, which is first converted to the oxazolidinone **70** by $LiAlH_4$-reduction and treatment with phosgene. Complexation of **70** leads to a diastereomeric mixture, which can be separated at the stage of **71**, i.e., after selective TBDPS-protection of the indole nitrogen and prenylation of the oxazolidinone nitrogen. In the key step of the sequence, deprotonation of **71** with LDA triggers intramolecular nucleophilic addition to afford the cyclized product **72** after oxidative quench with iodine. Finally, deprotection of the indole nitrogen and hydrolysis of the oxazolidinone gives clavicipitic alcohol (**73**).

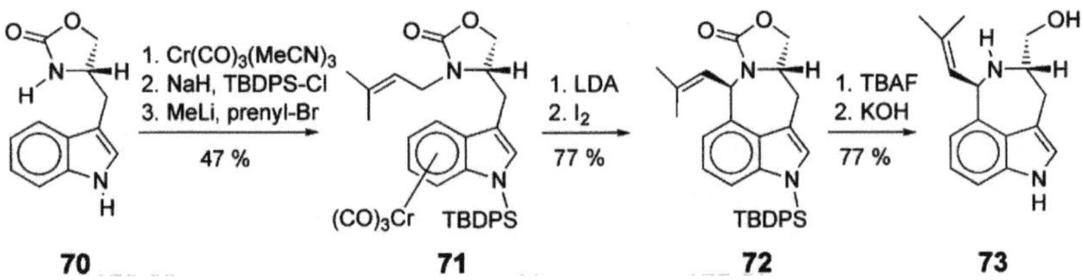

Scheme 13 Synthesis of clavicipitic alcohol (73) according to Semmelhack

An enantioselective total synthesis of (+)-ptilocaulin (**79**), a marine alkaloid with high antimicrobial and cytotoxic activity, was reported by Schmalz (Scheme 14) [40, 41]. The synthesis starts from anisole-$Cr(CO)_3$, which is converted to the planar-chiral building block **74** with ≥99% *ee* by enantioselective deprotonation/silylation [42] and subsequent recrystallization. After attachment of a 2-butenyl side-chain (→**75**) [43], nucleophilic addition of 2-lithio-1,3-

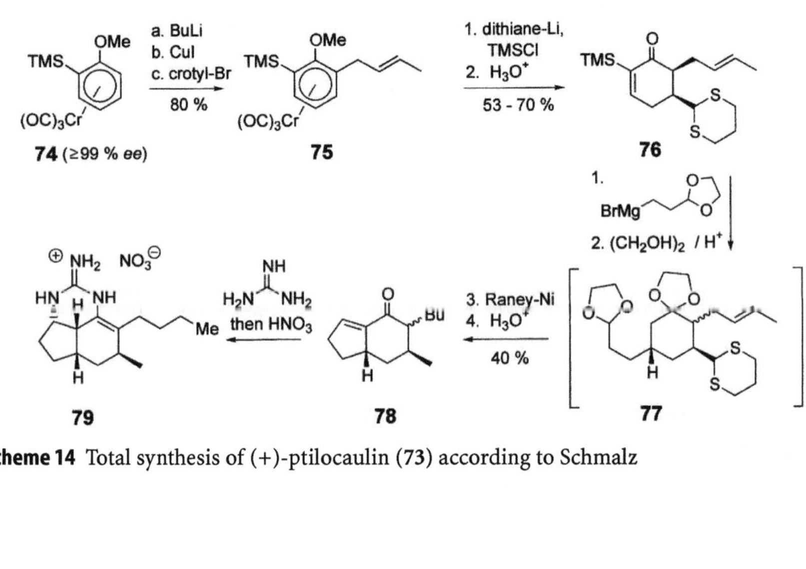

Scheme 14 Total synthesis of (+)-ptilocaulin (73) according to Schmalz

dithiane leads to the cyclohexenone **76** in good yield, but only if the crude reaction mixture is treated with TMSCl prior to hydrolysis. This remarkable dearomatization reaction proceeds with complete chirality transfer. The enantiopure Michael acceptor **76** is then transformed into the known ptilocaulin precursor **78** by a 4-step sequence consisting of diastereoselective 1,4-addition, ketal formation (→**77**), ultrasound-assisted desulfurization/hydrogenation (Raney-Ni) and aldol cyclization. In this synthesis, the $Cr(CO)_3$-unit assists in two crucial C-C bond-forming steps while the absolute stereochemical information stored in the planar-chiral building block (**74**) is efficiently exploited for the set-up of the lasting stereocenters.

In 1999, Kündig et al. disclosed a highly efficient synthesis of (-)-lasubine (**85**), a phenylquinolizidine from the *Lythraceae* family of alkaloids (Scheme 15) [44]. This synthesis is remarkable because mayor parts of it are carried out at the $Cr(CO)_3$-complexed ligand and several new bonds are formed with high diastereoselectivity under the stereochemical influence of the planar-chiral π-complex moiety. The chiral building block **80** is obtained in enantiomerically pure form either by resolution (via imine formation with l-valinol) or by diastereoselective complexation of a chiral cyclic aminal according to the procedure of Alexakis [45]. After preparation of the imine **81**, a highly diastereoselective aza-Diels-Alder reaction using Danishefsky's diene affords dihydropyridinone **82**. The alcohol function is converted into a bromide (→**83**) in order to prepare for the key radical cyclization, which gives rise to the quinolizidinone **84** in high yield. Diastereoselective reduction, desilylation and decomplexation finally affords the target molecule **85** in a straightforward manner. The Kündig lasubine synthesis demonstrates the high stereodirecting power of the $Cr(CO)_3$ unit as well as the compatibility of arene-$Cr(CO)_3$ complexes with a variety of reaction conditions.

There are a number of additional examples where arene chromium chemistry has been successfully employed in the synthesis of organic nitrogen compounds related to natural products[46–49]. For instance, Davies has demonstrated that

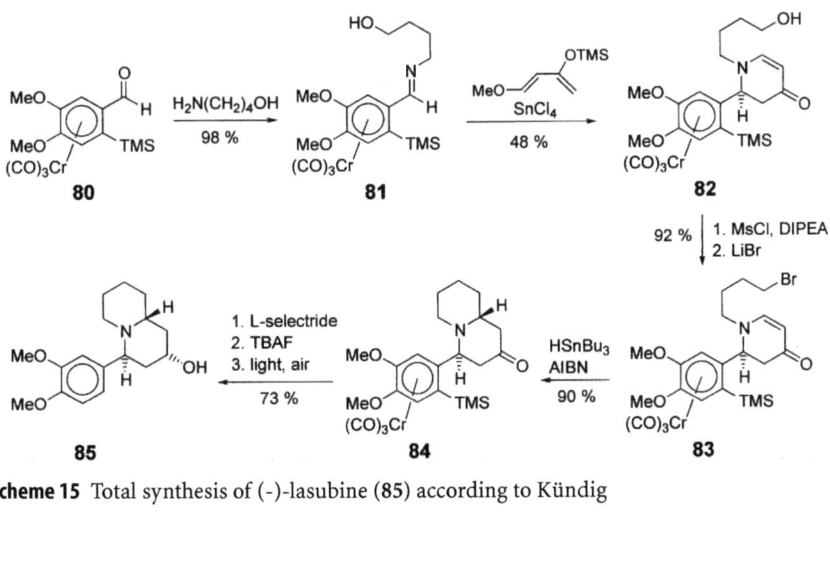

Scheme 15 Total synthesis of (-)-lasubine (**85**) according to Kündig

the stereospecific formation and configurational stability of $Cr(CO)_3$-complexed benzylic cations (such as **87**) offers valuable entries towards pharmacologically relevant "benzazepine alkaloids" [50]. The example given in Scheme 16 shows the high efficiency of the HBF_4-initiated cyclization of enantiopure **86** to afford **88** with virtually complete retention of configuration.

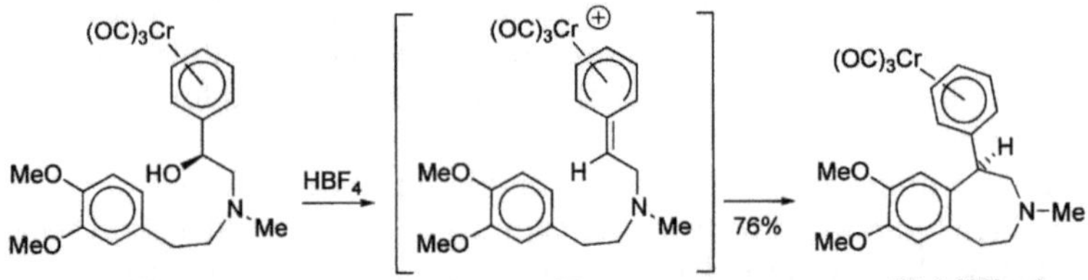

Scheme 16 Stereospecific synthesis of the benzazepine derivative **88** according to Davies

5 Synthesis of Compounds with Axial Chirality

Pioneering work of Uemura and coworkers has demonstrated that the formation of biaryl systems by Suzuki-coupling reactions of planar-chiral halobenzene-$Cr(CO)_3$ complexes with "arylmetals" may proceed with high diastereoselectivity, thus opening a very attractive method for the stereocontrolled synthesis of axially chiral biaryls [51, 52]. By proper choice of reaction conditions, both biaryl atropisomers may selectively be accessed from the same planar-chiral starting complex. These opportunities have inspired two research groups to develop a synthesis for the anti-leucaemic lignane (-)-steganone (**89**), based on the general strategy sketched in Scheme 17, where a chiral complex of type **92** is coupled to a boronic acid **91** to give **90** as a key intermediate.

Following this concept, Uemura and coworkers reported in 1995 a formal synthesis of (-)-steganone (**89**) (Scheme 18) [53, 54]. First, complex **93** is obtained from 3,4,5-trimethoxybenzaldehyde by protection with (*S*)-(-)-butane-1,2,4-triol, methylation of the free OH group and complexation. Diastereoselec-

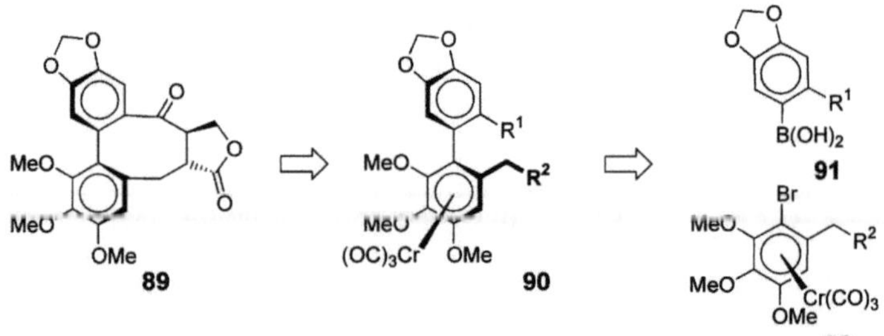

Scheme 17 Retrosynthetic analysis for steganone (**89**) according to Uemura and Molander

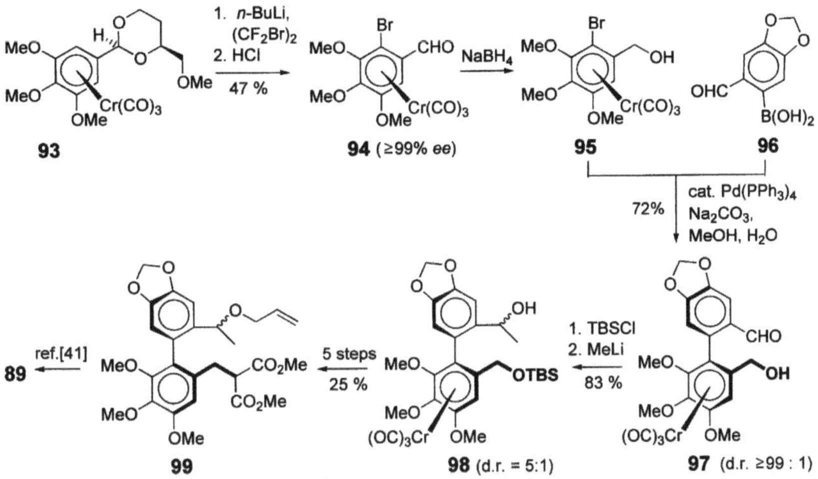

Scheme 18 Formal total synthesis of (-)-steganone (**89**) according to Uemura

tive *ortho*-lithiation using *n*-BuLi in a non-coordinating solvent (toluene) followed by electrophilic bromination and subsequent removal of the chiral auxiliary by acidic hydrolysis then affords the planar-chiral building block **94** with >99% enantiopurity after recrystallization. Borohydride reduction gave the alcohol **95**, which was coupled with the *o*-substituted arylboronic acid **96** to give the mono $Cr(CO)_3$-complexed biaryl **97** as a pure diastereomer. After protection of the OH-group, addition of methyl lithium proceeds with acceptable diastereoselectivity to give **98**. To complete the formal synthesis, **98** is converted into the malonate **99**, a compound which had previously served as an advanced intermediate in a synthesis of (-)-steganone (**89**) [55].

Five years later, another total synthesis of (-)-steganone (**89**) exploiting arene-$Cr(CO)_3$ chemistry was disclosed by Molander and coworkers[56]. This synthesis (Scheme 19) applies Uemura's method for the stereoselective construction of the diphenyl derivative **97** from the planar-chiral complex **95** but then provides a highly elegant and efficient solution for the conversion of **97** into the target molecule in only four further steps. After preparation of the bromide **100**, a butenolide unit is attached by Stille coupling to give complex **101**. Treatment of the latter with samarium(II) iodide in the presence of a proton source affords the hydroxy lactone **102** in good yield as a single diastereomer in a remarkable 8-*endo* radical cyclization. Oxidation/decomplexation of **102** and base-induced epimerization finally gives (-)-steganone (**89**).

Another important natural product containing an axially-chiral biaryl substructure is the antibiotic vancomycin **103**, which represents a tough challenge for chemical synthesis (Fig. 5) [57]. Again, arene chromium chemistry offers interesting opportunities for the stereoselective construction of the biaryl part, i.e., actinoidinic acid (**104**).

Recently, Uemura and coworkers reported a synthesis of the A-B ring system of vancomycin (Scheme 20) [58]. The planar-chiral building block **106** is pre-

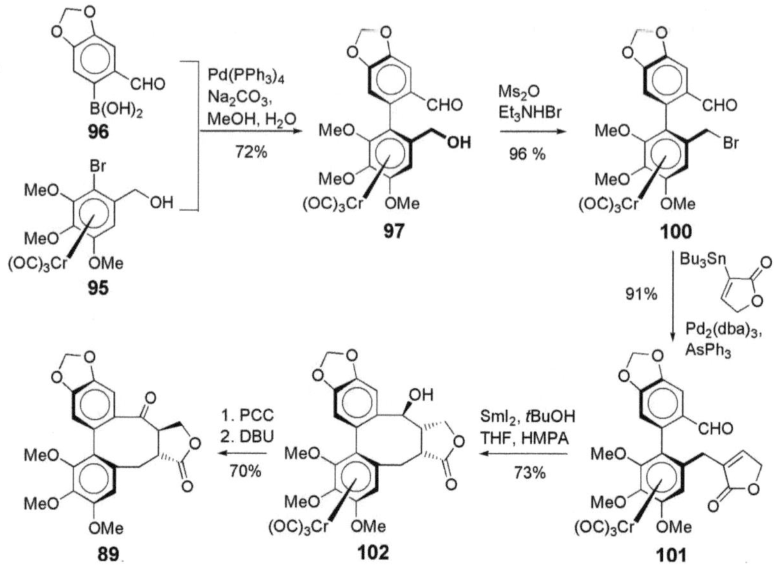

Scheme 19 Total synthesis of (-)-steganone (**89**) according to Molander

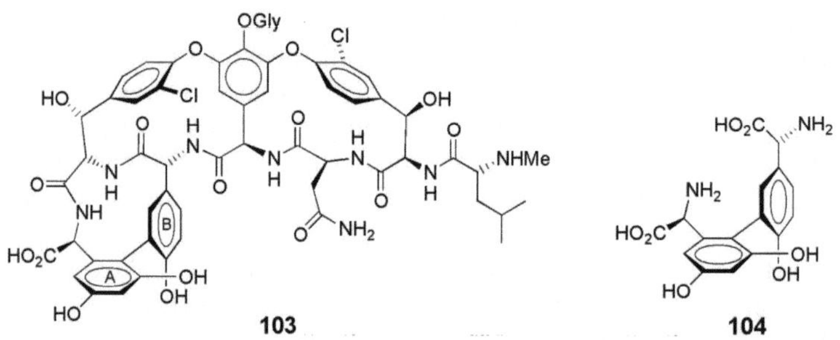

Fig. 5 Structure of vancomycin (**103**) and its axially-chiral AB-subunit actinoidinic acid (**104**)

pared (similar to **94**) by diastereoselective deprotonation/bromination of the chiral acetal **105**. Complex **106** is then first transformed into the derivative **107**, before diastereoselective Suzuki coupling with the boronic acid **108** affords the biaryl derivative **109**. Decomplexation and S_N2 substitution of the hydroxyl for a nitrogen functionality (azide) gives **110**, which is converted into the known vancomycin intermediate **111** [59] in a few further steps. Remarkably, the enantiomeric starting complex (*ent*-**106**) could be transformed into the same product enantiomer following a modified route.

An alternative approach toward the vancomycin AB system was described by Wilhelm and Widdowson in a synthesis of the *ent*-actinoidinic acid derivative **118** (Scheme 21) [60]. These authors start from 3,5-dimethoxy benzaldehyde (**112**) to prepare first complex **113**. The latter serves as a substrate for an enantio-

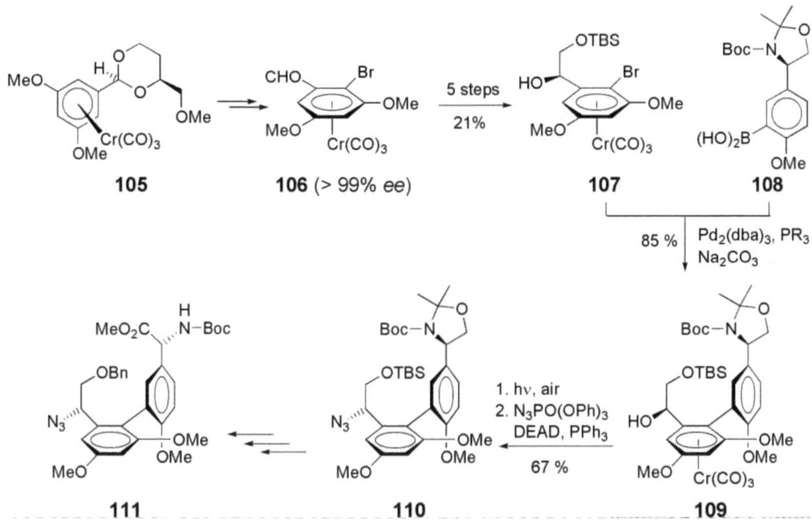

Scheme 20 Synthesis of the actinoidic acid derivative **111** according to Uemura

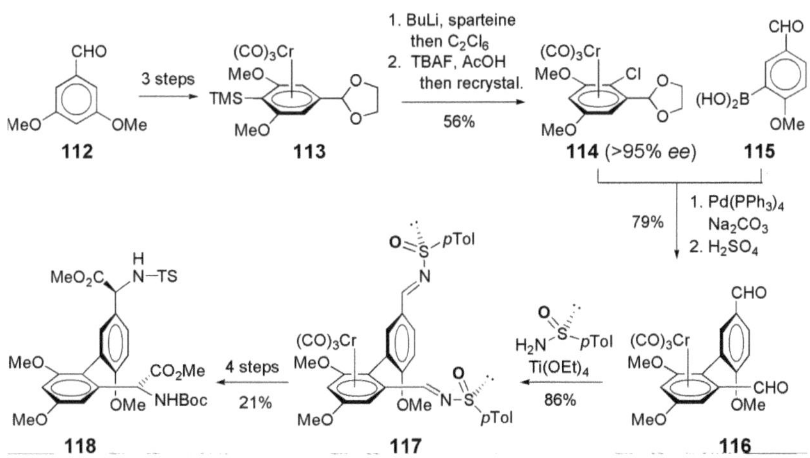

Scheme 21 Synthesis of the *ent*-actinoidinic acid derivative **118** according to Widdowson

selective deprotonation/chlorination reaction affording the planar-chiral complex **114** with >95% *ee* after recrystallization. The biraryl bond is then set-up again by Suzuki coupling to give the dialdehyde **116**, which is converted into the bis-sulfoximine **117**. These additional chiral groups are then exploited for the (diastereoselective) establishment of the benzylic stereocenters by means of Strecker reactions.

Uemura also applied his methodology in the stereoselective synthesis of korupensamines (Scheme 22) [54, 61, 62], biaryl isoquinoline alkaloids exhibiting

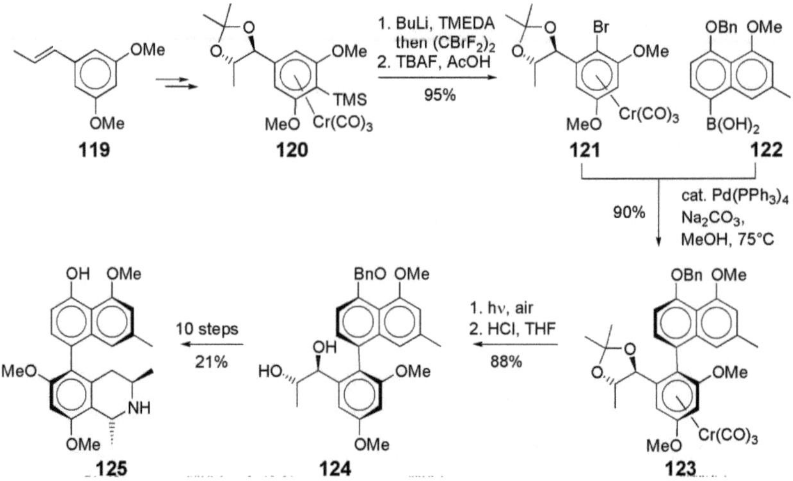

Scheme 22 Synthesis of *O,O*-dimethylkorupensamine **125** according to Uemura

promising antimalarial and anti-HIV activities. Complex **120**, obtained in non-racemic form via Sharpless asymmetric dihydroxylation of **119** followed by ketal formation, complexation and silylation undergoes diastereoelective *ortho*-lithiation to allow the preparation of the brominated derivative **121**. While *o*-formyl phenylboronic acids generally give rise to thermodynamically more stable *anti*-coupling products (see Schemes 18–21), the Suzuki coupling of **121** with the 1-naphthyl boronic acid derivative **122** proceeds under kinetic control to give the twisted biaryl **123** as a sole stereoisomer, in which the naphthyl unit and the metal fragment are oriented *syn* to each other. Decomplexation and subsequent transformations complete the synthesis of *O,O′*-dimethylkorupensamine (**125**), which is obtained in respectable 15% overall yield (16 steps) from the building block **120**.

6 Additional Aspects and Conclusions

So far, quite a number of natural product syntheses have been discussed, which all are *strategically* based on arene chromium chemistry. Of course, in certain cases, arene metal chemistry also offers superior *tactical* solutions for the realization of more or less "classical" strategies, which do not per se depend on the metal moiety. An example is the Iwata synthesis of macrocarpal C (**131**), an interesting acylphloroglucinol derivative from an *Eucalyptus* species (Scheme 23) [63, 64]. Here, the $Cr(CO)_3$-complexed building block **128** proved superior for the stereospecific alkylation of the terpenoid building block **129** under S_N1-type conditions (retention of configuration) to give **130** after oxidative decomplexation using ceric ammonium nitrate (CAN). For the preparation of **128** the facilitated deprotonation of complexed arenes could additionally be exploited (**126**→**127**).

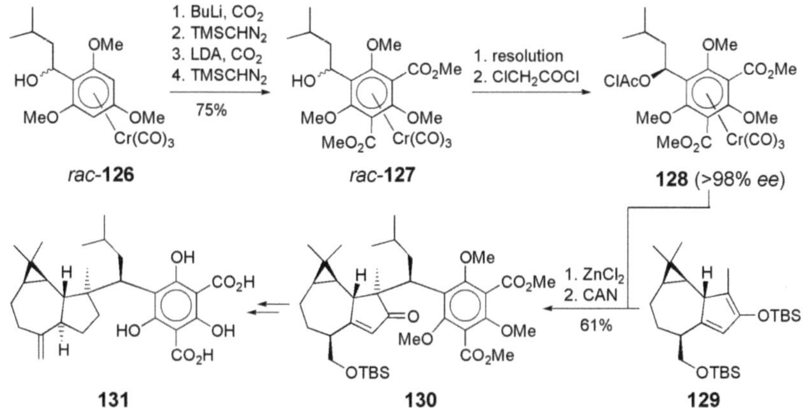

Scheme 23 Synthesis of macrocarpal C (**131**) according to Iwata

Chromium complexes of *ortho*-substituted benzaldehydes (and derivatives) are frequently used as chiral equivalents for the corresponding (achiral) aromatic building blocks, because addition reactions to such compounds usually proceed with a high degree of diastereoselectivity (compare Scheme 15). A chiral substitute for benzaldehyde itself is the *ortho*-silylated complex **132** which, for instance, undergoes highly stereoselective aldol addition reactions. The optically active aldol product **133** was used by Hanaoka and coworkers as a building block in the total synthesis of (+)-goniofufurone (**134**) and the taxol side chain analog **135** (Scheme 24) [65–68].

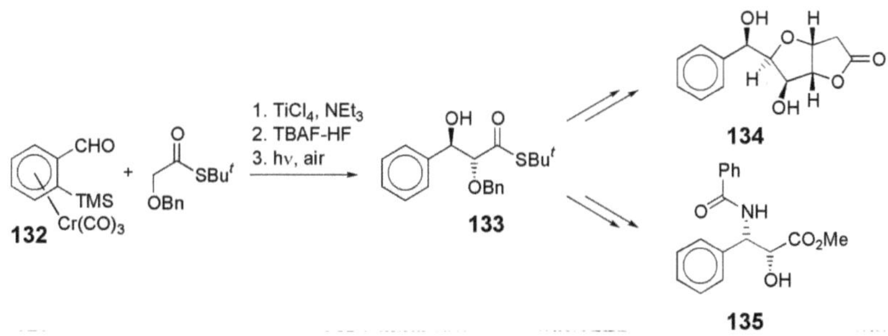

Scheme 24 Use of **132** as a chiral building block for the synthesis of (+)-goniofufurone (**134**) and the taxol side chain analog **135** according to Hanaoka

The synthetic potential of such *ortho*-silylated benzaldehyde complexes was recently further demonstrated by Moser et al. who developed an elegant and surprisingly efficient entry to spirocyclic compounds related to the antitumoral natural product fredericamycin A (**136**) (Fig. 6).

Reaction of **132** with the esterenolate **137** results in diastereoselective aldol addition, which is followed by a Brook rearrangement and cyclization to yield

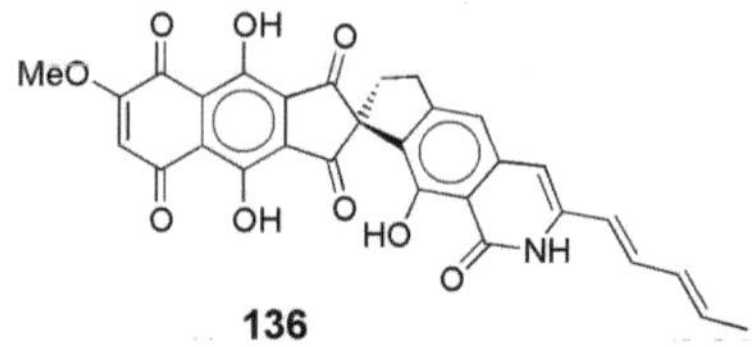

136

Fig. 6 Structure of fredericamycin A (**136**)

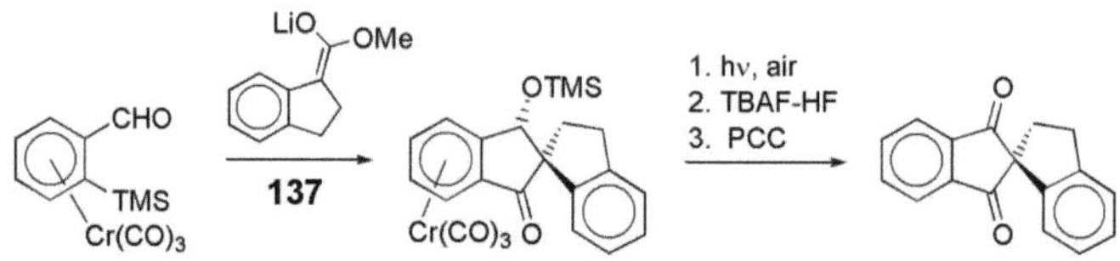

132 **138** **139**

Scheme 25 Synthesis of spiro-compounds related to fredericamycin according to Moser

the spirocyclic product **138** as a single diastereomer (Scheme 25) [69]. Thus, a key problem of any synthesis of fredericamycin has been solved in an astonishingly easy fashion applying arene chromium chemistry.

As mentioned above, glycopeptide antibiotics such as vancomycin (Fig. 5) pose some particularly difficult challenges for organic synthesis [70]. Besides controlling the stereochemistry of the biaryl unit, the formation of the diarylether bond(s) represents a severe difficulty [71]. Exploiting the finding that chloroarenes are strongly activated towards nucleophilic substitution by complex formation with electrophilic transition metal fragments [1,5], Pearson [72–74], Rich [75–76] and others [77] have developed highly useful techniques for diarylether synthesis using cationic ruthenium complexes.

An example is the synthesis of the fully functionalized DEF ring system of ristocetin A (**144**) reported by Pearson and Heo (Scheme 26) [78]. After complexation of the arylglycine derivative **140** the resulting product (**141**) [79] is fused with **142** by peptide coupling. On treatment of **143** with cesium carbonate as a base, the diarylether bond is smoothly formed by intramolecular S_NAr reaction. Decomplexation is finally achieved by light-induced ligand exchange to give the ristocetin DEF building block **144** in good overall yield.

In conclusion, η^6-arene transition metal complexes have demonstrated their unique potential for organic synthesis. In particular, planar-chiral η^6-arene-$Cr(CO)_3$ complexes are valuable building blocks for the diastereo- and enantioselective synthesis of complex natural products and related bioactive compounds. Highly original and competitive overall syntheses of various classes of natural products have been developed. The expenditure spent for the introduction of the metal fragment pays off especially in those cases, where the various chemical and stereochemical effects of the metal unit can be exploited in several subsequent transformations. Besides arene-$Cr(CO)_3$ complexes, cationic arene-RuCp complexes have also been applied in synthesis, especially as they allow for efficient arylether formation under mild conditions.

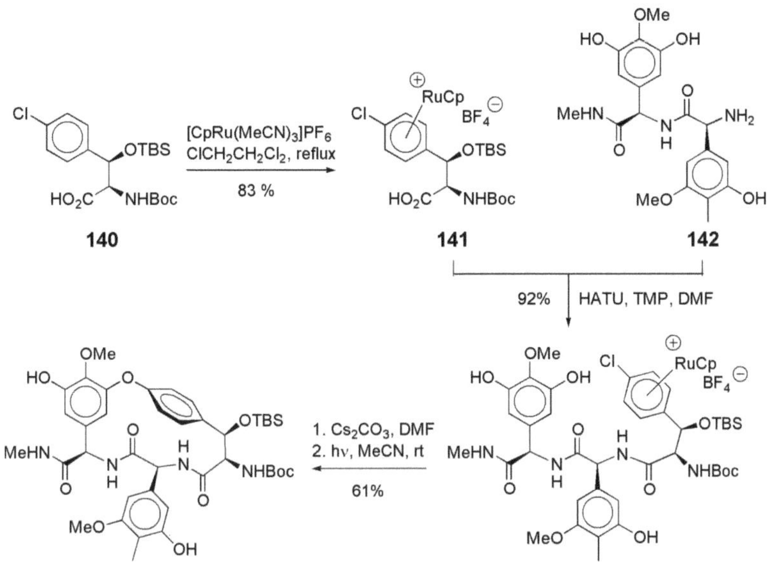

Scheme 26 Synthesis of the DEF ring system of ristocetin A (**144**) according to Pearson

It can be expected that arene metal chemistry will find many more applications in natural product synthesis in the future. Besides the (stereo-) selective functionalization of aromatic compounds, methods involving dearomatization reactions [4] possess a particular and not yet fully exploited potential.

References

1. Hegedus LS (1999) Transition metals in the synthesis of complex organic molecules, 2nd edn. University Science Books, Sausalito, CA, chap 10
2. Schmalz HG, Siegel S (1998) Chromium-arene complexes. In: Beller M, Bolm C (eds) Transition metals for organic synthesis, vol 1. Wiley-VCH, Weinheim, p 550
3. Uemura M (1991) Tricarbonyl(η^6-arene)chromium complexes in organic synthesis. In: Liebeskind LS (ed) Advances in metal organic chemistry, vol 2. Jai Press, London, p 195
4. Pape AR, Kaliappan KP, Kündig EP (2000) Chem Rev 100:2917
5. Semmelhack MF (1995) Transition metal arene complexes: nucleophilic addition. In: Abel EW, Stone FGA, Wilkinson G (eds) Comprehensive organometallic chemistry II, vol 12. Pergamon Press, New York, p 979
6. Davies SG, Donohoe TJ (1993) Synlett 323
7. Davies SG, Coote SJ, Goodfellow CL (1989) Synthetic applications of chromium tricarbonyl stabilized benzylic carbanions. In: Liebeskind LS (ed) Advances in metal organic chemistry, vol 2. Jai Press, London, p 1
8. Davies SG, McCarthy TD (1995) Transition metal arene complexes: side-chain activation and control of stereochemistry. In: Abel EW, Stone FGA, Wilkinson G (eds) Comprehensive organometallic chemistry II, vol 12. Pergamon Press, New York, p 1039

9. Semmelhack MF (1995) Transition metal arene complexes: ring lithiation. In: Abel EW, Stone FGA, Wilkinson G (eds) Comprehensive organometallic chemistry II, vol 12. Pergamon Press, New York, p 1017
10. For a review see: Carpentier JF, Petit F, Mortreux A, Dufand V, Basset JM, Thivolle-Cazat J (1993) J Mol Catal 81:1
11. Pfletschinger A, Dargel TK, Bats JW, Schmalz HG, Koch W (1999) Chem Eur J 5:537 (and references cited therein)
12. Schlögl K (1989) Stereochemistry of arenetricarbonylchromium complexes—useful intermediates for stereoselective synthesis. In: Werner H, Erker G (eds) Organometallics in organic synthesis 2. Springer, Berlin Heidelberg New York, p 63
13. Solladié-Cavallo A (1989) Chiral arene-chromium-carbonyl complexes in asymmetric synthesis. In: Liebeskind LS (ed) Advances in metal organic chemistry, vol 1. Jai Press, London, p 99
14. Semmelhack MF, Yamashita A (1980) J Am Chem Soc 102:5924
15. Uemura M, Kobayashi T, Minami T, Hayashi Y (1986) Tetrahedron Lett 27:2479
16. Uemura M, Isobe K, Take K, Hayashi Y (1983) J Org Chem 48:3855
17. Uemura M, Take K, Isobe K, Minami T, Hayashi Y (1985) Tetrahedron 41:5771
18. Schmalz HG, Hollander J, Arnold M, Dürner G (1993) Tetrahedron Lett 34:6259
19. Schmalz HG, Arnold M, Hollander J, Bats JW (1994) Angew Chem 106:77; Angew Chem Int Ed Engl 33:109
20. Schmalz HG, Millies B, Bats JW, Dürner G (1992) Angew Chem 104:640; Angew Chem Int Ed Engl 31:631
21. The term "CBS-reduction" refers to the enantioselective, oxazaborolidine-catalyzed borane reduction; for a review, see: Corey EJ, Helal C (1998) Angew Chem 110:2093; Angew Chem Int Ed Engl 37:1986
22. Schmalz HG, de Koning CB, Bernicke D, Siegel S, Pfletschinger A (1999) Angew Chem 111:1721; Angew Chem Int Ed Engl 38:1620
23. Roussis V, Wu Z, Fenical W, Strobel SA, Van Duyne G, Clardy J (1990) J Org Chem 55:4916
24. Look SA, Fenical W (1987) Tetrahedron 43:3363
25. Majdalani A, Schmalz HG (1997) Synlett 1303
26. Schmalz HG, Majdalani A, Geller T, Hollander J, Bats JW (1995) Tetrahedron Lett 36:4777
27. Schmalz HG, Siegel S, Bernicke D (1998) Tetrahedron Lett 39:6683
28. Majdalani A, Schmalz HG (1997) Tetrahedron Lett 38:4545
29. Tanaka JI, Ogawa N, Liang J, Higa T, Gravalos DG (1993) Tetrahedron 49:811
30. Geller T, Schmalz HG, Bats JW (1998) Tetrahedron Lett 39:1537
31. Geller T, Jakupovic J, Schmalz HG (1998) Tetrahedron Lett 39:1541
32. Höstermann D, Schmalz HG, Kociok-Köhn G (1999) Tetrahedron 55:6905
33. Dehmel F, Schmalz HG (2001) Org Lett 3:3579
34. Dehmel F, Lex J, Schmalz HG (2002) Org Lett 4:3915
35. Uemura M, Nishimura H, Minami T, Hayashi Y (1991) J Am Chem Soc 113:5402
36. Uemura M, Nishimura H, Hayashi Y (1990) Tetrahedron Lett 31:2319
37. Semmelhack MF, Rhee H (1993) Tetrahedron Lett 34:1399
38. Muratake H, Natsume M (1987) Tetrahedron Lett 28:2265
39. Semmelhack MF, Knochel P, Singleton T (1993) Tetrahedron Lett 34:5051
40. Schellhaas K, Schmalz HG (1996) Angew Chem 108:2277; Angew Chem Int Ed Engl 35:2146
41. Schellhaas K, Schmalz HG, Bats JW (1998) Chem Eur J 4:57
42. Ewin RA, MacLeod AM, Price DA, Simpkins NS, Watt AP (1997) J Chem Soc Perkin Trans 1:401
43. Semmelhack MF, Zask A (1983) J Am Chem Soc 105:2034
44. Ratni H, Kündig EP (1999) Org Lett 1:1997; (2000) Org Lett 2:1983
45. Alexakis A, Mangeney P, Marek I (1992) J Am Chem Soc 114:8288

46. Kündig EP, Xu LH, Romanens P, Bernardinelli G (1996) Synlett 270
47. Schinzer D, Abel U, Jones PG (1997) Synthesis 632
48. Davies SG (1990) J Organomet Chem 400:223
49. Sainsbury M, Mahon MF, Williams CS, Naylor A, Scopes DIC (1991) Tetrahedron 47:4195
50. Coote SJ, Davies SG, Middlemiss D, Naylor A (1990) Tetrahedron: Asymmetry 1:33
51. Kamikawa K, Uemura M (2000) Synlett 938
52. Kamikawa K, Watanabe T, Uemura M (2001) J Synth Org Chem Jap 59:1078
53. Uemura M, Daimon A, Hayashi Y (1995) Chem Commun 1943
54. Kamikawa K, Watanabe T, Daimon A, Uemura M (2000) Tetrahedron 56:2325
55. Meyers AI, Flisak JR, Aitken RA (1987) J Am Chem Soc 109:5446
56. Monovich LG, Le Huérou Y, Rönn M, Molander GA (2000) J Am Chem Soc 122:52
57. Nicolaou KC, Cho SY, Hughes R, Winssinger N, Smethurst C, Labischinski H, Endermann R (2001) Chem Eur J 7:3798
58. Kamikawa K, Tachibana A, Sugimoto S, Uemura M (2001) Org Lett 3:2033
59. Nicolaou KC, Natarajan S, Li H, Jain NF, Hughes R, Solomon ME, Ramanjulu JM, Boddy CNC, Takayanagi M (1998) Angew Chem Int Ed 37:2708
60. Wilhelm R, Widdowson DA (2001) Org Lett 3:3079
61. Watanabe T, Uemura M (1998) Chem Commun 871
62. Watanabe T, Shakadou M, Uemura M (2000) Synlett 1141
63. Tanaka T, Mikamiyama H, Maeda K, Ishida T, In Y, Iwata C (1997) Chem Commun 2401
64. Tanaka T, Mikamiyama H, Maeda K, Iwata C, In Y, Ishida T (1998) J Org Chem 63:9782
65. Mukai C, Kim IJ, Hanaoka M (1993) Tetrahedron Lett 34:6081
66. Mukai C, Kim IJ, Hanaoka M (1993) Tetrahedron: Asymmetry 3:1007
67. Mukai C, Kim IJ, Furu E, Hanaoka M (1993) Tetrahedron 49:8323
68. Mukai C, Hirai S, Hanaoka M (1997) J Org Chem 62:6619
69. Moser WH, Zhang J, Lecher CS, Frazier TL, Pink M (2002) Org Lett 4:1981
70. Nicolaou KC, Boddy CNC, Bräse S, Winssinger N (1999) Angew Chem Int Ed 38:2097
71. Theil F (2003) Synthesis of diaryl ethers: a long-standing problem has been solved. In: Schmalz HG, Wirth T (eds) Organic synthesis highlights V. Wiley-VCH, Weinheim, p 15
72. Pearson AJ, Park JG (1992) J Org Chem 57:1744
73. Pearson AJ, Bignan G, Zhang P, Chelliah MV (1996) J Org Chem 61:3940
74. Pearson AJ, Chelliah MV (1998) J Org Chem 63:3087
75. Janetka JW, Rich DH (1995) J Am Chem Soc 117:10585
76. Janetka JW, Rich DH (1997) J Am Chem Soc 119:6488
77. Venkatraman S, Njoroge FG, Girijavallabhan V, McPhail AT (2002) J Org Chem 67:3152
78. Pearson AJ, Heo JN (2000) Org Lett 2:2987
79. Pearson AJ, Heo JN (2000) Tetrahedron Lett 41:5991

Topics Organomet Chem (2004) 7: 181–204
DOI 10.1007/b12821

Arene Complexes as Catalysts

James H. Rigby · Mikhail A. Kondratenko

Department of Chemistry, Wayne State University, Detroit, MI 48202, USA
E-mail: jhr@chem.wayne.edu, mkondratenko@discoverypartners.com

Abstract An overview of organic reactions catalyzed by π-arene complexes of transition metals is presented. The chapter summarizes only those transformations that take place at the metal atom, which is coordinated to an arene ligand. Two different mechanisms of formation for catalytically active species have been distinguished: (1) via dissociation of arene-metal bonds and (2) with the η^6-arene ligand remaining on the metal. The topic is reviewed with emphasis on the types of organic reactions promoted by arene-metal complexes. The description of each reaction is provided with representative figures or schemes. Catalyst structures and key reaction intermediates are given to explain the reaction outcome. The use of π-arene complexes in the synthesis of complex molecules and industrial applications of arene-metal catalysts are discussed.

Keywords Arene complexes · Transition metal · Catalysis

1 Introduction

Transition-metal arene complexes have been known to serve as catalysts for a variety of organic transformations [1, 2]. Many of these processes (e.g., stereoselective hydrogenation, olefin metathesis, cycloaddition reactions) have routinely been used in the synthesis of complex organic molecules, including pharmacologically potent natural products. Arene metal catalysts have also found broad application in materials science. For example, transition-metal assisted cationic and radical photopolymerizations have been used extensively in industry for the production of polymers [3]. An attractive feature of this process follows from the fact that a slight variation in reaction conditions allows for the production of polymers possessing certain molecular weights, structures and physical properties. Arene metal photoinitiators have also been used worldwide for making polymeric films and protective coatings [4].

The current overview presents literature highlights on the catalytic applications of arene complexes and chiefly covers publications from 1990 through early 2002. In order to avoid overlaps with other chapters, we focused only on the transformations that occur at the metal center coordinated to an arene ring. The chapter is organized into two sections, which deal with two different ways of formation of catalysts from π-arene-metal precursors. The first part reviews processes where coordinatively unsaturated catalytically active species are generated by cleavage of arene-metal bonds (Fig. 1). The second section provides an account of reactions in which creation of free sites of coordination on the metal doesn't cause the dissociation of arene ligands. Some notable examples of employing π-arene catalysts in the synthesis of complex molecules are also presented.

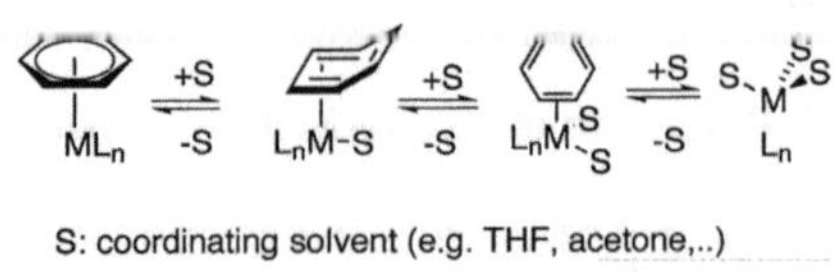

Fig. 1 General mechanism of dissociation of arene-metal bonds

2
Catalysis via Dissociation of Arene-Metal Bonds

Dissociation of the arene ligand has been a routine procedure for generating coordinatively unsaturated fragments [5]. This strategy was brought to practice by employing π-arene complexes to catalyze organic transformations. The detailed studies of catalytic reactions promoted by arene metal complexes have revealed that π-arene complexes are only precatalysts. Catalytically active species bearing free sites of coordination on the metal are usually generated in situ. In many cases, dissociation of arene ligands has been a method of choice for creating open sites of coordination on the metal center (Fig. 1).

Since the catalyst formation involves the cleavage of arene-metal bonds, the catalytic activity of the arene complex depends on the ease of dissociation of the arene ring. Arene dissociation can be facilitated by introducing coordinating substituents onto the benzene ring, by using polyaromatic compounds (naphthalene, anthracene) as 6π-ligands or by employing donor solvents. A powerful method for arene activation is to employ ligands (L) which bear a second coordinative site [6, 7]. To date, numerous approaches for labilizing arene-metal bonds have been developed with a view to practical synthesis of organic substances [2, 5, 7].

2.1
Hydrogenation

Hydrogenation of conjugated dienes is a frequently used model reaction for evaluating catalytic activities of arene complexes [2, 7, 8]. Normally, this reaction requires elevated temperatures >120 °C and high pressure of hydrogen (~70 atm). (η^6-naphthalene) $Cr(CO)_3$ and (η^6-benzene)(η^2-methyl acrylate)$Cr(CO)_2$ complexes convert dienes into alkenes at room temperature under an atmospheric pressure of hydrogen gas [2, 7]. The presumed mechanism of the diene hydrogenation involves initial complexation of diene and hydrogen to the 12-electron $Cr(CO)_3$ fragment **4** to afford the reactive intermediate **5**. Subsequent 1,4-hydrogen addition followed by the olefin release regenerates the $Cr(CO)_3$ moiety **4**, which can be involved in the next catalytic cycle (Scheme 1).

Notably, in the presence of (arene)$Cr(CO)_3$ complexes, dienes are exclusively converted into *cis*-olefins [2, 7–9]. These catalysts are superior in stereo- and chemoselectivity to ordinary catalysts including the Lindlar catalyst, cationic

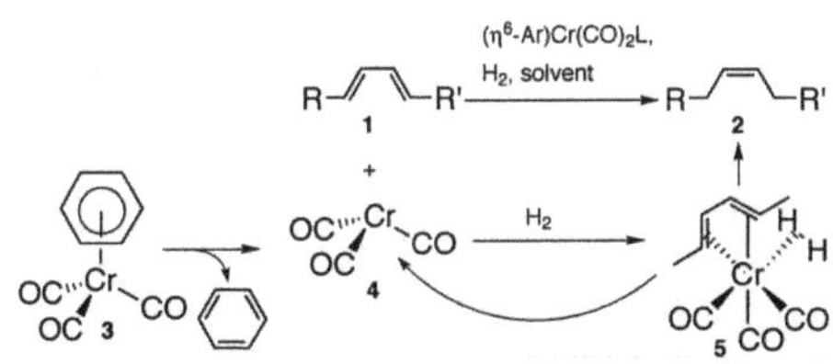

Scheme 1

rhodium complexes, and others. Catalytic amounts of arene-metal complexes also promote hydrogenation of α,β-unsaturated esters, ketones and imines [2, 10]. Alkyne hydrogenation results in *cis*-alkenes [11, 12] (Scheme 2). Mechanistically, these reactions are similar to the diene hydrogenation described above. Other metals such as ruthenium, molybdenum, and tungsten have also been used to promote hydrogenation of the carbon-carbon double and triple bonds [2, 10]. However, regio-and stereoselectivity of the hydrogenation was low when (arene)$Mo(CO)_3$ or (arene)$W(CO)_3$ were employed.

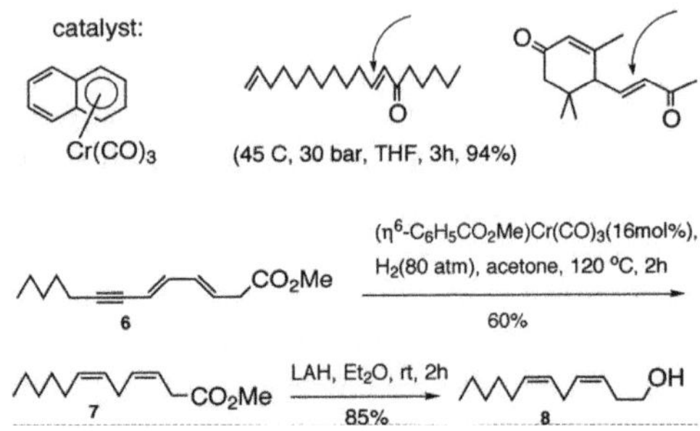

Scheme 2

The utility of the metal-assisted hydrogenation reaction has been nicely displayed in the elegant synthesis of biologically active compounds [2, 9, 11]. For instance, (η^6-$C_6H_5CO_2Me$)$Cr(CO)_3$ enables the simultaneous, stereoselective hydrogenation of alkyne- and conjugated diene fragments in **6** to give the skipped (*Z,Z*)-diene **7**, which is a precursor for the subterranean termites' pheromone mimic **8** (Scheme 2) [11].

Mono- and dinuclear ruthenium arene complexes ([(η^4-anthracene) $Ru(PCy)_2$] (**9**), [(η^6-9,10-dimethylanthracene)$RuH(PPh_3)_2$]$^+$ (**10**), etc.) and water-soluble tetranuclear clusters [(η^6-benzene)Ru_4H_4][BF_4]$_2$ (**11**), [(η^6-benzene)Ru_4H_4]Cl_2 (**12**), etc.) serve as catalysts for arene hydrogenation to give saturated hydrocarbons [2, 13]. Hydrogenation of arenes is an important industrial process for the generation of cleaner diesel fuel. Use of **11** and **12** allows for conducting the hydrogenation of aromatics under so-called ionic liquid-organic biphasic conditions, which eliminate the environmental problem associated with water contamination [14].

2.2 Hydrosilylation

Harrod and co-workers demonstrated that (arene)chromium tricarbonyl complexes are active catalysts for the thermal hydrosilylation of 1,3-dienes (Scheme 3) [15].

Similar to the hydrogenation reaction, addition of triethoxysilane **14** to *trans*-1,3-pentadiene **13** proceeded in 1,4-fashion yielding two isomeric olefins **15** and

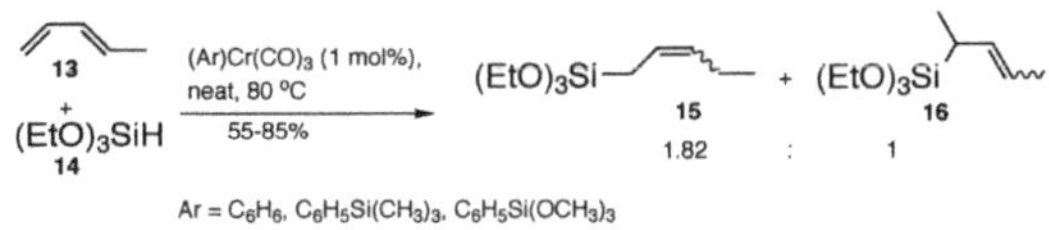

Scheme 3

16 in a ratio of 1.82:1. The catalytic activity was not greatly affected by the nature of the arene ligand attached to the metal.

2.3
Addition of Halocarbons to Alkenes

A range of (η^6-arene)Cr(CO)$_3$ complexes have been examined as mediators of the reaction between CCl_4 and alkenes [16]. Only terminal alkenes **17** and cyclic olefins **18** undergo CCl_4 addition in the presence of THF to give 1,1,1,3-tetrachloroalkanes **19** and **20** (Scheme 4). Interestingly, carbon tetrachloride does not react with conjugated dienes, but non-conjugated ones containing a terminal C=C bond do react. In spite of the fact that the terminal C=C bond in the substrate is necessary for the reaction to occur, addition of CCl_4 to dienes takes place at the non-terminal olefin fragment leaving the terminal double bond untouched.

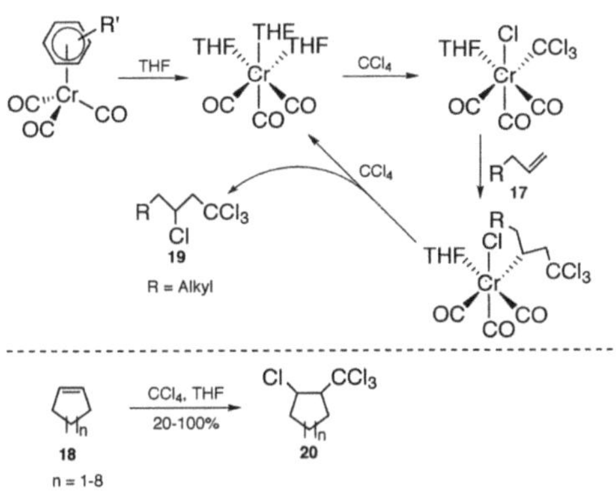

Scheme 4

It is believed that the mechanism of haloalkylation involves oxidative addition of CCl_4 to the metal followed by alkene insertion into the Cr-CCl_3 bond (Scheme 4). The cycle is then completed by reaction with another molecule of CCl_4.

2.4
Dehydrohalogenation

The Group VIB arene-metal tricarbonyl complexes are effective homogeneous catalysts for the dehydrohalogenation of aliphatic alkyl halides to create an olefin function [17]. Often mixtures of several isomers occur in these reactions with terminal alkenes being formed as the major product (Scheme 5). This suggests that the proportion of products depends on their relative rate of formation (kinetic control).

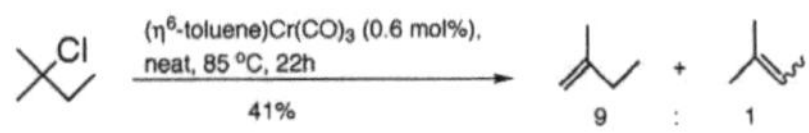

Scheme 5

The catalytic activity was found to depend on the nature of the metal catalyst. Higher efficiency was obtained with Mo catalysts as compared to Cr and W ones. The nature of the ligand attached to the metal has some control over the conversion. In addition to dehydrogenation double bond migration also takes place.

2.5
Isomerization

Arene chromium complexes exhibit very high catalytic activity for the isomerization of dienes [2, 18]. Thus, in the presence of (η^6-naphthalene)$Cr(CO)_3$ 1,5-hydrogen shift proceeds smoothly at room temperature (Scheme 6) [18]. The mild reaction conditions make it possible to avoid undesired side reactions, which might occur under normal thermal conditions. The presumed mechanistic pathway involves the preliminary coordination of a diene substrate **21** to the $Cr(CO)_3$ fragment to give the intermediate **22**, which undergoes oxidative addition to produce the η^5-pentadienylhydridochromium species **23** (Scheme 6). Subsequent reductive elimination followed by decomplexation complete the hydrogen shift from the C-1 to C-5 position providing the diene **24**. It is noteworthy that the hydrogen migration proceeds in a stereospecific manner.

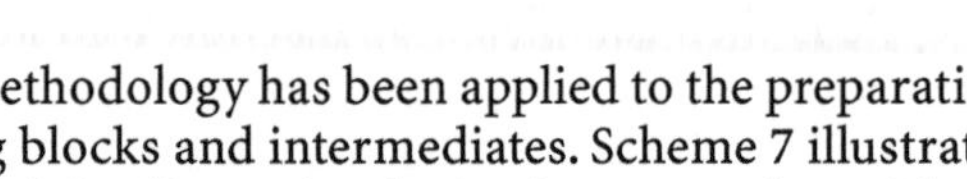

Scheme 6

This unique methodology has been applied to the preparation of synthetically useful building blocks and intermediates. Scheme 7 illustrates the stereocontrolled synthesis of the dienamine derivative **26** starting with the readily available butadiene **25**. Compound **26** undergoes intramolecular Diels-Alder reaction to afford the *cis*-octahydroquinoline **27** as a single diastereomer [18].

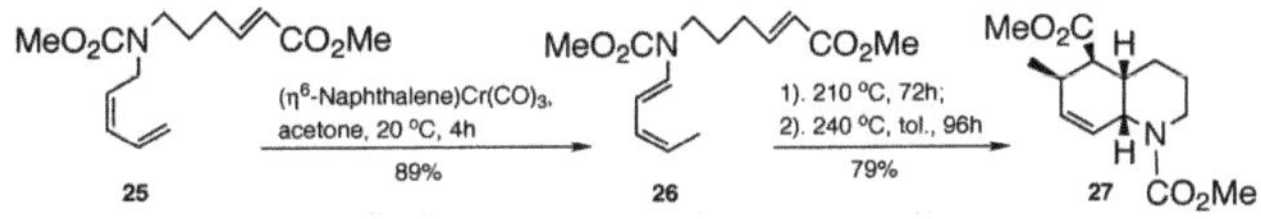

Scheme 7

The chromium-assisted isomerization of 1,4-dienes to their conjugated 1,3-isomers has also been observed, but no isomerization has been detected with 1,5-dienes [2].

(η^6-Arene)NiR_2 (R=SiCl_3, SiF_3, C_6F_5) complexes display very high catalytic activity for the isomerization of 1-butene to produce a mixture of 2-butenes **29** and **30** (Scheme 8) [19]. It is remarkable that the mechanism of this reaction is one of the few that does not involve the complete dissociation of an arene ligand, but appears to follow $\eta^6 \rightarrow \eta^4$ slippage pathway. From the detailed mechanistic study the catalysis appears to proceed via a hydride intermediate **28**. Complexation of **27** with 1-butene with a concomitant $\eta^6 \rightarrow \eta^4$ arene slippage followed by a series of rearrangements furnishes a mixture of **29** and **30**. The *cis/trans* ratio of products depends on the arene, the R-ligand, solvent and 1-butene/Ni proportion.

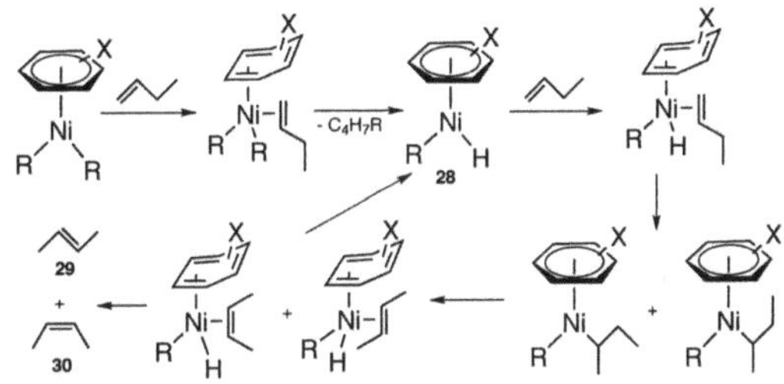

Scheme 8

2.6 Furan Alkylation

So far only one example of furan alkylation assisted by arene complexes has been reported. D.J. Milner has shown that furan reacts with *tert*-butyl chloride in the presence of (η^6-mesitylene)Mo$(CO)_3$ to afford a mixture of 2-*tert*-butylfuran and 2,5-di-*tert*-butylfuran in 65–80% yields [20]. Both mono- and disubstituted furans can be prepared as the major product by changing the furan/*t*-BuCl ratio. Although there is only one report of this reaction, the ability to synthesize mono- and disubstituted furans in a simple manner reflects the potential of this methodology which can be used for preparing substituted furans, that are valuable synthetic materials.

2.7 Borylation

An interesting rhodium-mediated process has recently been reported in which the Hartwig's catalyst (η^4-C_6Me_6)Rh(η^5-C_5H_5) (**31**) promotes highly selective borylation of aryl C-H positions and alkane methyl groups [21, 22]. Arylboronate esters and acids are versatile reagents in organic chemistry providing a straightforward methodology for the formation of carbon-carbon bonds (Suzuki coupling). They are also easily converted to amines, alcohols, alkanes and other classes of functionalized molecules in a single step. Typically, areneboron reagents are prepared from aryllithium or haloarene precursors. A direct borylation of arene rings recently proposed by Smith and co-workers is very attractive since it substantially simplifies and shortens the route to arylboranes [21]. A key for converting aryl C-H bonds into C-B bonds is the use of the arene-rhodium catalyst **31** (Scheme 9).

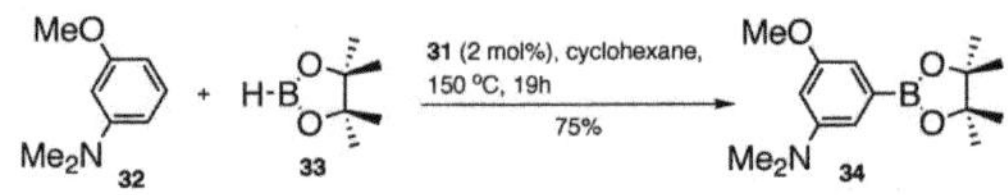

Scheme 9

The catalytic borylation offers unique selectivity when compared to traditional aromatic substitution reactions. Thus, mono- and 1,3-disubstituted (type **32**) arenes react predominantly at *meta*-positions. This high *meta*-selectivity is a sterically controlled process. In the case of 1,3-difunctionized substrates borylation occurs exclusively at the 5-position even where functionalizations at the 5-position are difficult to achieve by other methods (for instance, dimethyl resorcinol, 1,3-bis[dimethylamino]benzol, etc.). Published evidence suggests that the catalytic borylation tolerates most heteroatom substituents, which enhances the value of this methodology.

Complex **31** also catalyses the high-yield formation of linear alkylboranes from alkanes under thermal conditions [22]. Scheme 10 depicts the dehydrogenative coupling of pinacolborane **33** with n-octane **35.** In this reaction, 1-octylpinacolborane **36** was the only product to be detected. No borane derivatives from functionalization at the internal positions were observed. 2-Methylheptane and methylcyclohexane provided only alkylboranes from primary C-H bond functionalization. The high selectivity for the termini of alkanes is kinetically determined and results from a steric preference for formation of a linear metal-alkyl complex.

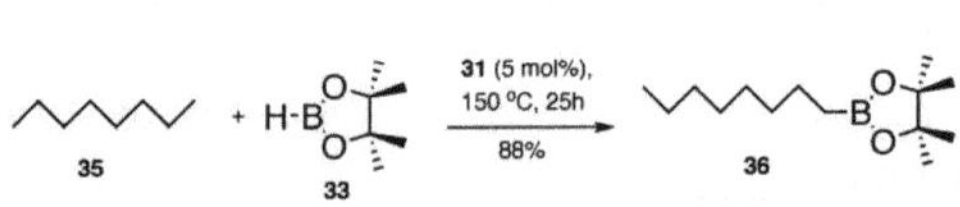

Scheme 10

Undoubtedly, the synthetic value of the Rh-catalyzed alkyl borylation reaction follows from the fact that it can selectively deliver a functional group, such as an alcohol, an amine or an alkene, at the terminus of an alkyl chain during the synthesis of more complex structures. With the proper substituent at boron, this process can extrude the same borane reagent that was used in the alkane functionalization step to allow recycling of the main group reagent for large-scale synthesis.

To date the detailed mechanism for this process has not been defined. However, data obtained with **31** and similar transition-metal complexes indicate that the obligatory step in the catalytic pathway is the formation of the (η^5-C_5H_5)Rh moiety, which results from the dissociation of the hexamethylbenzene ligand.

2.8 Methoxydefluorination

The cationic (η^6-arene)(η^5-cyclopentadienyl)rhodium(III) complexes (type **37**) were found to be active in catalyzing methanolysis of fluoroarenes yielding the corresponding methoxyarenes [23]. Thus, *para*-fluorobenzene **38** undergoes a methoxydefluorination reaction to afford the corresponding *para*-methoxybenzene **39** (Scheme 11). Other fluoroarenes are also amenable to halogen substitution. This reaction represents an example of catalytic nucleophilic substitution on the arene ring. Normally, nucleophilic arene substitution requires previous activation of the arene by anion-stabilizing groups or the use of π-arene metal complexes as starting materials.

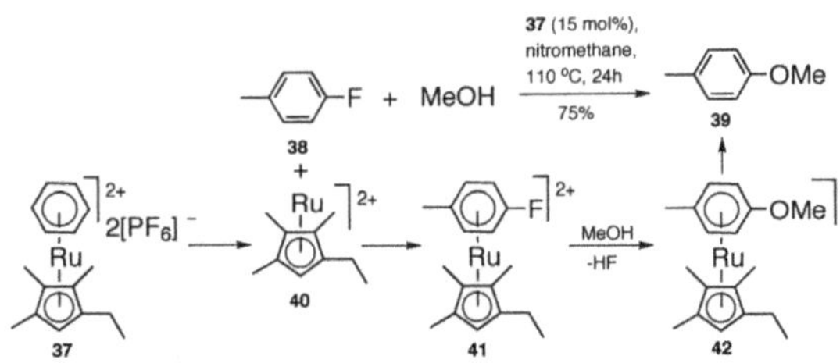

Scheme 11

Gorunov and co-workers have performed a detailed examination of this reaction. A mechanistic pathway consistent with their observations is depicted in Scheme 11. The observations firmly established the initial formation of the coordinatively unsaturated dicationic species **40**, which complexes with **38** to form the [(η^6-*para*-fluorobenzene)(η^5-1-ethyl-2,3,4-trimethylcyclopentadienyl) Ru(III)][PF_6]$_2$ intermediate complex **41**. The fluorine atom in **41** readily exchanges for the methoxy group yielding **42**. Finally, the cleavage of methoxyarene **39** regenerates the catalyst **40** or reactive intermediate **41**.

2.9 Cycloaddition

First reported in the last decade, catalytic higher-order cycloaddition reactions illustrate some of the unique synthetic potential of processes promoted by arene-metal complexes. The Cr(0)-mediated cycloaddition reactions are of particular significance since they afford versatile, and often difficult to prepare, bicyclic [4.2.1] **45** and [4.4.1] **50** systems in diastereomerically homogeneous form [24]. Extensive studies by Rigby and co-workers have established much of what is known about the effect of metals, solvents, reagents and reaction conditions for improving the cyclization process [24–27]. Reaction between cycloheptatriene **43** and ethyl acrylate **44** is a typical example of the chromium-induced [6+2] reaction (Scheme 12). The reaction proceeds with great facility and with no complications to provide the bridged adduct **45.** A limitation of this process is the need to employ only electron deficient olefins.

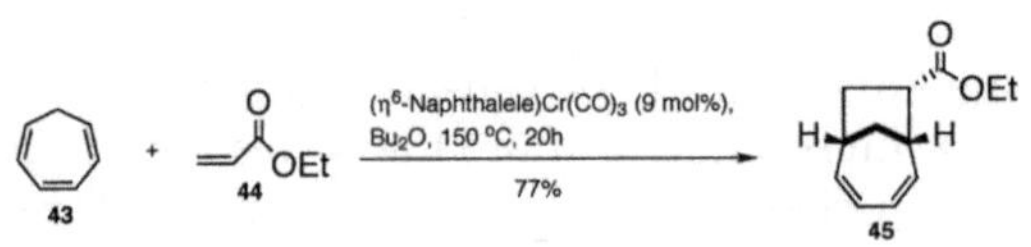

Scheme 12

Very recently Kündig and coworkers reported that the use of (η^6-benzene)(η^2-methyl acrylate)$Cr(CO)_2$ or (η^4-cyclohexadiene)$_2$ $Cr(CO)_2$ as catalysts allows [6+2] cycloadditions to be carried out at 20 °C [28].

An attractive result from the point of view of "green" chemistry is that a [6+2] bicyclic system can be prepared in water media in the presence of magnesium metal [25]. The magnesium additive presumably serves as reducing agent that returns oxidized chromium back to the active Cr(0) species.

A critical advance in the development of metal-promoted [6π+2π] cycloaddition as a synthetic tool occurred with the implementation of the heterogeneous catalyst **46** (Scheme 13) [26]. Polymeric complex **46** is easily accessible from commercial starting materials. It was established that the resin-supported catalyst **46** displays essentially the same efficiency as its homogeneous counterparts.

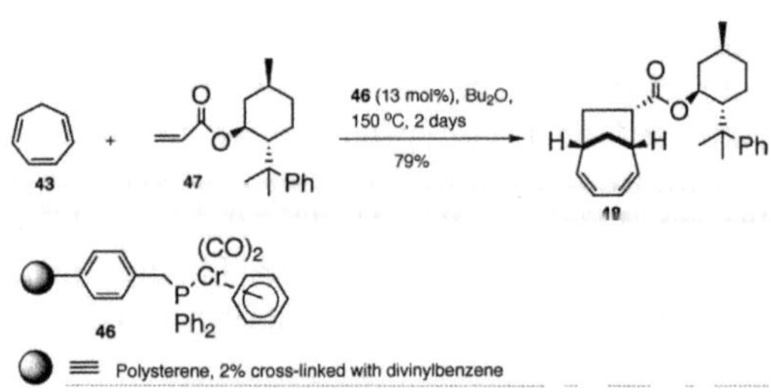

Scheme 13

The use of chiral esters as 2π-substrates permit the production of diastereomerically enriched cycloadducts. For example, the menthyl ester **48** was obtained as a 4:1diastereomeric mixture. The most intriguing feature of polystyrene-bound catalyst **46** is the fact that it can be removed from the reaction mixture by simple filtration and repeatedly reused. This means that the two basic problems of homogeneous catalysis, separation and recycling of the catalyst, can be solved by using the solid-supported complex **46**. Additionally, the environmental problems associated with chromium can also be effectively eliminated.

Another very appealing Cr-assisted cycloaddition procedure, the [6+4] combination, affords bicyclo[4.4.1]undecane systems, that are otherwise difficult to make [27]. Thus, in the presence of (naphthalene)$Cr(CO)_3$, cycloheptatriene **43** underwent clean cyclization with 2,4-*trans,trans*-hexadiene **49** to yield compound **50** in good yield (Scheme 14). Unlike the previously described [6π+2π] cycloaddition, the catalytic [6+4] process requires Mg-powder to facilitate re-reduction to the active Cr(0). The catalytic [6+4] cyclization has proven to be quite general and tolerates protected hydroxyl and carboxyl functionalities.

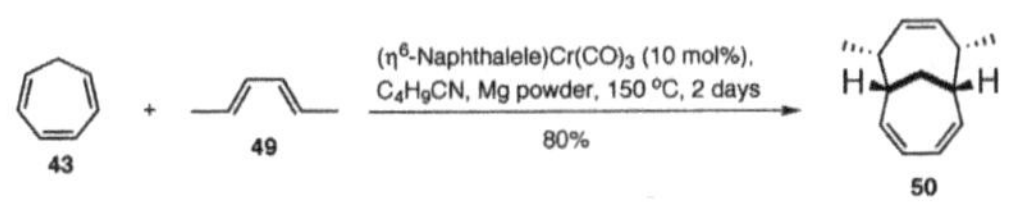

Scheme 14

Obviously, metal-assisted higher-order cycloadditions offer unique and efficient entries into the intricate molecular structures of a number of natural product targets. For example, the BC ring system of ingenol **51** can be readily assembled by employing [6+4] reaction (Scheme 15) [29].

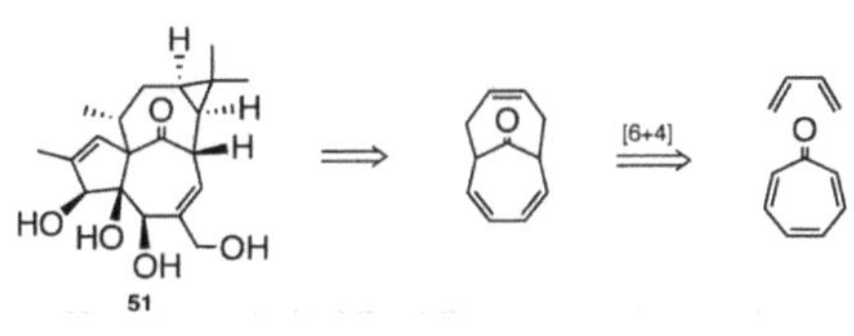

Scheme 15

A number of observations have suggested that both [6+2] and [6+4] reactions are closely related in terms of mechanism. The catalytic pathways involve the complexation of both addends to the metal center to form similar complexes **52** and **53** (Scheme 16). The formation of the putative intermediates **52** and **53** is in good accord with the endo selectivity observed with the Cr-assisted cycloaddition reactions.

Recently, Lapido's laboratory reported on one more type of arene complex, which can promote cycloaddition reactions [30]. It was demonstrated that [(DMSC)Ti{η^6-1,2,4-$C_6H_3(SiMe_3)_3$}] **54** and [(DMSC)Ti{η^6-1,3,5-$C_6H_3Bu^t{}_3$}] **55**

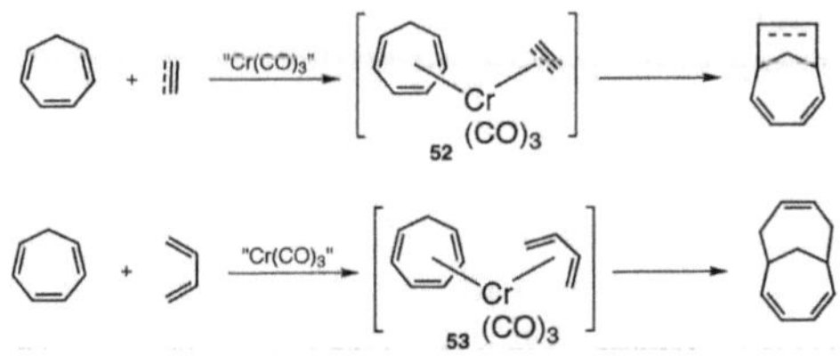

Scheme 16

(DMSC=1,2-alternate Me_2Si-bridged *p-tert*-butylcalix[4]arene) are highly efficient catalysts for [2+2+2] cycloaddition of terminal alkynes producing 1,2,4-substituted benzenes with excellent regioselectivity and in excellent yields. The efficient cyclization of trimethylsilylacetylene **56** and propargyl sulfide **58** depicted in Scheme 17 shows that both mono- and dialkynes are amenable to catalytic trimerization under the established conditions. It is noteworthy that both aliphatic and aromatic terminal alkynes cyclotrimerize efficiently in a rapid and exothermic fashion. However, internal alkynes such as 3-heptyne and 1-trimethylsilyl-1-propyne do not react under these conditions.

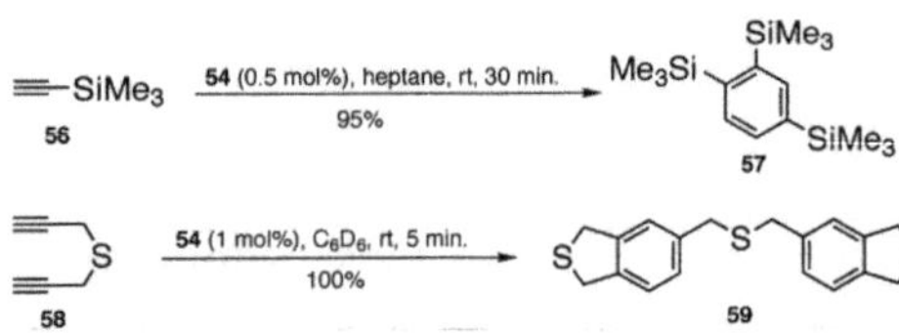

Scheme 17

Although the exact catalytic pathway for this process has not been established, extensive mechanistical study confirms that catalysis occurs on the [(DMSC)Ti] moiety, which is formed by the dissociation of an arene ligand.

2.10
Olefin Dimerization

Several reports have been published on olefin dimerization reactions [31, 32]. For example, Ohgomori and co-workers showed that (η^6-arene)(η^4-cyclooctadiene)Ru(0) **60** complexes (arene=benzene, toluene, anisole, etc.) promote the tail-to-tail dimerization of acrolein to produce (*E*)-2-hexene-1,6-diol **61** (Scheme 18) [31]. The coordinatively unsaturated Ru(cyclooctadiene) species appears to play a key role in the catalytic process.

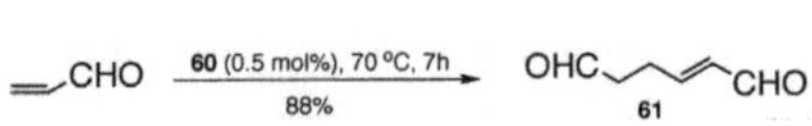

Scheme 18

Another report from the Klabunde laboratory shows the catalytic dimerization of ethylene and propylene in the presence of arene nickel complexes [32].

2.11 Polymerization

The catalytic activity of arene metal complexes in polymerization processes is a widely studied area and an immense amount of work has appeared in the literature dealing with photopolymerization reactions mediated by arene complexes [3,33,34]. Many of these processes have been shown to be useful routes to a large variety of products of scientific and commercial interest. Over the last decade most of the work has been performed in the area of iron-arene complex catalysis. Iron-arene salts such as **62** can photoinitiate polymerization of epoxides, pyrroles, olefins, etc. in either cationic or radical modes. Scheme 19 depicts the general principle of cationic polymerization of epoxides with iron-arene initiators. Irradiation of **62** initiates ligand exchange of the arene by several epoxides yielding the triepoxy complex **63**. The Lewis acidic property of the iron species induces ring opening reactions in the ligand sphere to produce intermediate **64** which initiates the epoxide polymerization process [33].

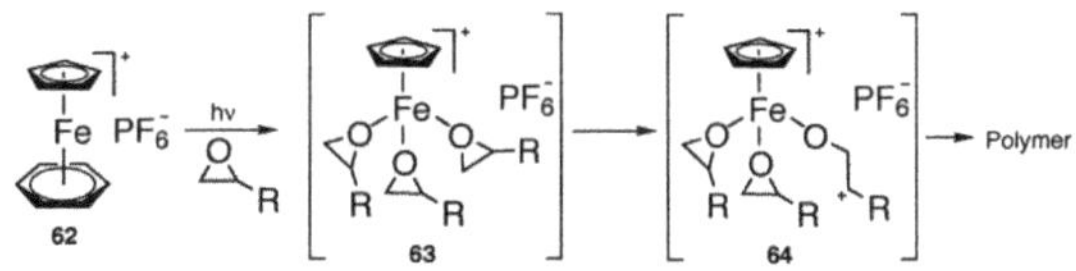

Scheme 19

[Arene(Cp)Fe]$^+$ complexes **62** are also found to be very efficient catalysts for polymerization of pyrroles and aromatic dicyanate esters [34,35].

Interestingly, arene iron salts **62** have also proven to be effective catalysts for the free radical polymerization of epoxides, acrylates, etc. [3,36]. Similar to cationic polymerization, radical processes are also photochemically initiated reactions but are initiated in the presence of radical precursor additives (halogenated solvents, bis(*p*-*N*,*N*-dimethylaminobenzylide)cyclopentanone, etc.).

Another type of polymerization promoted by arene complexes is based on the well-known olefin metathesis reaction. Olefin and alkyne metathetical polymerizations have been observed with catalytic amounts of Group VI arene metal carbonyls under refluxing conditions [37]. The same process takes place at ambient temperatures when electron-transfer-chain catalysis is invoked [37].

3 Catalysis with Retention of Arene-Metal Coordination

An alternative approach to arene complex catalysis involves generation of catalytically active species without breaking arene-metal bonds. In the event, open sites of coordination on the metal center are created by the dissociation of ligands other than arene. The aromatic rings remain as integral parts of catalytically active units. Most of the synthetic applications associated with this type of catalysis have been effected using ruthenium complexes.

3.1 Hydrogenation

Ruthenium(II) complexes have shown excellent catalytic performance in transfer hydrogenation of ketones and aldehydes [38]. A combined system of [{$RuCl_2$(η^6-arene)}$_2$], a bidentate ligand, and a base in 2-propanol converts carbonyl compounds into alcohols, furnishing acetone as a bi-product. An important feature of this strategy is the ability to exploit the stereocontrol afforded by chiral ligands. The use of chiral amino alcohols or amino acids as bidentate ligands leads to enantiomerically enriched alcohols. An example of such an enantioselective ketone reduction is depicted in Scheme 20. In this case, ethyl benzoylacetate **65** undergoes asymmetric transfer hydrogenation with a mixture of [{$RuCl_2$(η^6-arene)}$_2$] **67** and (1S,2R)-*N*-methylnorephedrine **68** in isopropanol to afford the hydroxylester **66** in a yield of 80% with 94% ee.

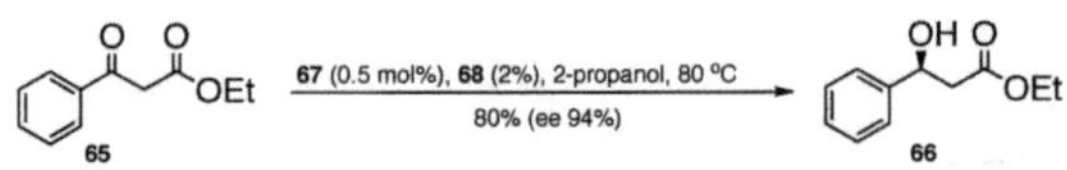

Scheme 20

The catalysis process starts from the reaction between bis[(η^6-cymene) dichlororutheniun(II)] **67** with amino alcohol **68** yielding 18-electron complex **69**. In the presence of base, **69** gives 16-electron species **70**, which is the true catalyst (Scheme 21). Subsequent interaction of **70** with 2-propanol provides the 18-electron reducing agent **72**. The hydridic Ru-H and protic N-H in **72** are transferred simultaneously to the C=O bond in **65** to form β-hydroxyester **66**

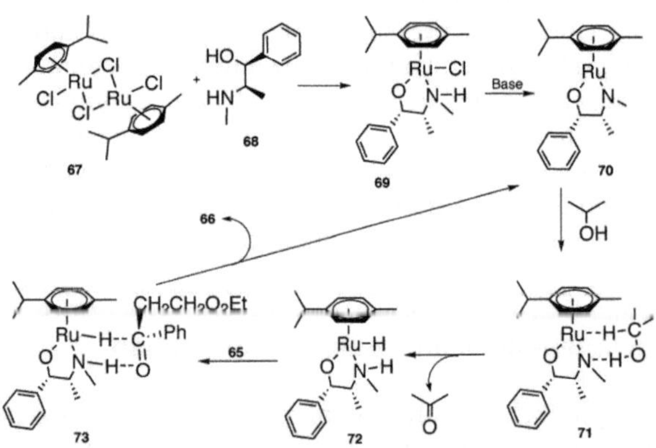

Scheme 21

and regenerate the complex **70**. This catalytic pathway is in good accord with the fact that intermediate complexes **69**, **70**, and **72** have been isolated and characterized by NMR, ESI-MS, as well as by X-ray in the case of **69**.

Transfer hydrogenation of ketones is quite general with regard to the molecular complexity of substrates and tolerates many functional groups. Recently, enatioselective transfer hydrogenation has been successfully applied to imine reduction [39]. From a synthetic perspective the principal virtue of this approach is the ease of generation of alcohols and amines in excellent yields with high enantiomeric excess.

In addition to transfer hydrogenation reactions, arene ruthenium complexes also display excellent activity in the catalytic hydrogenation of olefins and alkynes including asymmetric reduction [40]. Remarkably, this process occurs under milder conditions, than required for catalysis with the dissociation of arene-metal bond. Lately, arene iridium complexes have also been found to be effective hydrogenation catalysts [41]. It is noteworthy that iridium can also promotes addition to the carbon-nitrogen double bond.

3.2
Hydroaminomethylation

Originally discovered by Reppe and co-workers, hydroaminomethylation of olefins has attracted attention as an economical, one-pot synthesis of amines. Alper and co-workers have demonstrated that zwitterionic rhodium complex [$Rh^+(\eta^4$-cyclooctadiene)(η^6-$C_6H_5BPh_3$)-] **74** is an excellent catalyst for hydroaminomethylation of phenylalkenes **75** to afford the corresponding branched methylated amines **76** with high regioselectivity (Scheme 22) [42].

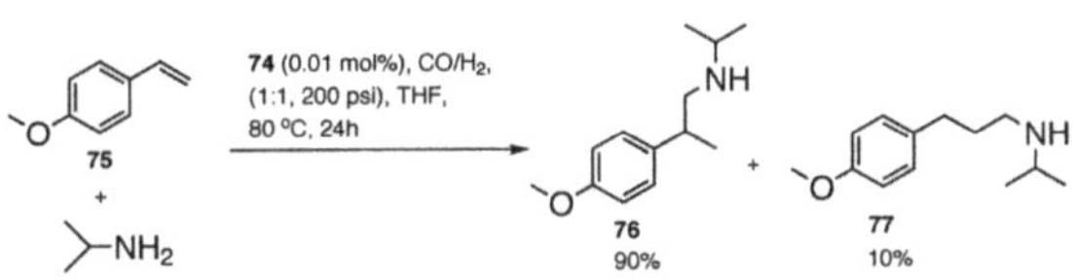

Scheme 22

The hydroaminomethylation reaction catalyzed by **74** has been found to be rather general with regard to amines and possesses many advantages that stem from the relatively mild reaction conditions.

3.3
Enantioselective Desymmetrization

So far, only one example of enantioselective desymmetrization of olefins with (η^6-arene)Ru(II) half-sandwich complex **78** has been reported [43]. It was found, that isomerization of 2-*n*-butyl-4,7-dihydro-1,3-dioxepin **79** with a plane of symmetry, leads to the (+)-enantiomer of 2-*n*-butyl-4,5-dihydro-1,3-dioxepin

80 (Scheme 23). A number of observations have shown that the catalytic efficiency of **78** depends slightly on the reaction conditions themselves.

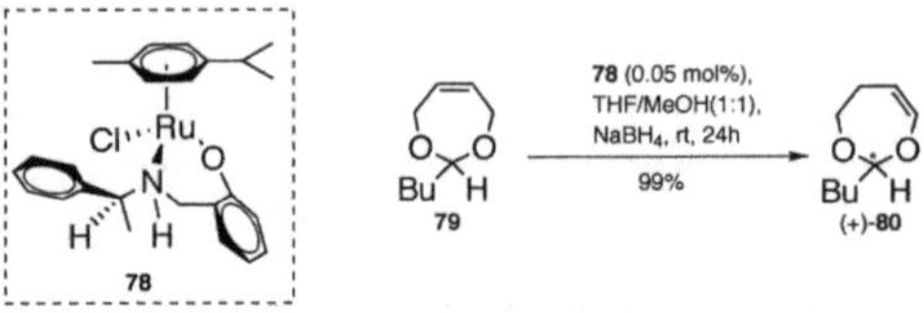

Scheme 23

3.4 Cycloaddition

In the last few years, cationic (arene)ruthenium complexes containing chiral bidentate ligands have attracted considerable attention as enantioselective Lewis acid catalysts for Diels-Alder reactions [44]. Thus, Davies and co-workers have demonstrated the Ru-complex **81** promotes the enantioselective [4+2] cycloaddition between methacrolein **82** and cyclopentadiene **83** to furnish 2-methybicyclo[2.2.1]hepta-5-ene-2-carbaldehyde **84**, with a high *exo*-selectivity (>95%) (Scheme 24). Catalyst **81** is normally prepared in situ from its neutral precursor. Remarkably, the reaction proceeds rapidly at room temperature. Extensive modifications employing different types of ligands and reaction conditions have began to reveal the considerable potential of arene ruthenium complexes for Diels-Alder type transformations.

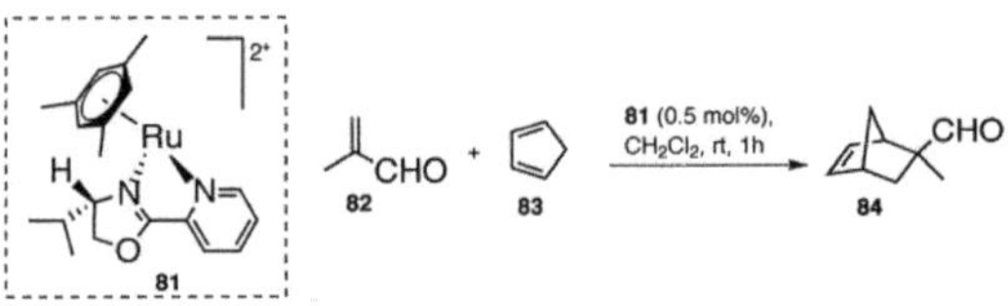

Scheme 24

A fascinating approach to the synthesis of pyridine heterocycles has been discovered by using (arene)Fe(0) complexes to catalyze [2+2+2] cycloaddition reactions. The proposed method consists of a cross cyclization of alkynes and nitriles [45]. An example of such a reaction is depicted in Scheme 25. In the presence of (η^4-1,5-cyclooctadiene)(η^6-phosphinine)Fe(0) **85**, butyronitrile **86** combines with two molecules of methyl propargyl ester **87** yielding a mixture of isomeric pyrroles **88** and **89**. The reaction proceeds at ambient temperature, but relatively long reaction times are required to obtain good yields. The symmetric 2,4,6-substituted pyridine **88** is produced in excess relative to the alternative issue, the 2,3,6-derivative **89**. Surprisingly, only benzene derivatives were found to be formed as side products, and no nitrile trimerization was observed.

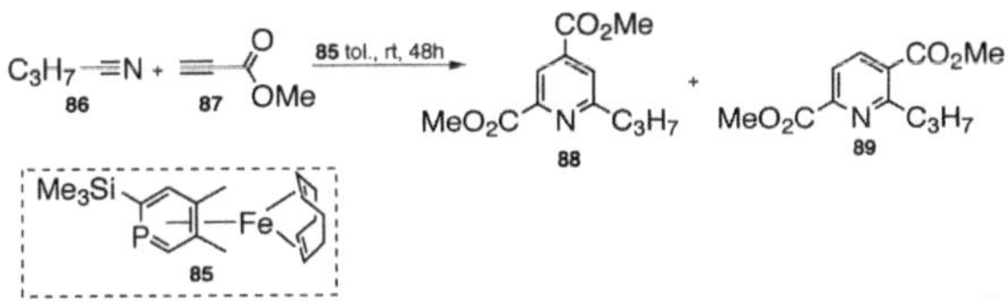

Scheme 25

Cyclodimerization of 1,3-butadiene via arene metal-mediation has received relatively little attention. To date, the only example of such a dimerization has been described by Zenneck and co-workers [46]. This group has shown that (η^6-arene)(η^4-diazadiene)iron complexes catalyze 1,3-butadiene dimerization to afford a mixture of 3-vinylcyclohexene and 1,5-cyclooctadiene. The initial observations showed a strong dependence of [4+4]/[4+2] product ratio on the nature of the arene coordinated to the metal center. These early results offer good opportunities for the further improving on the chemoselectivity of this cyclization.

3.5 Cyclopropanation

Although cyclopropanation can be viewed as a [2+1] cycloaddition reaction, this transformation is considered separately in this review due to the unique specificity of this process and its importance in synthetic chemistry.

Ruthenium-mediated cyclopropanation is a frequently used process in contemporary organic synthesis. Demonceau and co-workers developed an interesting Ru-catalyzed cyclopropanation with (arene)Ru(phosphine) complexes **90** and **92** (Fig. 2) [47, 48].

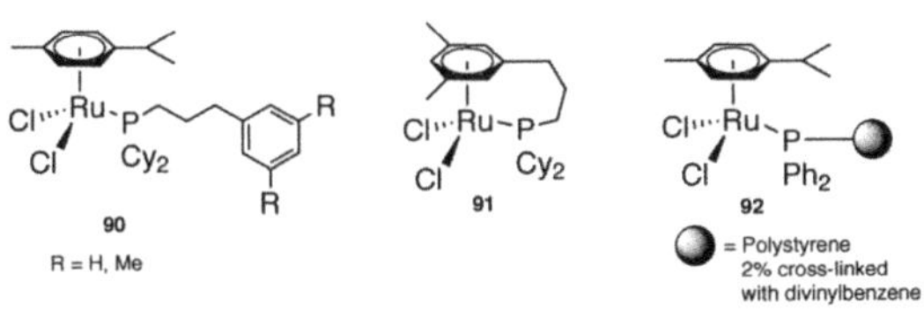

Fig. 2 Ruthenium-based cyclopropanation catalysts

Satisfactory results have been obtained in the reaction of styrenes **93** with ethyl diazoacetate **94** in the presence of **90** to afford the cycloadduct **95** (Scheme 26) [47]. Preliminary studies have shown reasonable catalytic efficien-

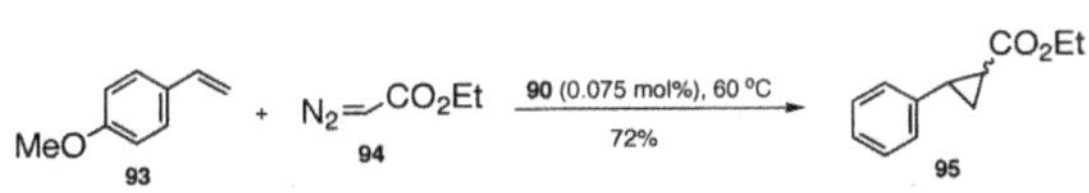

Scheme 26

cy for both complexes **90** and **91**. The inherent problem of this process is the low *endo/exo* selectivity of the reaction.

Recent interest in the development of environmentally benign synthesis has driven the development of the polymer-bound cyclopropanation catalyst **92** [48]. It is noteworthy that the immobilized catalyst **92** is as active as its homogeneous analogs **90** and **91** and has the advantage that it can be reused numerous times.

3.6 Cycloisomerization

A newly described intramolecular cycloisomerization of dienes catalyzed by arene ruthenium complexes is an attractive method for the construction of medium-sized rings [49]. Interestingly, cycloisomerization catalysts have also been found to promote alkene metathesis under the same conditions. The unique feature of these processes is that either of the two competitive reactions can be carried out selectively by choosing an appropriate substrate or a co-catalyst. For example, the in situ generated cation **96** cycloisomerized the nitrogen-containing diene **97** to the aminoolefine **98** with only traces of the metathesis product **99** (Scheme 27). However, a similar diene without the nitrogen atom exclusively underwent the metathesis reaction.

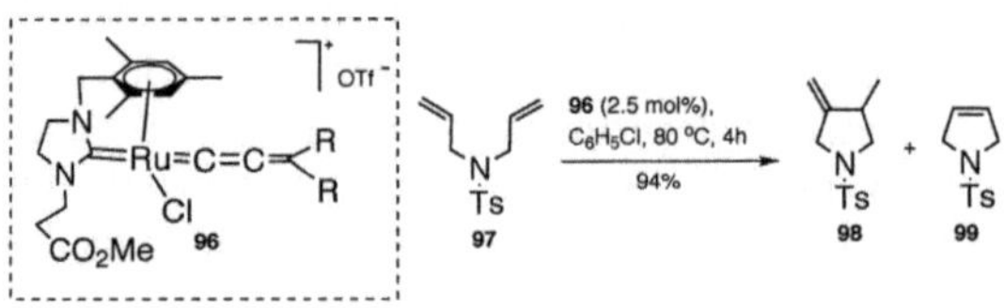

Scheme 27

3.7 Ring Closing Metathesis

First described by Grubbs and Schrock, ring closing metathesis reactions have been of interest for many years as they have been particularly effective for the construction of macrocycles. Several research groups have recently shown high catalytic activity of (arene)ruthenium carbene cations **96**, **100** in intramolecular metathesis reactions [49, 50]. This is not surprising, since complexes **96** and **100** are structurally reminiscent of the well-known Grubbs catalyst. They effect the cyclization of various functionalized dienes and enynes with good to excellent yields and show a great tolerance towards an array of functional groups.

This process is of immense importance and has been applied in a range of synthetic protocols. Thus, in the presence of **100**, the diene **101** was effectively converted into cyclic olefin **102**, which upon deprotection affords the potent insect alkaloid, azamacrolide epilachnene **103**, which is isolated from the pupae of the Mexican beetle *Epilachnar varivestis* (Scheme 28).

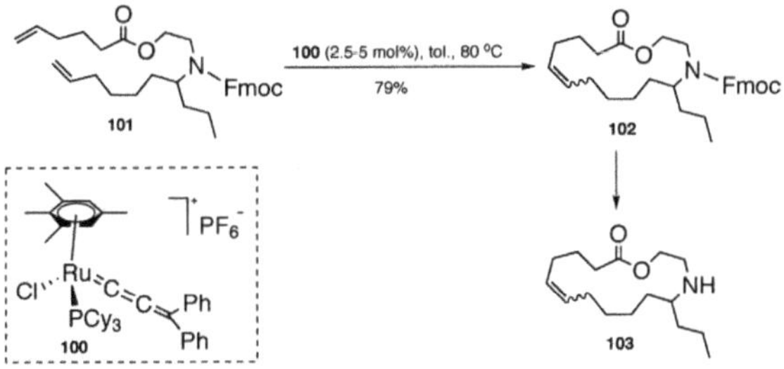

Scheme 28

3.8
Metathesis Polymerization

The development of Ru-based catalysts has made ring-opening olefin metathesis polymerization a valuable tool for the preparation of synthetic target. Noels and co-workers have reported a highly efficient ring-opening metathesis polymerization of alkenes in the presence of arene-ruthenium complexes bearing phosphine or carbene ligands. Such complexes spontaneously promote the ring-opening metathesis polymerization of norbornene and cyclooctene, when activated by addition of a diazo compound or by irradiation with visible light [51]. As a rule, catalytically active species have been formed in situ from commercially available [(η^6-cymene)$RuCl_2$] **67**. The observed functional group tolerance of this catalyst looks very promising for promoting polymerization of a variety of functionalized cyclic olefins.

3.9
Hydration of Terminal Alkynes

Several arene-ruthenium complexes containing bidentate phosphine ligands have been shown to be useful catalyst precursors for the hydration of terminal aryl alkynes **105** to afford acetophenones **106** (Scheme 29) [52]. For example, the cationic complex **104**, when activated by $AgSbH_6$, promoted addition of water to a carbon-carbon triple bond. It was found that such reactions proceeded slowly but in good to excellent yields. It is remarkable, that the water in commercial acetone was sufficient to achieve complete conversion to the product.

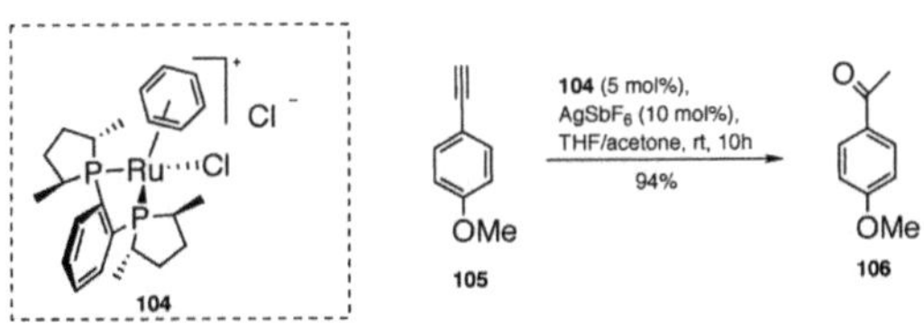

Scheme 29

3.10 Enol-Ester Formation

The electrophilic activation of terminal alkynes by arene-ruthenium(II) catalysts has provided selective access to enol esters. Enol esters are much more reactive than alkyl esters and have been used in a variety of reactions. In the past decade, Dixneuf and co-workers have developed selective approaches to the Markovnikov and *anti*-Markovnikov addition of carboxylic acids across alkynes by employing different arene-ruthenium(II) catalysts [48, 53, 54]. Of special interest is the synthesis of *N*-Boc-protected l-alanine isopropenyl ester **110** from *N*-Boc-l-alanine **108** and propyne **109** mediated by (η^6-cymene)$RuCl_2(PPh_3)$ complex **107** (Scheme 30) [53]. Addition of the amino acid **108** to the propyne **109** proceeded exclusively in the Markovnikov sense and without accompanying racemization of the substrate.

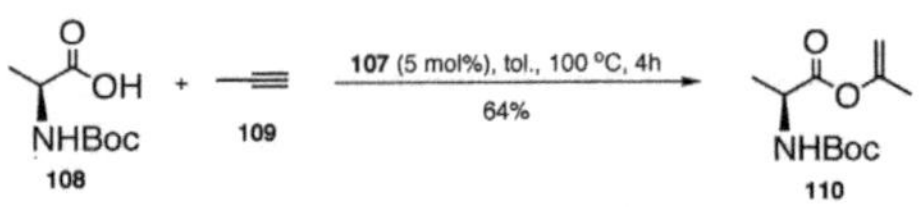

Scheme 30

Scheme 31 depicts an example of the *anti*-Markovnikov addition of nucleophiles to alkynes. (η^6-Cymene)$RuCl_2(PMe_3)$ complex **111** mediated the reaction between diethylamine **112**, CO_2, and phenylacetylene **113** to provide a mixture of *trans*- and *cis*-carbamates **114** and **115** respectively in the ratio of 5:1 [53].

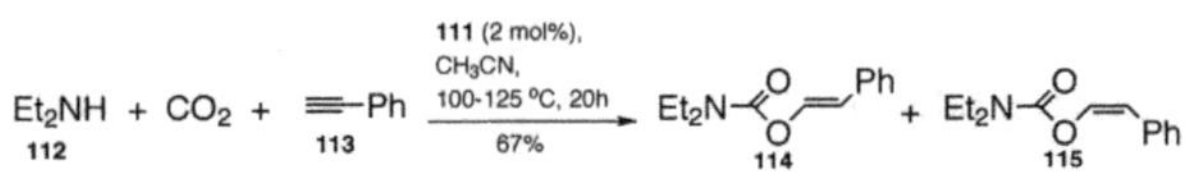

Scheme 31

Employing a resin-bound catalyst **92** for enol ester synthesis is an important extension of this strategy [48]. It is noteworthy that the solid-supported catalyst **92** displays the same activity as its homogeneous analog **107** and is amenable for recycling.

3.11 Furan Synthesis

In a series of fascinating studies, Dixneuf's group has demonstrated the capability of arene-ruthenium complexes for promoting the intramolecular cyclization of (*Z*)-3-methylpent-2-en-4-yn-1-ols, such as **116** and **118**, to furan derivatives **117** and **119** (Scheme 32) [53, 55–57]. This reaction corresponded to the intramolecular Markovnikov addition of a hydroxyl group to a non-activated terminal triple bond. It was found that the *cis*-geometry of double bond is crucial for the

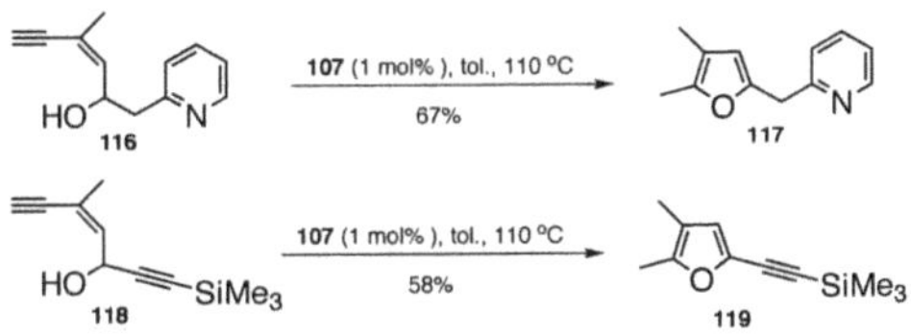

Scheme 32

cyclization to occur, and particularly noteworthy feature of this process is that it makes possible the cyclization of enynols containing functional groups [56].

Interesting results have been obtained with (η^6-arene)(η^1-pyridine)$RuCl_2$ catalysts anchored on an inorganic matrix. Preliminary results from Seçkin's laboratory demonstrated that such solid-supported complexes were easily prepared and displayed high catalytic activities towards the intramolecular cyclization of enynols to furans [57].

3.12 Dehydrogenative Coupling of Silanes

Several reports on the ability of arene complexes to mediate silicon-carbon bond forming processes are particularly attractive since silanes are important intermediates in organic synthesis [58–60]. In particular, it was shown that arene-ruthenium complexes are active catalysts for dehydrogenative transfer coupling of arene C-H and silane Si-H bonds to produce arylsilanes. For instance, (η^6-C_6Me_6)$Ru(H_2)(SiEt_3)_2$ complex **120** mediates the coupling of triethylsilane with trifluorotoluene **121** to afford the arylsilane **122** along with some amount of carbosilane dimer **123** (Scheme 33) [58].

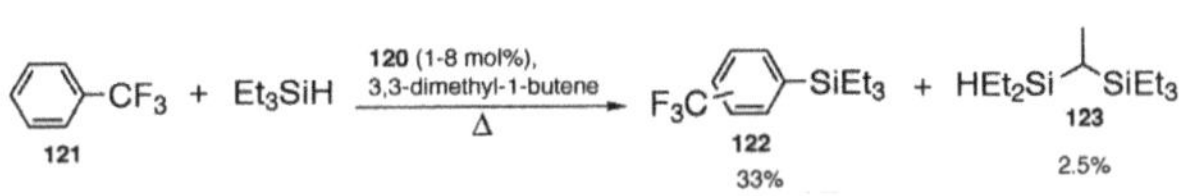

Scheme 33

Metal-assisted dehydrogenative coupling of silanes with nucleophiles can also lead to the silicon-heteroatom bond formation. Thus, in the presence of catalytic amount of (η^6-benzene)(η^2-$HSiHPh_2$)$Cr(CO)_2$ **124** diphenylsilane **125** reacts smoothly with either aniline or thiophenol to afford the products **126** or **127** respectively (Scheme 34) [60]. This process represents a rare example of arene-chromium complex catalysis, which proceeds without breaking arene-metal bonds.

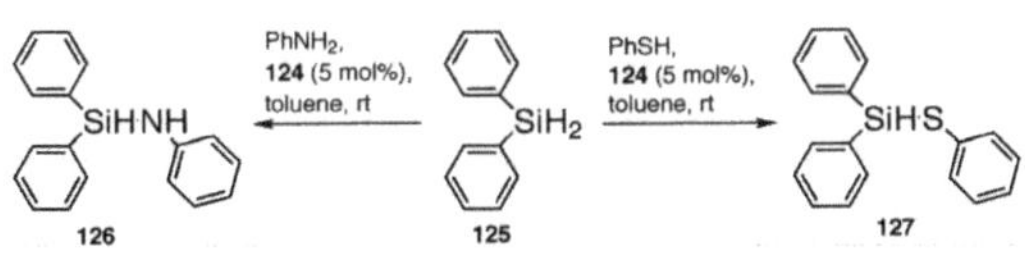

Scheme 34

4
Summary

Catalysis with transition-metal arene complexes has provided a new manifold of synthetic options. The broad features of accessible reaction pathways are well established, and organic chemists can feel comfortably about incorporating such methodology into their armory. We are sure that many new synthetically useful reactions using arene metal catalysts will be developed in the nearest future.

References

1. For recent reviews see: (a) Brunner H (1996) In: Cornis B, Herrmann WA (eds) Applied homogeneous catalysis with organometallic compounds, vol 1. VCH, Weinheim, p 201; (b) Semmelhack MF (1995) In: Hegedus LS (ed) Comprehensive organometallic chemistry II, vol 12. Pergamon, Oxford, pp 979, 1017; (c) Bennett MA (1997) Coord Chem Rev 166:225; (d) Brait S, Deabate S, Knox SAR, Sappa E (2001) J Cluster Sci 12:139
2. (a) Sodeoka M, Shibasaki M (1993) Synthesis 643; (b) Frankel EN, Selke E, Grass CA (1968) J Am Chem Soc 90:2446; (c) Cais M, Frankel EN, Rejoan (1968) Tetrahedron Lett 1919; (d) Cais M, Frankel D, Weidenbaum K (1975) Coord Chem Rev 16:27
3. (a) Rabek JF, Lucki J, Zuber M, Qu BJ, Shi WF (1992) Pure Appl Chem A29–297; (b) Wang P, Shen Y, Wu S, Adamczak E, Linden LA, Rabek JF (1995) Pure Appl Chem A32:1973
4. Dietliker K (1991) In: Oldring PKT (ed) Chemistry and technology of UV and EB formulation for coatings, inks and paints, vol III. SITA Technology, London
5. Kündig EP, Perret C, Spichiger S, Bernardinelli G (1985) J Organomet Chem 286:183
6. (a) Semmelhack MF, Chlenov A, Wu L, Ho D (2001) J Am Chem Soc 123:8438; (b) Tobita H, Hasegawa K, Minglana JJG, Luh LS, Okazaki M, Ogino H (1999) Organometallics 18:2058
7. Kündig EP, Kondratenko MA, Romamens P (1998) Angew Chem Int Ed 37:3146
8. Chandiran T, Vancheesan S (1994) J Mol Catal 88:31
9. (a) Vasili'ev AA, Engman L, Serebryakov EP (2000) J Chem Soc Perkin Trans 1 2211; (b) Vasili'ev AA, Engman L, Serebryakov EP (2000) Mendeleev Commun 101; (c) Serebryakov EP, Vasili'ev AA, Titskii DY, Beletskaya IP (2000) Mendeleev Commun 168; (d) Vasili'ev AA, Engman L, Serebryakov EP (1999) Acta Chem Scand 53:611; (e) Vasili'ev AA, Serebryakov EP (1996) Russ Chem Bull 45:2232; (f) Vasili'ev AA, Vlasyuk AL, Kryshtal GV, Serebryakov EP (1995) Russ Chem Bull 44:1946; (g) Vasili'ev AA, Kryshtal GV, Serebryakov EP (1995) Mendeleev Commun 41
10. Bennett MA, Neumann H, Thomas M, Wang X (1991) Organometallics 10:3237
11. Vasili'ev AA, Vlasyuk AL, Gamalevich GD, Serebryakov EP (1996) Bioorg Med Chem 4:389
12. Sodeoka M, Shibasaki M (1985) J Org Chem 50:1147
13. (a) Süss-Fink G (2002) In: Herrmann WA (ed) Synthetic methods in organometallic and inorganic chemistry, vol 10. Thieme, Stuttgart, p 122; (b) Borowski AF, Sabo-Etienne S, Chaudret B (2001) J Mol Catal A Chem 174:69; (c) Fidalgo EG, Plasseraud L, Süss-Fink G (1998) J Mol Catal A Chem 132:5; (d) Plasseraud L, Süss-Fink G (1997) J Organomet Chem 539:163; (f) Lin Z, Halpern J (1991) J Organomet Chem 417:C24
14. Dyson PJ, Ellis DJ, Packer DG, Welton T (1999) Chem Commun 25
15. Harrod JF, Shaver A, Tucka A (1985) Organometallics 4:2166
16. (a) Bland WJ, Davis R, Durrant JLM (1985) J Organomet Chem 280:95; (b) Le Maux P, Jaouen G, Saillard JY (1981) Organomet Chem 212:193
17. Tsonis CP, Hwang JS (1984) J Mol Catal 26:219

18. Yamada H, Sodeoka M, Shibasaki M (1991) J Org Chem 56:4569
19. Kanai H, Choe SB, Klabunde KJ (1986) J Am Chem Soc 108:2019
20. Milner DJ (1981) J Organomet Chem 217:199
21. (a) Tse MK, Cho JY, Smith MR III (2001) Org Lett 3:2831; (b) Cho JY, Iverson CN, Smith MR III (2000) J Am Chem Soc 122:12868
22. Chen H, Schlecht S, Semple TC, Hartwig JF (2000) Science 287:1995
23. (a) Goryunov LI, Nikitin YM, Steingarts VD (1998) Russian J Org Chem 34:1646; (b) Goryunov LI, Steingarts VD (1998) Russian J Org Chem 34:1640
24. (a) Rigby JH (1999) In: Harmata M (ed) Advances in cycloaddition, vol 6. JAI Press, Greenwich, p 97; (b) Rigby JH, Krueger AC (1995) In: Coxon JM (ed) Advances in detailed reaction mechanisms, vol 4. JAI Press, Greenwich, p 1; (c) Rigby JH (1993) Acc Chem Res 26:579
25. Rigby JH, Mann LW, Myers BJ (2001) Tetrahedron Lett 42:8773
26. Rigby JH, Kondratenko MA, Fiedler C (2000) Org Lett 2:3917
27. Rigby JH, Fiedler C (1997) J Org Chem 62:6106
28. Kündig EP, Robvieux F, Kondratenko M (2002) Synthesis 2053
29. (a) Rigby JH, De Sainte Claire V, Cuisiat SV, Heeg MJ (1996) J Org Chem 61:7992; (b) Rigby JH, Hu J, Heeg MJ (1998) Tetrahedron 39:2265
30. Ozerov OV, Patrick BO, Lapido FT (2000) J Am Chem Soc 122:6423
31. Ohgomori Y, Ichikawa S, Sumitani N (1994) Organometallics 13:3758
32. Choe SB, Kanai H, Klabunde KJ (1989) J Am Chem Soc 111:2875
33. Chen X, Chen Y (1977) J Appl Polym Sci 66:2551
34. Kotch TG, Lees AJ, Fuerniss SJ, Papathomas KI (1995) Chem Mater 7:801
35. Rabek JF, Lucki J, Zuber M, Qu BJ, Shi WF (1992) Polymer 33:4838
36. Bowser R, Davidson RS (1994) J Photochem Photobiol A Chem 77:269
37. (a) Mula MB, Beaumont AJ, Doyle KO, Gallagher ML, Rooney AD (1999) J Mol Catal A Chem 148:23; (b) Vijayaraj TA, Sundararajan G (1997) Organometallics 16:4940; (c) Ganesamoorthy S, Sundararajan G (1992) J Mol Catal 76:335; (d) Ganesamoorthy S, Sundararajan G (1992) Macromolecules 25:2060
38. (a) Ohta T, Nakahara S, Shigemura Y, Hattori K, Furukawa I (2001) Appl Organomet Chem 15:699; (b) Evaraere K, Mortreux A, Bulliard M, Brussee J, van der Gen A, Nowogrocki G, Carpentier JF (2001) Eur J Org Chem 275; (c) Yamakawa M, Yamada I, Noyori R (2001) Angew Chem Int Ed 40:2818; (d) Takehara J, Hashigushi S, Fujii A, Inoue S, Ikariya, T, Noyori R (1996) Chem Commun 233
39. Uematsu N, Fujii A, Hashiguchi S, Ikariya T, Noyori R (1996) J Am Chem Soc 118:4916
40. (a) Moldes I, de la Encarnacion E, Ros J, Alvarez-Larena A, Piniella JF (1998) J Organomet Chem 566:165; (b) Heinemann F, Klodwig J, Knoch F, Wündisch M, Zenneck U (1997) Chem Ber Recueil 130:123; (c) Bennett MA, Ennett JP (1992) Inorg Chim Acta 198/200:583
41. (a) Torres F, Sola E, Martin M, Lopez JA, Lahoz FJ, Oro LA (1999) J Am Chem Soc 121:10632; (b) Torres F, Sola E, Martin M, Ochs C, Picazo G, Lopez JA, Lahoz FJ, Oro LA (2001) Organometallics 20:2716
42. Lin YS, El Ali B, Alper H (2001) Tetrahedron Lett 42:2423
43. Brunner H, Prommesberger M (1998) Tetrahedron Asymm 9:3231
44. (a) Davies DL, Fawcett J, Garratt SA, Russell DR (2001) Organometallics 20:3029; (b) Davenport JD, Davies DL, Fawcett J, Garratt SA, Russell DR (2000) J Chem Soc Dalton Trans 4432; (c) Davies DL, Fawcett J, Garratt SA, Russell DR (1997) Chem Commun 1351
45. (a) Knoch F, Kremer F, Schmidt U, Zenneck U (1996) Organometallics 15:2713; (b) Schmidt U, Zenneck U (1992) J Organomet Chem 440:187
46. Le Floch P, Knoch F, Kremer F, Mathey F, Scholz J, Scholz W, Thiele KH, Zenneck U (1998) Eur J Inorg Chem 119
47. Simal F, Jan D, Demonceau A, Noels AF (1999) Tetrahedron Lett 40:1653
48. Leadbeater NE, Scott KA, Scott LJ (2000) J Org Chem 65:3231

49. (a) Cetinkaya B, Demir S, Özdemir I, Toupet L, Sémeril D, Bruneau C, Dixneff PH (2001) New J Chem 25:519; (b) Miyaki Y, Onishi T, Ogoshi S, Kurosawa H (2000) J Organomet Chem 616:135
50. (a) Fürstner A, Liebl M, Lehmann CW, Picquet M, Kunz R, Bruneau C, Touchard D, Dixneuf P (2000) Chem Eur J 6:1847; (b) Fürstner A, Picquet M, Bruneau C, Dixneuf P (1998) Chem Commun 1315; (c) Jafarpour L, Huang J, Stevens ED, Nolan SP (1999) Organometallics 18:3760
51. (a) Delaude L, Demonceau A, Noels AF (2001) Chem Commun 986; (b) Stumpf AW, Saive E, Demonceau A, Noels AF (1995) Chem Commun 1127
52. Chen Y, Valentini M, Pregosin PS, Albinati A (2002) Inorg Chim Acta 327:4
53. Bruneau C, Dixneuf P (1997) Chem Commun 507
54. (a) den Reijer CJ, Drago D, Pregosin P (2001) Organometallics 20:2982; (b) Kabouche A, Kabouche Z, Bruneau C, Dixneuf P (1999) J Chem Research (S) 249; (c) Neveux M, Bruneau C, Dixneuf P (1991) J Chem Soc Perkin Trans 1:1197
55. (a) Cetinkaya B, Seçkin T, Özdemir I, Alici B (1998) J Mater Sci 8:1835; (b) Cetinkaya B, Özdemir I, Bruneau C, Dixneff PH (1997) J Mol Catal A Chem 118:L1; (c) Kücükbay H, Cetinkaya B, Guesmi S, Dixneuf P (1996) Organometallics 15:2434
56. Seiller B, Bruneau C, Dixneuf PH (1995) Tetrahedron 51:13089
57. Cetinkaya B, Özdemir I, Bruneau C, Dixneff PH (2000) Eur J Inorg Chem 29
58. Ezbiansky K, Djurovich PI, LaForest M, Sinning DJ, Zayes R, Berry DH (1998) Organometallics 17:1455
59. Djurovich PI, Dolich AR, Berry DH (1994) Chem Commun 1897
60. Matarasso-Tchiroukhine E (1990) Chem Commun 681

Topics Organomet Chem (2004) 7: 205–223
DOI 10.1007/b12822

Planar Chiral Arene Chromium(0) Complexes as Ligands for Asymmetric Catalysis

Kilian Muñiz

Kekulé-Institut für Organische Chemie und Biochemie, Universität Bonn, Gerhard-Domagk-Strasse 1, 53121 Bonn, Germany
E-mail: kilian.muniz@uni-bonn.de

Abstract Bidentate (η^6-arene) tricarbonylchromium(0) complexes can serve as efficient ligands and ligand precursors for various (transition) metals. Such complexes can be employed as catalysts in asymmetric catalytic synthesis such as chiral hydrogenations, reductions, heterofunctionalisations and C-C bond formations. Within this article, the stereochemical and electronic properties of these compounds are summarised and discussed in relation with the respective catalytic reaction for which they have been applied.

Keywords Chromium complexes · Chiral catalysts · Asymmetric synthesis · Chiral ligands · Chiral metal complexes

1 Introduction

The development of novel chiral metal complexes and chiral ligands is crucial for both progress and development of asymmetric catalytic synthesis [1–3]. Within this area, the appearance of planar-chiral ferrocenes as ligands in asymmetric catalysis has been an important advancement [4–7]. While most of these complexes bear side chains or atom groups with stereogenic centres, it is often the 1,2-disubstitution pattern of the π-complexed ring that creates an inherent planar chirality [8] and exercises efficient stereochemical control.

The origin of planar chirality is depicted in Fig. 1. For ACTCs, complexes displaying an unsymmetrical 1,2- or 1,3-disubstitution pattern (or related unsymmetrical higher substitution) are no longer superimposable with their mirror images (**A** vs **ent-A**). In case that the respective side chains are capable of exercising as donor moieties for chelation or coordination to transition metals, optically active complexes **B** result that have potential as asymmetric metal catalysts or catalyst precursors [9, 10].

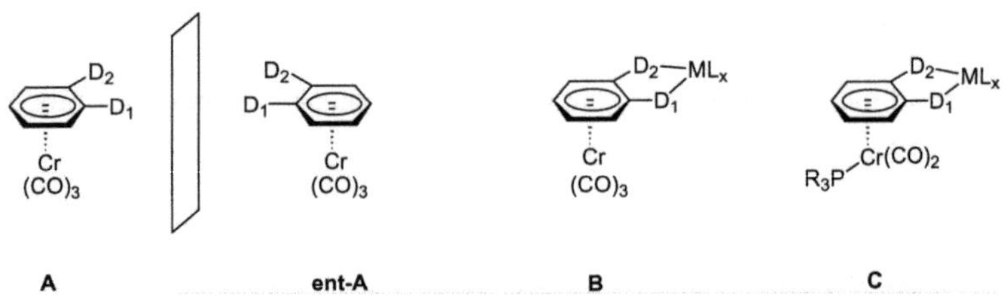

Fig. 1 Planar-chiral ACTCs and respective metal complexes

Contrary to their ferrocene counterparts, (η^6-arene) tricarbonylchromium(0) complexes (hereafter abbreviated to *ACTC*) have found only limited use in asymmetric catalytic synthesis. However, their number is steadily increasing and their further application might be of special interest since electronic and, to a certain extent, steric tuning of these ligands can be undertaken by replacing one of the carbonyl ligands of the ACTC moiety with more basic phosphorus groups (**C**) [11]. Due to the reduced π-acceptor quality of the phosphine groups, the overall electron density on the chromium, the coordinated arene ring and, finally, on the attached donor moieties themselves will increase. When appropriately devised, such a variation would create a new entry into efficient catalyst design.

2 Synthesis of Planar Chiral ACTC Ligands

The synthesis of planar-chiral arene chromium complexes has been reviewed several times and has also been discussed in an earlier article [9, 10]. Apart from stereoselective complexation reactions, diastereoselective and enantioselective *ortho*-functionalisations and use of chiral bases have been the main routes towards planar-chiral ACTCs [10, 12, 13].

3
Catalyses

It is evident that, due to the inherent oxidative lability of their carbon monoxide ligands and of the chromium(0) centre itself, ACTCs cannot serve as ligands in oxidation chemistry. However, they have been employed in other areas of asymmetric catalysis, namely in C-C-bond formation, in reductions of both C-C and C-O double bonds and in heterofunctionalisation reactions.

3.1
Asymmetric Catalytic Reductions

While the use of achiral ACTCs in catalytic hydrogenation and isomerisation reactions is well-studied [14], planar chiral derivatives for asymmetric catalysis have so far received relatively scant attention.

On the other hand, it is well understood that the binding properties of bisphosphines are of utmost importance in asymmetric catalysis in general, and the outstanding performance of some of their corresponding Rh and Ru complexes in asymmetric catalytic hydrogenation have caused particular attention [1]. Interestingly, there are no examples for ACTC ligands in Ru(II) hydrogenation chemistry. As an example of the classical Rh(I)/Rh(III) catalysed hydrogenation of dehydroamino acids, Salzer, Stürmer and co-workers have described the use of planar-chiral ACTCs **1** and **2** [15], Scheme 1.

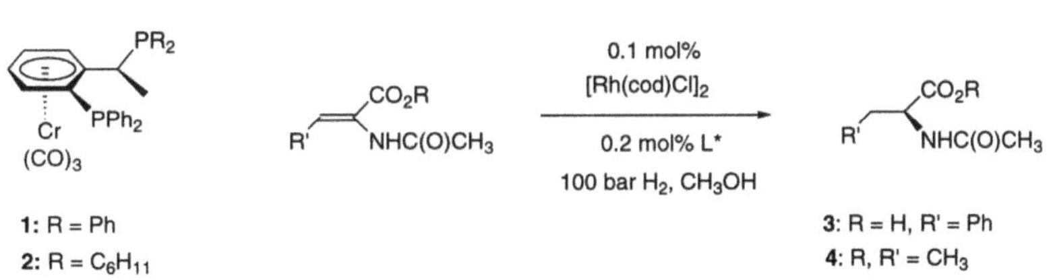

Scheme 1 Asymmetric catalytic hydrogenation (I)

At an S/C of 500 and at a hydrogen pressure of 100 bar, the phenyl-derived ACTC **1** gave 80% chemical yield and 66% enantiomeric excess in the synthesis of the phenyl alanine derivative **3**. This result could be increased to a quantitative yield and 78% ee with the more basic biscyclohexylphosphino complex **2** as ligand. The enantiomeric excess was further enhanced to a value of over 95% when methyl acetamidoacrylate was the substrate. Clearly, the promising potential of ACTCs as ligands in asymmetric hydrogenation has been established and, given the easy modification of these ligands as introduced by Salzer [13], one can expect more applications in this area, including Ru-catalysed hydrogenations, Scheme 2.

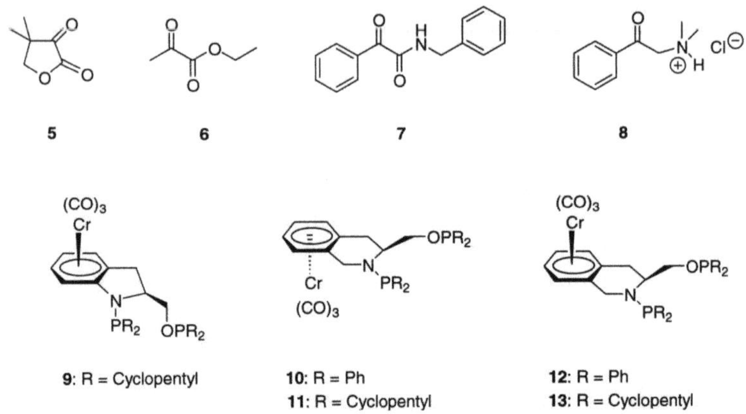

Scheme 2 Asymmetric catalytic hydrogenation (II)

An alternative approach has been undertaken by studying the related carbonyl reduction with molecular hydrogen. This reaction can be catalysed by a complex derived from a diphosphino ligand, bearing phosphite and phosphinite groups, and rhodium salts [16]. Here, ACTCs **9–13** have found application and were employed as ligands for an in situ complexation to rhodium sources such as $[Rh(cod)Cl]_2$ and $[Rh(cod)OCOCF_3]_2$. The complex obtained from $[Rh(cod)OCOCF_3]_2$ was initially used as catalyst for the asymmetric hydrogenation of dihydro-4,4-dimethyl-2,3-furandione and a product with an enantiomeric excess of more than 99% was obtained. However, the enantiomeric excesses for the two catalysts derived from either ACTC **9** or its uncomplexed arene were found to be equally high indicating that in this case the diphosphine structure of phosphite and phosphinite dominates the course of the catalysis. However, when the substrate was *N*-benzylbenzoylformamide, catalysts derived from **9** and the respective Rh(I)-source performed better than those prepared from the uncomplexed arene. This indicates that either the additional stereoelement or the electronic change resulting from the complexation has a significant influence on the catalysis.

A more detailed study with other functionalised ketone substrates revealed several further features of this system [17]. First, the *anti*-configurated diastereoisomer of **9** gave a significantly lower induction in the reduction of **5** (84% ee). When **7** was reduced by a catalyst prepared from either $[Rh(cod)Cl]_2$ or $[Rh(cod)OCOCF_3]_2$ and **9**, up to 97% ee at full conversion could be obtained. Catalysts derived from the diastereomeric complex or the non-coordinated ligand gave inferior results. A significant improvement was achieved when the ligand structure was changed to the tetrahydroisoquinolines **10**–**13** [18]. While only one of the phenyl derivatives, the *anti*-isomer **10**, was suitable for reduction of **5** (90% conversion and 72% ee at 50 bar and with $[Rh(cod)OCOCF_3]_2$ as catalyst precursor), the respective cyclopentyl derivatives showed good general performances in the reductions of the substrates **5–8**. As a general result, the *anti*-configurated ACTC was superior to its *syn*-diastereomer. In the reduction of **5**,

87% ee could be obtained; however, the uncomplexed ligand was superior (95% ee). This was not the case for reduction of **6** where ACTC **11** gave values of 86 (in toluene as solvent) and 95% ee (in methanol), respectively that are well above those obtained with the diastereomer **13** or the uncomplexed ligand. For substrates **7** and **8** no beneficial effect was found for the chromium moiety since a high asymmetric induction of more than 99% ee was reached with both **11** and its uncomplexed arene ligand.

These results indicate that the diastereomeric composition with regards to the planar chirality is of major importance in this kind of hydrogenations [19] while the complexation itself might in some cases have a deleterious effect on the enantiomeric excess of the product, Scheme 3.

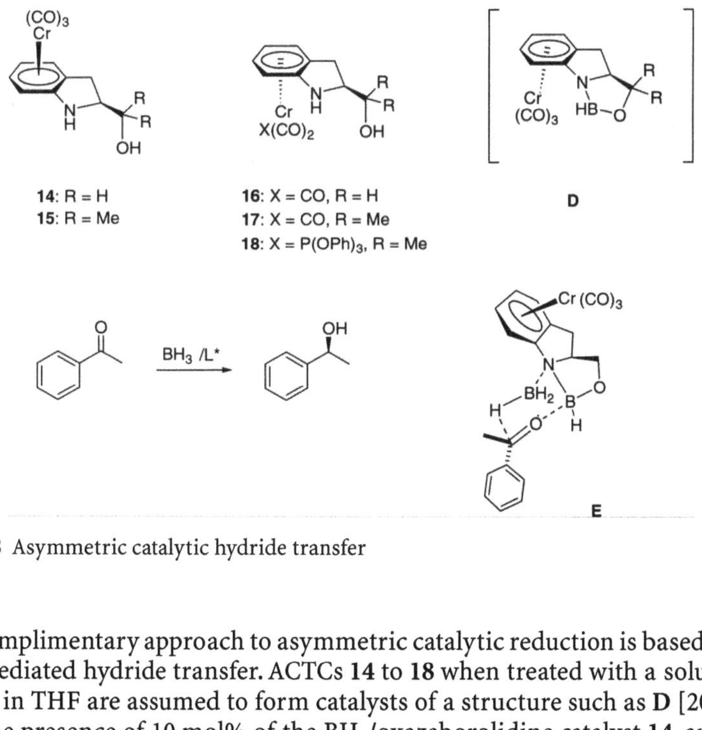

Scheme 3 Asymmetric catalytic hydride transfer

A complimentary approach to asymmetric catalytic reduction is based on borane mediated hydride transfer. ACTCs **14** to **18** when treated with a solution of borane in THF are assumed to form catalysts of a structure such as **D** [20].

In the presence of 10 mol% of the BH_3/oxazaborolidine catalyst **14**, an enantiomeric excess of 50% was obtained in the reduction of acetophenone. Use of a bulkier aromatic substituent such as 9-acetylphenanthrene led to a rise in enantioselectivity (62% ee). For the reduction of acetophenone, it was further found that catalyst precursor **16** with opposite absolute planar chirality but identical central chirality gave a drastic drop in enantioselectivity (25% ee).

Interestingly, for the other two diastereomers **15** and **17**, which in their side chain bear two additional methyl groups, the reaction outcome was found to be reverse. Here, a higher enantiomeric excess was obtained in catalyses with the diastereomer having the opposite absolute planar configuration (39% ee for ACTC **17** compared to 20% ee for **15**). All these results have been explained by an assumed transition state **E** which is depicted for ACTC **14** representing the most efficient catalyst precursor within this set. In **E**, the bulky tricarbonylchromium(0) moiety enables an appropriate distinction between the methyl and the phenyl group by minimising steric interactions with the carbonyl ligands after coordination of acetophenone and borane onto the oxazaborolidine catalyst. For ACTC **16**, the chromium moiety is located on the opposite side of the arene leading to a lower selectivity in the coordination step. For the other pair of diastereomers, it was suggested that in the case of **15** the additional two methyl groups would cause unfavourable interactions with the chromium moiety leading to a highly strained overall complex. Thus, **17** which does not suffer from such interactions is a better catalyst precursor.

The related ACTC **18** with a phosphinite unit yielded a product with only 18% ee in acetophenone reduction. This result was explained assuming unfavourable steric interactions which result in partly reversed coordination preferences. However, electronic reasons also have to be taken into account because an increase in electron density on the complexed arene ring will undoubtedly affect the electronic nature of the amine and thereby slightly decrease the Lewis acidity of the complex as a whole.

3.2 Asymmetric Catalytic Heterofunctionalisations

3.2.1 *Nitrogen-Transfer Reactions*

The use of racemic planar-chiral ACTCs in Hartwig/Buchwald coupling [21–23] of aryl bromides with secondary amines has been described by Uemura [24], Scheme 4. A variety of ACTC ligands (**19–28**) were investigated for an in situ ligation to palladium. The coupling reaction between 4-bromo toluene and piperidine was chosen as a test reaction in order to evaluate the influence of electronic tuning in ACTC ligands. Since amine coupling reactions are known to be strongly dependent on the steric and electronic structure of the ligands, the observed differences in chemical yield are not unexpected. Apparently, ligand **20** served best indicating that replacement of one CO by a more basic triphenylphosphine represents the optimum electronic tuning. As a general trend, amino-substituted ACTCs give higher yields than their oxygenated counterparts. It is noteworthy that the corresponding uncomplexed aromatic compounds gave lower conversion. When further coupling reactions between aryl bromides and secondary acyclic amines were investigated **20** was again the ligand of choice yielding the expected coupling products in 53–95% chemical yield. For some substrates bearing electron-withdrawing substituents, however, the more weakly coordinating phosphino alkoxy complexes **1** and **2** gave slightly superior yields. Thus, **20, 24** and **25** represent the first efficient ACTC ligands for Hartwig/Buchwald amination reactions.

Even though the coupling reactions as investigated are only of achiral nature, there is an important precedence in asymmetric catalysis: Rossen and Pye at Merck employed the planar-chiral bisphosphine PHANEPHOS to attempt a ki-

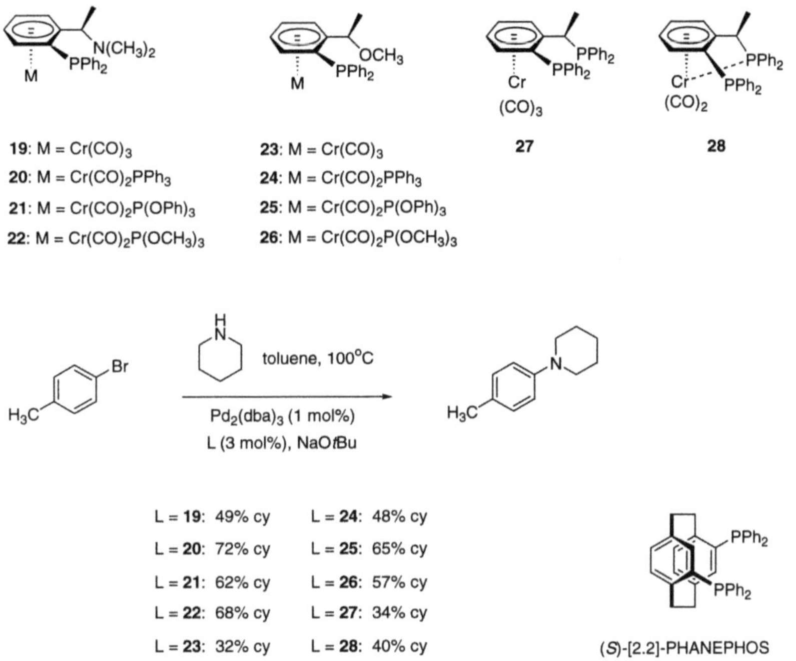

Scheme 4 Catalytic amination of aryl bromides

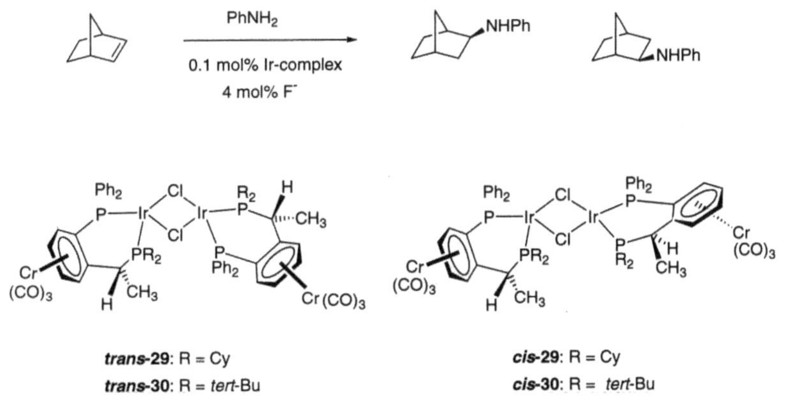

Scheme 5 Asymmetric catalytic hydroamination

netic resolution of racemic 4,12-dibromo-[2.2]paracyclophane via an enantioselective Hartwig/Buchwald coupling of aryl precursors and amines [25]. When racemic 4,12-dibromo-[2.2]paracyclophane was aminated with benzylamine and thallium hexafluorophosphate in the presence of 2 mol% of a Pd/(*S*)-PHANEPHOS complex a reasonable kinetic resolution was achieved: at 79% conversion an enantiomeric excess of 93% was obtained for the remaining (*R*)-

enantiomer, and at about 90% conversion this was increased to 99.9% ee. Future efforts in this area to address the use of these or related ACTC complexes for kinetic resolution must appear promising, Scheme 5.

As a different type of amination, the intermolecular hydroamination of norbornene with aniline was carried out with preformed Ir complexes **29** and **30** bearing bidentate ACTC ligands [15]. As it is commonly the case for the generation of such complexes, the catalyst precursors were isolated as an inseparable mixture of the respective *cis*- and *trans*-isomers. At 60 °C, only traces of the desired hydroamination product could be detected, and moderate enantioselectivities of 51 and 70% ee were determined for catalysts derived from the cyclohexyl derivative **29** and the *tert*-butyl derivative **30**, respectively. Still, a significant improvement in rate is required to transform the present chiral catalyst system into a truly useful synthesis.

3.2.2 *Hydroboration and Hydrosilylation Reactions*

The hydroboration of some styrene derivatives by means of a chiral rhodium complex has been investigated [26]. The active chiral catalyst was prepared in situ from a reaction between an achiral Rh complex and the chiral ACTC compound **31**. In order to prevent formation of achiral catalysts, a slight excess amount of ligand was added. Employing such in situ formation, a 1.2/1-ratio for ACTC/Rh was found to be sufficient. For example, 4-methoxy styrene was smoothly hydroborated under these conditions to give, after standard basic oxidative work-up, the desired secondary alcohol in 87% chemical yield and 62% ee. Enhancing the amount of ligand to an ACTC/Rh-ratio of 2/1 had only a slight beneficial effect. Under otherwise unchanged reaction conditions, several styrene derivatives were submitted to hydroboration from which products with up to 81% ee could be obtained. The results indicate a strong correlation between the electronic substitution pattern and the induced enantioselectivity, although this effect has not been quantified. The stereochemical reaction outcome can be rationalised by assuming a bidentate *P,N*-coordination mode for the ACTC resulting in a three-dimensional scenario in which three quadrants of the transition state are blocked leaving just one preferred side for face-selective olefin approximation. The favourable transition state is depicted below and is in agreement with the absolute configuration as isolated in the products, Scheme 6.

An extension of this work was achieved when the aminal donor moiety in **31** was replaced by a pyridine group [27]. Three different pyridinyl phosphino ACTCs **32**–**34** were synthesised differing only in the protection group at the hydroxyl position. In comparison to the first generation, these compounds now give reasonably high asymmetric induction. For example, 4-methoxy styrene could now be converted with up to 84% ee. Ligand **33** which offered the best result with regards to chemical yield and enantioselectivity was employed for subsequent reactions. Here, no dependence on electronic substitution was observed, and even 4-bromostyrene gave 93% chemical yield and 84% ee. The highest enantiomeric excess was obtained for 2,4-dimethylstyrene (86%). The

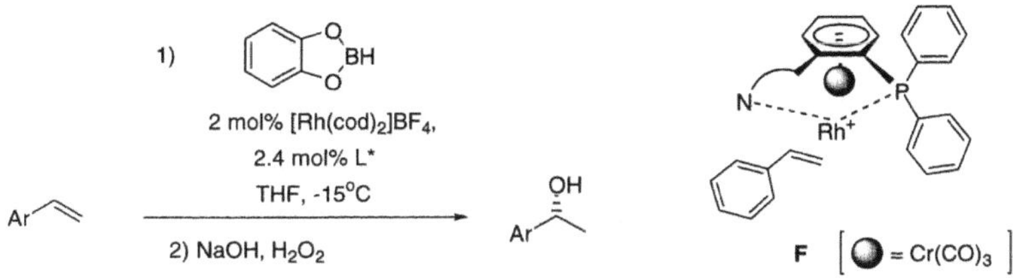

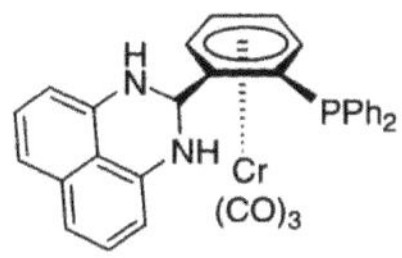

31

Ar = 4-CH_3O-C_6H_4:	87% cy, 62% ee
Ar = 4-Br-C_6H_4:	80% cy, 19% ee
Ar =H:	95% cy, 53% ee
Ar = 3,4-$(CH_3O)_2$-C_6H_3:	65% cy, 75% ee
Ar =2,4-$(CH_3)_2$-C_6H_3:	94% cy, 81% ee

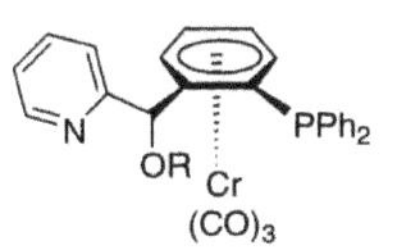

32: R = CH_3
33: R = $CH_2C_6H_5$
34: R = 4-CH_3O-C_6H_4

R = **32**, Ar = 4-CH_3O-C_6H_4:	72% cy, 84% ee
R = **33**, Ar = 4-CH_3O-C_6H_4:	86% cy, 82% ee
R = **34**, Ar = 4-CH_3O-C_6H_4:	39% cy, 61% ee
R = **33**, Ar = 4-Br-C_6H_4:	94% cy, 84% ee
R = **33**, Ar = 3,4-$(CH_3O)_2$-C_6H_3:	31% cy, 73% ee
R = **33**, Ar =2,4-$(CH_3)_2$-C_6H_3:	56% cy, 86% ee

Scheme 6 Asymmetric catalytic hydroboration

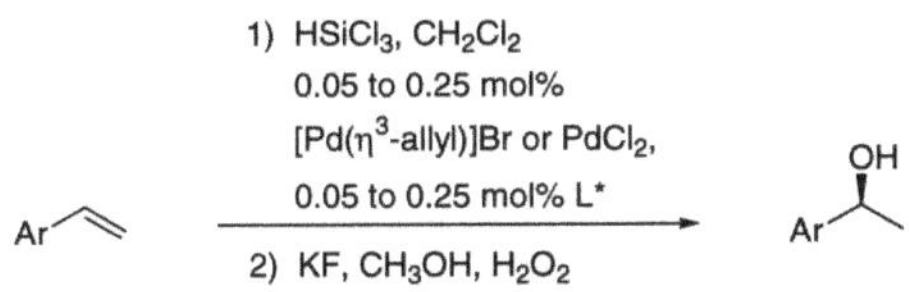

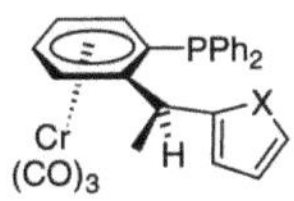

35: X = O
36: X = S

X = S, Ar = C_6H_5:	- 50°C, 92% ee
X = O, Ar = 4-CH_3-C_6H_4:	- 25°C, 83% ee
X = O, Ar = 4-CH_3O-C_6H_4:	- 25°C, 77% ee
X = O, Ar = 4-F-C_6H_4:	- 25°C, 82% ee
X = S, Ar = 4-Cl-C_6H_4:	- 25°C, 83% ee
X = O, Ar = 4-CF_3-C_6H_4:	- 25°C, 86% ee

Scheme 7 Asymmetric catalytic hydrosilylation

working model for the assumed transition state was generally the same as the one discussed above for the related aminal derived ACTCs, Scheme 7.

An important example of hydrosilylation was reported by Weber and Jones [28]. In contrast to most other work in this area, palladium instead of rhodium was chosen as the metal component of the catalyst. Since it had been known that

Pd-catalysts are much more active in hydrosilylations, this allowed for a much lower catalyst loading. As the authors predicted correctly that suitable ligands must devise one weakly and one strongly coordinating group, ACTCs with labile thiophene and furane groups were devised. The actual catalyst was prepared in situ from an allyl palladium precursor or from palladium dichloride and 1 to 3 equivalents of chiral ACTC. Hydroboration of styrene was thus accomplished in the presence of 0.05 mol% of the metal component and trichlorosilane as reagent to give the expected secondary alcohol in 47% ee. Subsequent lowering of the reaction temperature to -50 °C enhanced the product ee to 87% which could be further optimised to 92% ee when 0.25 mol% of the catalyst were used. Other substrates gave generally good to high enantiomeric excesses in the range of 77–86%. In these reactions either **35** or **36** were used complementary as chiral ligand depending on whether a palladium allyl precursor or palladium dichloride was chosen. Interestingly, a catalyst system from (η^3-allyl)palladium bromide, ligand **35** and trichlorosilane allowed for solvent-free hydrosilylation of styrene. This reaction yielded 52% ee and was complete within 2 min, which equivalents to a turnover frequency of 60,000 per hour making the present system to the most efficient asymmetric catalysis based on ACTC ligands so far.

3.3
C-C Bond Formation

3.3.1
Asymmetric Catalytic Diels-Alder Reactions

With regards to the importance of stereoselective C-C bond formation, the asymmetric Diels-Alder cycloaddition is certainly the most important [29, 30]. In the field of ACTC ligands, Jones and Guzel have introduced the first example of enanantiopure 1,3-disubstituted planar-chiral derivatives for a chiral Diels-Alder reaction [31]. Methacroleine and cyclopentadiene were chosen as the reaction partners. For the ligand structure, a vicinal diol group which generally can be conveniently generated via Sharpless AD reaction was predisposed for metal chelation. Main group III elements were screened for catalytic activity due to their inherent Lewis acidity, however, boronates were found to be only of moderate catalytic efficiency, and the best result was obtained for a combination of **38** and $BHCl_2$ (91:9 exo:endo-selectivity, 44% ee in favour of (***R***)-**37**). Better selectivities could be achieved with aluminium-based catalysts and especially with dihaloaluminates which led to enantiomeric excesses of up to 61% (95:5 exo:endo-selectivity). However, the absolute stereochemistry was altered and the major enantiomer of **37** was now of *S*-configuration. The diastereoisomer **39** gave a slightly inferior reaction outcome (96:4 exo:endo-selectivity, 53% ee). This has also been the first example of a comparison of diastereomeric planar chiral catalysts displaying a 1,3-substitution pattern.

An amplification of the stereodirecting effect of the planar chirality was achieved by modification of the ligand framework [32, 33]. Thus, replacing the flexible diol side chain by a more rigid ring annulation led to the ligand structures **40**, **41** and **42**. Again, the Diels-Alder reaction between methyl acrylate and cyclopentadiene was chosen for evaluation, and again, aluminium catalysts were the most efficient ones. In all cases, diethylaluminium chloride was more effective than its monoalkyl derivative. For example, with ligand **41** up to 90% ee were obtained at a catalyst loading of 20 mol%, while this value could be increased to 95% when the reaction temperature was lowered to -95 °C. At 10 mol%, the ee value reached 84%. Diastereomer **40** proved to be essentially less efficient and gave only 25% ee. On the other hand, catalysis with the uncomplexed ligands proceeded in an even more inefficient manner, Scheme 8.

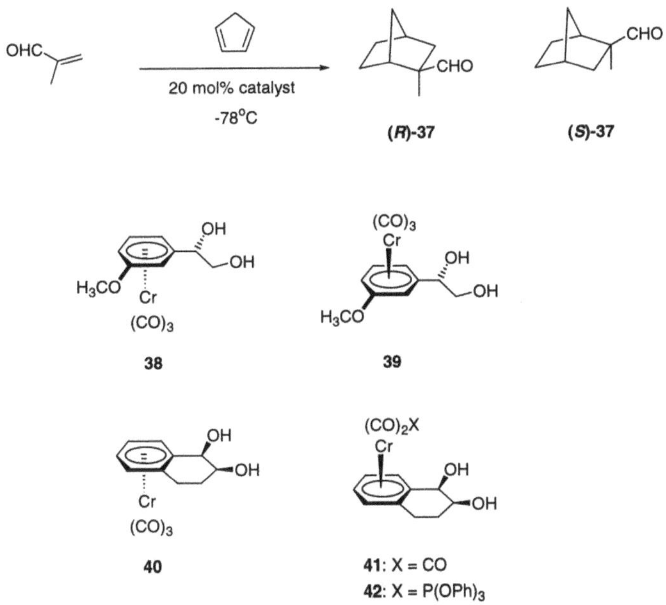

Scheme 8 Asymmetric catalytic Diels-Alder cycloaddition

In order to investigate a potential electronic tuning of the coordinated metal moiety CO displacement by more basic phosphorous ligands was accomplished. Out of a total of eight different ligands, ranging from triphenylphosphine to triethylphosphite, only $P(OPh)_3$ gave a convenient result (95:5 exo:endo-selectivity, 86% ee for catalysis with **42**). Thus, such electronic tuning has no beneficial effect since all these values are still below than the ones for the parent ACTC **41**.

3.3.2
Asymmetric Allylic Alkylation

The asymmetric alkylation of allylic systems by means of palladium catalysis, the so-called Tsuji-Trost reaction, is one of the most investigated asymmetric catalytic reactions [34, 35]. It is therefore no surprise that it has also caused interest in the area of ACTC ligands.

In the initial work by Uemura [36] ACTC complexes **43** to **46** were employed as ligands and, again, the catalytically active species was formed in situ. For the standard allyl system, the reaction of 1,3-diphenylacetoxy propene with sodium dimethyl malonate an enantiomeric excess of 94% in favour of the (*S*)-enantiomer was achieved with **44**, while ACTC **43** gave a closely related value of 92% ee. Under the same conditions, a catalysis relying on ligands **ent-27** and **46** led to a moderate ee of only 61 and 48%, respectively, and the reaction had to be carried out at a much lower temperature (-78 °C) in order to obtain at least 86% ee. Sub-

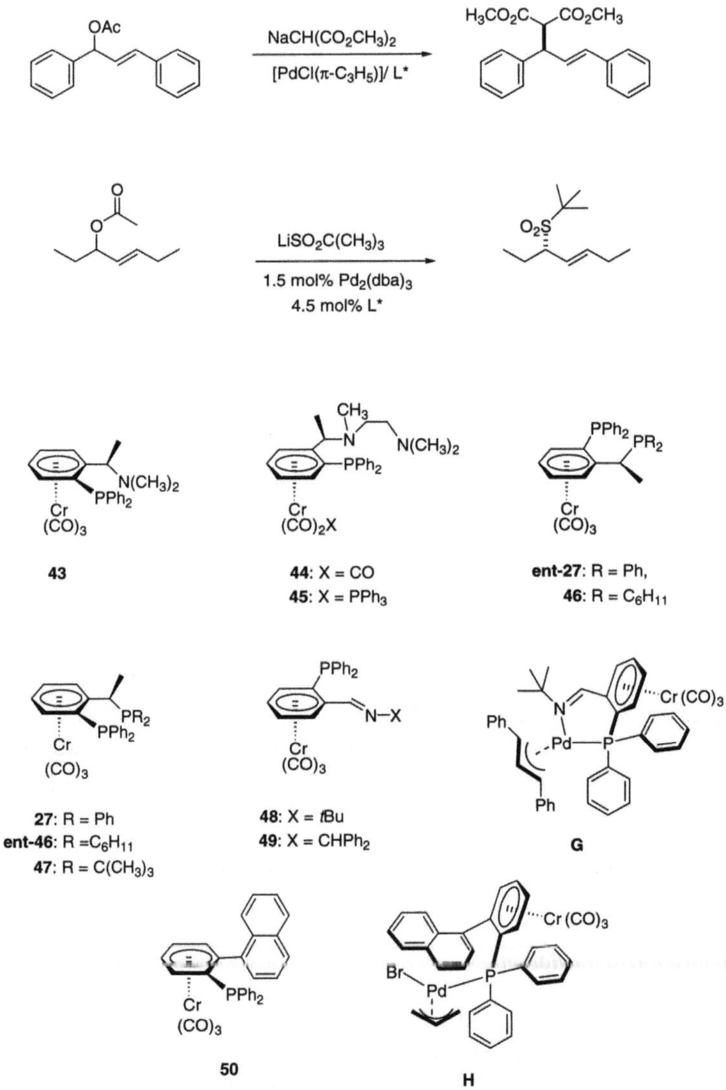

Scheme 9 Asymmetric catalytic allylic substitution

stituting one CO moiety in the best ligand **44** for PPh_3 gave a lower enantioselectivity of 76% ee which indicates that for this reaction the tricarbonylchromium(0) group is best, Scheme 9.

Related work by Salzer and Gais is based on the use of the three ACTCs **27**, **ent-46** and **47** [15]. This work is of interest since it does not investigate the mentioned standard reaction, but instead deals with a sulfur nucleophile. While the first two mentioned ligand candidates gave enantiomeric excesses of 57 and 14% ee, respectively, product formation was accompanied with formation of 2,4-heptadiene, the achiral elimination product. This was overcome by using the *tert*-butyl substituted complex **47**, albeit the enantiomeric excess now dropped to 11%.

A series of chiral imine/phosphine bidentate ACTCs such as **48** and **49** were synthesised by Chung. In this series, planar chirality was sufficient to render the catalysis highly enantioselective. Thus, ACTC **49** catalysed an asymmetric Tsuji-Trost reaction with diethyl malonate to yield the corresponding *S*-configurated product with 98% ee. The related complex **48** gave a lower enantiomeric excess of 79% ee (*S*-configuration). A crystal structure of **48-[Pd(η3-1,3-diphenylallyl)]BF4** revealed a preferential cationic structural conformation **G** that is in agreement with the observed absolute configuration of the product. Extensive nmr studies including noe experiments confirmed that this solid state structure corresponds to the solution behaviour of **48-[Pd(η3-1,3-diphenylallyl)]BF4** [37]. Interesting monodentate phosphino-substituted ACTCs have been reported by Nelson and Hilfiker [38]. Enantioselective complexes such as **50** are derived from stereoselective Suzuki-coupling between appropriately constructed planar-chiral 2-bromo carbamates and aryl boronic acids followed by nucleophilic ipso-substitution with lithium diphenylphosphine. These ACTCs form palladium complexes that are excellent catalyst precursors for asymmetric Tsuji-Trost reactions. For example, standard allylic substitution in the presence of 5 mol% yields the product in 90–92% ee depending on ligand stoichiometry. A working model **H** for the transition state has been deduced from the X-ray structure of a related, non-prochiral palladium allyl complex **50-[Pd(η3-allyl)]Br.** Based on this evaluation, the chromium moiety does not exercise strong influence over the Pd-P bond, however, the stereochemical environment originates from the chirality of the ACTC moiety. This overall arrangement is dominated by the perpendicular arrangement of both the naphthalene system and the phenyl rings of the phosphino group. As a consequence the ligated allyl system is not accessible from three sides and the two allyl termini are thus differentiated efficiently.

3.3.3
Catalytic Cross Coupling Reactions

Another classical reaction in the area of asymmetric catalytic C-C bond formation is given by Grignard cross-coupling [39], Scheme 10.

For the purpose of an efficient coordination to palladium, Hayashi and Uemura synthesised three bidentate ACTCs [40]. Using **43** as ligand in palladium catalysed cross-coupling reactions between 1-phenyl ethyl zinc chloride

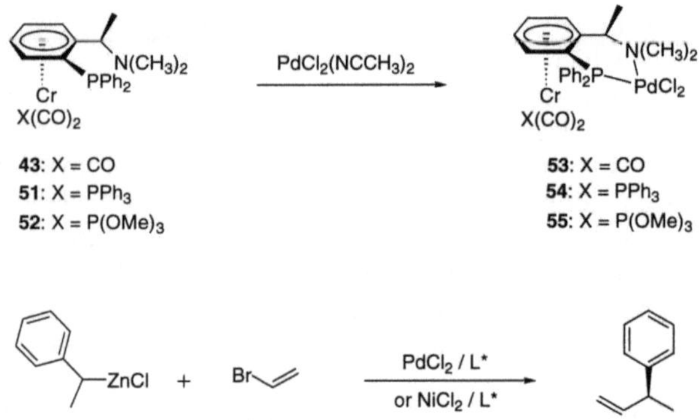

Scheme 10 Asymmetric catalytic cross-coupling

and vinyl bromide the coupling product was obtained with up to 61% ee. Enantioselectivities were lower for the corresponding magnesium Grignard complex and for the use of 2-phenyl vinyl bromide. Alternatively, nickel dichloride could be used as the metal source and the complex was then generated in situ, albeit without any change in enantioselectivity and at the expense of slightly lower chemical yields. When the electronic properties of the coordinated chromium fragment were altered via CO replacement, catalysts from **51** and **52** led to significantly lower enantiomeric excesses (37 and 17% ee, respectively). Importantly, the free arene ligand gave 40% ee, indicating that both the planar chirality induced by the chromium fragment as also the electronic tuning by the tricarbonyl moiety exercise the optimum influence. Importantly, the respective palladium catalyst precursors **53–55** were isolated from the complexation reactions. These have been the first examples of planar-chiral ACTC complexed catalyst precursors and, together with the mono-ligated complexes **48-[Pd(η3-1,3-diphenylallyl)]BF4** and **50-[Pd(η3-allyl)]Br** still remain the only examples in the area.

3.3.4
Asymmetric Hydrovinylation

Finally, the preparation and subsequent use of α-phosphino substituted ACTCs has been reported by Salzer and co-workers [41]. The three complexes **56–58** were investigated as ligands for a palladium-catalysed hydrovinylation employing styrene and ethylene as substrates. The active catalyst was formed in situ from an achiral palladium source and the respective ligand, Scheme 11.

The best value was obtained for ligand **58** which yielded a very active system that within 15 min gave a 62% yield of the desired chiral monomer (out of a total conversion of 67%). The enantiomeric excess which at this point was of 78.5%. However, a significant isomerisation takes place upon longer reaction time which increases the amount of the undesired achiral copolymer. Still, this isomerisation occurs with a kinetic resolution, thus increasing the enantiomeric excess of the remaining chiral compound to 92%.

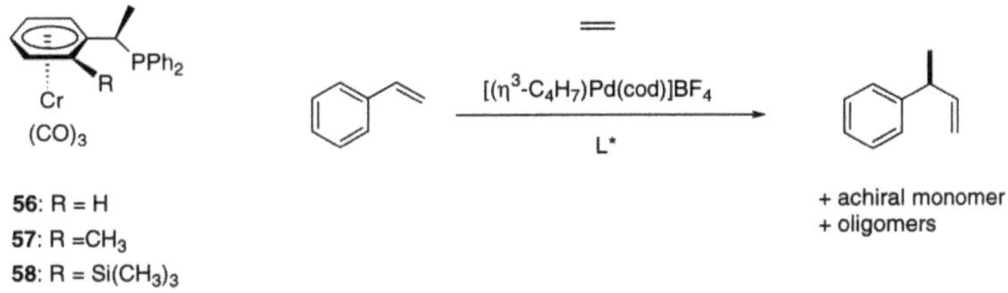

Scheme 11 Asymmetric catalytic hydrovinylation

3.3.5
Catalytic Asymmetric Dialkylzinc Additions

A complementary reaction towards the construction of secondary alcohols consists of a dialkylzinc addition to aldehydes in the presence of substoichiometric amounts of suitable chiral ligands, mostly amino alcohols [42–44]. For this reaction, which already is one of the classical asymmetric catalytic syntheses, three structurally related catalyst precursors have been described.

The first series of **59–64** was developed by Uemura and represents the most successful ACTCs in this reaction to date [45]. For example, 5 mol% of ACTCs **59** or **60** in the diethylzinc addition to benzaldehyde led to the formation of (*S*)-1-phenyl propanol with very high enantioselectivities. Higher functionalisation at the hydroxymethyl side chain was found to be essential since the unsubstituted hydroxymethyl compound **59**, that lacks sterically demanding substituents in the benzylic position, led to only 15% ee. When additional stereocenters were introduced in the hydroxymethyl side chain there was no beneficial effect and the enantiomeric excesses of products from catalyses with ACTCs **62** and **63** were identical or slightly lower. However, a change of absolute configuration at this stereogenic centre from (*S*) to (*R*) led to a dramatic decrease in enantioselectivity. Also, switching planar chirality by the use of a complex in which the chromium moiety was attached to the other face of the arene ring than in **60** decreased the enantiomeric excess of the product to 29% ee. This is another important example that in order to achieve excellent enantioselectivities an efficient internal cooperation of all stereoelements is essential [19]. For the present system, ACTC **60** evidently contains the best combination of planar and central chirality, as can be further deduced from the proposed transition state **I** with its optimum arrangement of seven- and six-membered chelate rings. This stereochemical situation could be further optimised by introducing phosphino groups into the chromium carbonyl moieties. The respective complexes **64** with either a triphenylphosphino or a triphenylphosphite ligand acted as superior ligands in diethylzinc additions. Since their electronic properties are far from being similar, it must be concluded that steric bulk from these groups is the obvious explanation for the increase in stereoinduction.

A different ligand type was presented with ACTC **63** [46]. From the four structurally related ligands, complex **65** performed best yielding the (*S*)-configurated alcohol with 70% ee from the reaction of benzaldehyde with diethyl zinc. In this ligand precursors the chromium moiety is essential for high asymmetric induction, since the arene ligand alone catalysed the formation of a product with only 10% ee.

Finally, ACTCs **66** and **67** were employed in the asymmetric addition of dimethyl or diethylzinc to benzaldehyde and ferrocenecarbaldehyde [47]. In order to obtain acceptable conversion and enantioselectivities the catalyst loading had to be 10 mol%. In such circumstances, enantiomeric excess of up to 86% could be obtained. Complex **67** with its diphenylhydroxymethyl substituent performed much better than **66**, which is in accord with the observation made for the series of ACTCs **59–64.**

A closely related asymmetric reaction is the nickel catalysed conjugate addition of dialkylzincs to Michael acceptors such as chalcone. Here, ACTC **63** showed the best performance [45]. A catalyst loading of 5 mol% of nickel salt was sufficient, albeit a high ligand loading of 50 mol% was necessary in order to suppress catalysis by achiral species, and the product was obtained in 90% chemical yield and with an enantiomeric excess of 62%, Scheme 12.

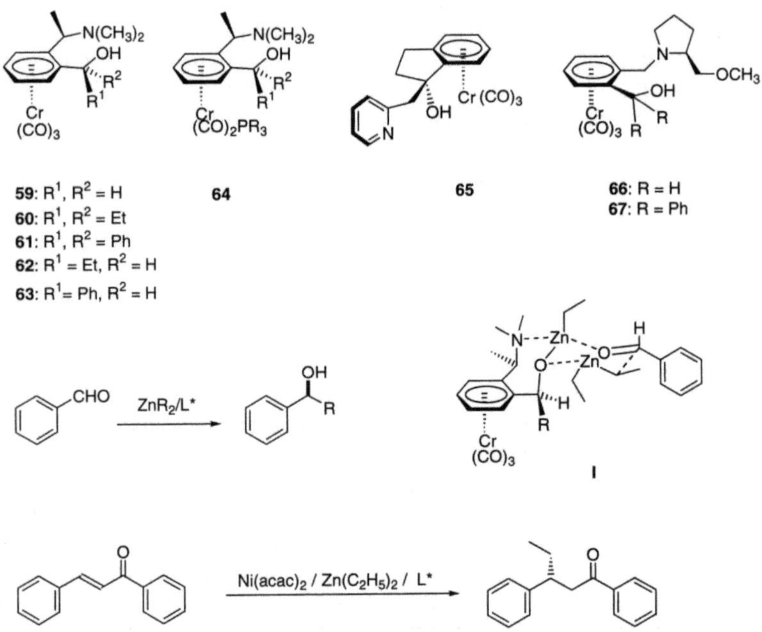

Scheme 12 Asymmetric catalytic dialkylzinc addition

4
Structural Information

The complexes **48-[Pd(η3-1,3-diphenylallyl)]BF4** and **50-[Pd(η3-allyl)]Br** are the only direct catalyst precursors described in this article that have been characterised by X-ray crystal structure. Additionally, several of the ACTC ligands have been characterised in the solid state (i.e. compounds **ent-27, 28, 31, 32, 35, 56** and **63**), and nmr data has been obtained for the preformed palladium catalyst precursors **53–55**. However, the available data is not sufficient to allow for any accurate discussion on preferential ligand conformation, and thus will not enable any ligand optimisation by rational design. It will be especially important to gain data on the question whether the preferential ligand conformation as undertaken in its uncomplexed state is maintained when complexed onto a transition metal. This has been investigated in depth for the corresponding ferrocene ligands [48] and has had dramatic impact on catalyst design in this area. Moreover, except for the series of **48** to **50** chiral ACTCs with planar chirality as the only stereogenic element are still virtually unknown. Future research should also deal with synthetic approaches towards this type of potential ligand.

5 ACTC Complexes for Asymmetric Catalysis Towards Natural Products

Despite the relative high number of ACTCs that have been employed in asymmetric catalysis, there has been no application of these catalysts for the construction of intermediates of interest for natural product synthesis. $Cr(CO)_3$-complexed ligands that have been described for such purposes lack an element of planar chirality. Thus, the importance of the chromium(0)-moiety relies on the creation of a suitable electronic and/or steric environment within the active structure of the catalyst. Natural products that have been synthesized by such route include (*R*)-(-)-rhododendrol, the aglycone of rhododendrin, a hepatoprotective agent, the macrolides (*R*)-(-)-phorcantholide I and (*R*)-(+)-lasiodipolin and related bifunctional synthons [49–53], Scheme 13.

For example, methylation of trityl-protected ω-hydroxyl nonanone (**69**) with the aid of 5 mol% of the ACTC **68** gave rise to the desired secondary alcohol with 88% ee, which can be further elaborated to (*R*)-(-)-phorcantholide I (**70**). Use of the related non-complexed amino alcohol gave only 59% ee at the higher catalyst loading of 20 mol%. The superiority of the complexed ligands had already been

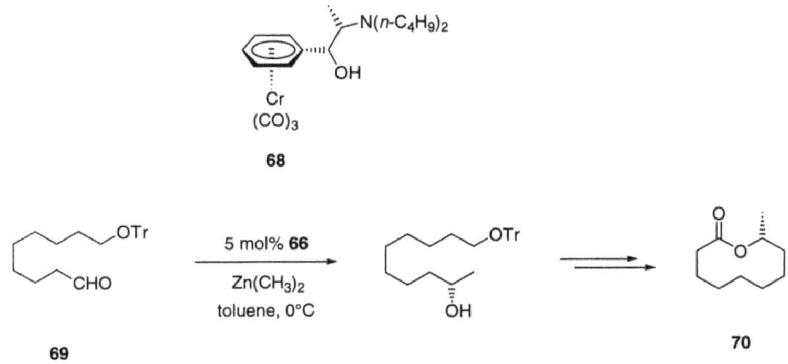

Scheme 13 Asymmetric diethylzinc addition for natural product synthesis

demonstrated for the standard dimethylzinc addition to benzaldehyde. Here, complex **68** gave a 99% enantiomeric excess, while the simple tertiary norephedrine led to a product with only 94% ee [54].

6 Conclusion

Optically active ACTCs with both planar and central chirality may serve as ligands for asymmetric catalytic synthesis. Since the first review on this topic has appeared [9], substantial progress in this area has been achieved. However, the full potential of such ligands has not yet been fully exploited. It is to be expected that further efforts will deal with new ligand structures and will pay special focus on electronic tuning, a possibility of ligand design that is nearly unique to ACTC structure. Further, it will be necessary to gain broader knowledge on the complexation ability of the ligands and their coordination behaviour in metal complexes.

Acknowledgements. The author is grateful to Prof. Dr. S. E. Gibson for her hospitality during an ERASMUS stay at Imperial College in 1993/4, to Prof. Dr. C. Bolm, RWTH Aachen, for his continuous support and encouragement and to Prof. Dr. K. H. Dötz, Universität Bonn, for his generous support and interest. Financial support from the Fonds der Chemischen Industrie (Liebig fellowship 2000–2003) is gratefully acknowledged.

References

1. Noyori R (1993) Asymmetric catalysis in organic synthesis. Wiley, New York
2. Ojima I (ed) (2000) Catalytic asymmetric synthesis, 2nd edn. Wiley-VCH, Weinheim
3. Jacobsen EN, Pfaltz A, Yamamoto H (eds) (1999) Comprehensive asymmetric catalysis. Springer, Berlin Heidelberg New York
4. Togni A (1996) Angew Chem 108:1581; Angew Chem Int Ed Engl 35:1475
5. Togni A (1996) Chimia 50:86
6. Hayashi T, Togni A (eds) (1995) Ferrocenes, VCH, Weinheim
7. For an outstanding example of a ferrocenyl ligand in asymmetric catalysis, see: Blaser H-U (2002) Adv Synth Catal 344:17 (and cited literature)
8. Schlögl K (1967) Topics Stereochem 1:39
9. The present article represents an update of an earlier review: Bolm C, Muñiz K (1999) Chem Soc Rev 28:51
10. During preparation of the present article, a related review on the synthesis of planar chiral ACTCs and their application in asymmetric catalysis appeared: Gibson SE, Ibrahim H (2002) Chem Commun 2465
11. For a more detailed study on electronic tuning in the tripod chromium tricarbonyl moiety, see: Jones GB, Chapman BJ, Mathews JE (1998) J Org Chem 63:2928
12. Recently, several additional publications have appeared that describe the use of chiral bases for the two-step construction of diastereomerically pure ligand candidates: Gibson (neé Thomas) SE, Ibrahim H (2001) Chem Commun 1070; Gibson SE, Ibrahim H, Pasquier C, Steed JW (2002) Tetrahedron 58:4617
13. An important new α-substitution reaction that has found decisive application in the introduction of coordinating phosphino groups has been discovered by Salzer: Englert U, Salzer A, Vasen D (1998) Tetrahedron: Asymmetry 9:1867
14. Sodeoka M, Shibasaki M (1993) Synthesis 643

15. Vasen D, Salzer A, Gerhards F, Gais HJ, Stürmer R, Bieler NH, Togni A (2000) Organometallics 19:539
16. Pasquier C, Naili S, Pelinski L, Brocard J, Mortreux A, Agbossou F (1998) Tetrahedron: Asymmetry 9:193
17. Pasquier C, Naili S, Mortreux A, Agbossou F, Pelinski L, Brocard J, Eilers J, Reiners I, Peper V, Martens J (2000) Organometallics 19:5723
18. Pasquier C, Pelinski L, Brocard J, Mortreux A, Agbossou-Niederkorn F (2001) Tetrahedron Lett 42:2809
19. Muñiz K, Bolm C (2000) Chem Eur J 6:2309
20. Jones GB, Heaton SB, Chapman BJ, Guzel M (1997) Tetrahedron: Asymmetry 8:3625
21. Hartwig JF (1998) Angew Chem 110:2154; Angew Chem Int Ed 37:2046
22. Hartwig JF (1998) Acc Chem Res 31:852
23. Wolfe JP, Wagaw S, Marcoux J-F, Buchwald SL (1998) Acc Chem Res 31:805
24. Kamikawa K, Sugimoto S, Uemura M (1998) J Org Chem 63:8407
25. Rossen K, Pye PJ, Maliakal A, Volante RP (1997) J Org Chem 62:6462
26. Son SU, Jang H-Y, Lee IS, Chung YK (1998) Organometallics 17:3236
27. Son SU, Jang H-Y, Han JW, Lee IS, Chung YK (1999) Tetrahedron: Asymmetry 10:347
28. Weber I, Jones GB (2001) Tetrahedron Lett 42:6983
29. Corey EJ (2002) Angew Chem 114:1724; Angew Chem Int Ed 41:1650
30. Nicolaou KC, Snyder SA, Montagnon T, Vassilikogiannakis G (2002) Angew Chem 114:1742; Angew Chem Int Ed 41:1668
31. Jones GB, Guzel M (1998) Tetrahedron: Asymmetry 9:2023
32. Jones GB, Guzel M (2000) Tetrahedron: Asymmetry 11:1267
33. Jones GB, Guzel M, Heaton SB (2000) Tetrahedron: Asymmetry 11:4303
34. Trost BM, Van Vranken DL (1996) Chem Rev 96:395
35. Pfaltz A, Helmchen G (2000) Acc Chem Res 33:336
36. Hayashi Y, Sakai H, Kaneta N, Uemura M (1995) J Organomet Chem 503:143
37. Jang H-Y, Seo H, Han JW, Chung YK (2000) Tetrahedron Lett 41:5083
38. Nelson SG, Hilfiker MA (1999) Org Lett 1:1379
39. For a recent general review on catalytic asymmetric arylation, see: Bolm C, Hildebrand JP, Muñiz K, Hermanns N (2001) Angew Chem 113:3382; Angew Chem Int Ed 40:3284
40. Uemura M, Miyake R, Nishimura H, Matsumoto Y, Hayashi T (1992) Tetrahedron: Asymmetry 3:213
41. Englert U, Haerter R, Vasen D, Salzer A, Eggeling EB, Vogt D (1999) Organometallics 18:4390
42. Noyori R, Kitamura M (1991) Angew Chem 103:34; Angew Chem Int Ed Engl 30:49
43. Soai K, Niwa S (1992) Chem Rev 92:833
44. Pu L, Yu H-B (2001) Chem Rev 101:757
45. Uemura M, Miyake R, Nakayama K, Shiro M, Hayashi Y (1993) J Org Chem 58:1238
46. Malfait S, Pélinski L, Brocard L (1996) Tetrahedron: Asymmetry 7:653
47. Bolm C, Muñiz K, Ganter C (1998) New J Chem 1371
48. Togni A (1997) In: Helmchen G, Dibo J, Flubacher D, Wiese B (eds) Organic synthesis via organometallics (OSM 5). Vieweg, Braunschweig Wiesbaden, p 185
49. Jones GB, Heaton SB (1993) Tetrahedron: Asymmetry 4:261
50. Jones GB, Chapman BJ, Huber RS, Beaty R (1994) Tetrahedron: Asymmetry 5:1199
51. Jones GB, Huber RS (1993) Synlett 367
52. Jones GB, Huber RS, Chapman BJ (1997) Tetrahedron: Asymmetry 8:1797
53. Jones GB, Guzel M, Chapman BJ (1998) Tetrahedron: Asymmetry 9:901
54. Heaton SB, Jones GB (1992) Tetrahedron Lett 33:1693

Author Index Volumes 1–7

The volume numbers are printed in italics

Subject Index

MIX
Papier aus verantwortungsvollen Quellen
Paper from responsible sources
FSC® C105338
www.fsc.org

If you have any concerns about our products,
you can contact us on
ProductSafety@springernature.com

In case Publisher is established outside the EU,
the EU authorized representative is:
Springer Nature Customer Service Center GmbH
Europaplatz 3, 69115 Heidelberg, Germany

Printed by Libri Plureos GmbH
in Hamburg, Germany